Milk

MILK

A Local and Global History

Deborah Valenze

Yale

UNIVERSITY PRESS

New Haven & London

Yale University Press books may be purchased in quantity
for educational, business, or promotional use.
For information, please e-mail sales.press@yale.edu
(U.S. office) or sales@yaleup.co.uk (U.K. office).

Set in Galliard Oldstyle type by Keystone Typesetting, Inc.,
Orwigsburg, Pennsylvania.
Printed in the United States of America by
Sheridan Books, Ann Arbor, Michigan.

Library of Congress Cataloging-in-Publication Data
Valenze, Deborah M., 1953–
Milk : a local and global history / Deborah Valenze.
p.   cm.
Includes bibliographical references and index.
ISBN 978-0-300-11724-0 (hardback)
1. Milk — History. 2. Milk — Social aspects. 3. Cooking
(Milk) — History. 4. Dairy products — History. 5. Food
habits — History. I. Title.
GT2920.M55V35   2011
641.3′7109 — dc22
2011002514

A catalogue record for this book is available from the
British Library.

This paper meets the requirements of ANSI/NISO
Z39.48–1992 (Permanence of Paper).

10 9 8 7 6 5 4 3 2 1

For Nancy Carney

. . .

CONTENTS

Preface, ix
Acknowledgments, xiii

Introduction, 1

Food is fun to think with. That's the mantra of scholars and students involved in food studies, which is fast becoming a vital interdisciplinary field of study in academe. Of course, food studies also involves people outside academe, a vast network of writers, cooks, vendors, policy advocates, health specialists, farmers, and those generally known as "foodies," who are eager to think and talk about food. Milk is often the focus of discussion, much of it passionate: should everyone consume milk? Is it safe? Should it be fed to children? And why are people all over the world so engaged and often emotionally invested in issues having to do with milk? The answers to all of these questions can be helped along by a perspective informed by history.

Modern milk has clear origins in the last one hundred and twenty years. When I began this project, I confess, I kept an eye out for a neat arc of change, a story of the rise and triumph of the contemporary carton of milk. After all, milk really *is* the consummate commodity, the virtual queen of today's supermarket. There is no question that the modern industrial complex of food production helped mass-produced milk come into being. The product itself shows clear signs of a makeover, having shed its natural origins in exchange for a sterile commercial identity. Yet it is the only food that is also produced by humans, adding another dimension to how we see its nature. Practically every controversy over contemporary milk, whether raw, organic, or "regular," has to do with wishing to recover the past, when milk was less a commodity and somehow more authentically natural. Wouldn't it be reasonable to write about what happened to make modern milk "milk"?

The following pages tell that story, but the more profound discovery about milk became clear to me when I looked at what historians call "the long durée," the sweep of hundreds and even a thousand years of the past. Though a fact of nature, milk is really a product of culture. From the moment of expression, milk enters a matrix of human contrivance. Simply defining what people believed milk to be in the past is a tricky task. Was it a magical bodily fluid or a bestial by-product? Was it sacred or profane? Medicinally useful or indigestibly dangerous? The food of infants or the nutrient of everybody? Nearly every field of knowledge has had something to say about this mystifying liquid: mythology, religion, natural philosophy, medicine, agronomy, culinary arts, chemistry, and even cosmetology have weighed in on what milk is. In fact, the dilemmas surrounding milk through the ages have been far more fascinating than its formulaic rise to nutritional supremacy.

These dilemmas bring to light the argument of my book. A classic modernizing story indeed comes into focus as soon as milk enters the market in earnest. But while technology, transport, and business play critical roles, the forces of modernity turn out to be contenders rather than victors in determining what milk would become in the twentieth century. Even the science of nutrition, arguably the most powerful force in its history, needed help from other constituencies (such as insistent mothers and wartime governments) in order to define milk as universally necessary. What we now think of as milk is really a shifting combination of many perspectives, some of them residing in the mentalities of local cultures, rooted in beliefs that have been around for a very long time.

Different cultures and regions have had distinctive ideas about milk, but one book can cover only so much. With the exception of the first and last chapters, this account examines the areas where milk first became crucial to diets and economies, namely Western Europe and the United States. I have discussed briefly developments affecting other parts of the world in order to give a sense of how commercial-minded milk producers transplanted milk to many places. In the case of India, one of the most ancient milk cultures and now the largest producer of milk in the world, I offer a short narrative of how the nation organized its dairy forces.

And finally, the historian's apology: I have not written a book about contemporary milk. My expertise and interest rest in the past, beginning with ancient times and moving through almost two millennia (with unabashed hubris) to the aftermath of the Second World War. A final chapter covers major issues of milk history since then, such as rBGH (recombinant Bovine Growth Hormone) and the burgeoning interest in raw milk. For readers interested in factual knowledge about modern-day milk, food encyclopedias and the Internet may supplement the narrative. What this book will provide is a sense of how milk acted as a potent force in history, participating in some of the major historical shifts in food production and consumption over the past several hundred years. Its unique, often inflammatory power to provoke the imagination and inspire action provides a fascinating lesson in how human desire, sometimes realistic and sometimes far beyond the power of any one commodity, lies embedded in the food we talk about every day.

ACKNOWLEDGMENTS

Milk, dairying, food, diets, breast milk, raw milk: the countless conversations I've had on these subjects over the years have made this book a collaborative effort from beginning to end. First thanks go to Emma Parry and Jon Schneer, who coaxed me into embarking on the project, and Maxine Berg, whether she knows it or not, who made the plan possible by getting me to write on world dairying history several years before that. Huge thanks to my editor, Jean Thomson Black, who has been the best of book midwives and an enormous inspiration in helping me carry out this project. At Barnard College, I am grateful to Provost Elizabeth Boylan for continuing support, along with all of my wonderful colleagues, especially Herb Sloan and Lisa Tiersten, who supplied me with much bibliographic help (I will never be as widely read as they are), and Sully Rios for her expert support. The Barnard and Columbia students in my history of food courses deserve big thanks for their engagement and ideas, as do Alhelí Alvarado-Diaz and Brian Karl, for many stimulating discussions. Various audiences have heard portions of this book and have offered excellent suggestions; in particular, I would like to thank the Northeast Conference on British Studies and the Boston chapter of the Barnard Alumni Association. On the other side of the Atlantic, Eric Hobsbawm inspired the *Past and Present* article of long ago, and continued to ask me hard questions about milk ever since; special thanks to him and Marlene for their energetic presence over the years. Derek Oddy advised me on research in

European food history in ways that proved indispensable in the writing and teaching of the subject. Jim Obelkevich directed me to invaluable resources and shared his knowledge of twentieth-century consumer culture with me. Many people took the time to send me articles, information, and advice; among them are Marc Abrahams, Peter Atkins, Virginia Berridge, Mariel Bossert, Mitchell Charap, Nancy Cott, Thomas Fenner, Alan Gabbey, Owen Gutfreund, Tim Hitchcock, Joel Hopkins, Kaia Huseby, Penny Ismay, Shelagh Johnston, Kathleen Keller, Rachel Manley, Andrew Mathews, Judy Motzkin, Keith Moxey, Walt Schalick, Judith Shapiro, Pat Thane, Chris Waters, and Nancy Woloch. And thanks to Christoph Grosjean-Sommer and Isidor Lauber of Emmi AG in Switzerland; Martin Tanner of Bern, Switzerland; Neil and Jane Dyson of Holly Green Dairy Farm, Bledlow, Buckinghamshire, England; and Philip and Teah Ranney of Putney, Vermont.

I am grateful to the following libraries: Widener Library and the Houghton Library, Harvard University; the Schlesinger Library of Radcliffe College; Butler Library, Columbia University; the British Library; the Colindale Newspaper Library; and the Wellcome Library and Rare Books Room.

Thanks to Embry Owen for help with the bibliography, Melissa Flamson for assistance with image permissions, and Jaya Chatterjee and Phillip King for editorial assistance. I am also grateful to Joan Winder, Joan Adler and the Straus Historical Society, Eric Fowell of Rushden, England, Kay Collins and the Rushden Heritage Project, Jessica Kaiser of the Mannheim University Library, and Debby Rose.

Several people gave me invaluable advice after reading chapters or all of the manuscript: I am indebted to three anonymous readers for Yale University Press for excellent critiques and suggestions at the penultimate stage of writing; thanks to Timothy Alborn and Marya Huseby for reading at an early stage; Nancy and Charlie Carney read the first draft and offered me suggestions, detailed and conceptual, some drawn from their own experiences in American dairies; Kim Hays offered her expert eye right up until the very last minute; and Janna Malamud Smith gave me helpful feedback on the entire manuscript. Kim also organized my visits to the Emmi plant, the Tanner farm, and my meeting with Christoph Grosjean-Sommer of the Swiss Milk Producers. My understanding of contemporary milk owes an enormous debt to those visits. Special thanks to my virtual writing partner and food studies guide, Merry White, and our hours of exchange over high-quality coffee. Her readings of many chapters changed my thinking in important ways.

A number of friends, in addition to those mentioned above, deserve mention for other kinds of support and inspiration: Laura Bossert, Irene Briggin, Martin Dineen, Phyllis Emsig and Mark Goldberg, Kitty Griffith, John Pratt, Janna and David Smith, Ruth Smith, Timea Szell, Donna Smith Vinter, Richard Vinter, Peter and Kathleen Weiler, and Lew Wurgaft and Carole Cosell.

I owe special and loving thanks to my husband, Michael Timo Gilmore, who has shared this project with me in every way and has probably endured more dinner-table conversation about milk than he ever believed possible. He was also a constant sounding board and an invaluable reader of every page. My daughters, Emma and Rosa Gilmore, have always been foremost in my thoughts as I worked on this book; suffice it to say that milk and motherhood are inseparable. They also supplied me with all kinds of help in the form of books, images, information, and encouragement. I would also like to thank their friends, whose genuine interest added to my own enthusiasm for the subject. And finally, profound thanks to Nancy Carney, my high school history teacher and the dedicatee of this book. She has taught me how to think about history as a subject full of humans, animals, plants, and nature. And if it hadn't been for her example and guidance many years ago, and her continuing presence today, this book never would have been written.

Introduction

Cows that have names give more milk than cows without names. This discovery won recognition at the 2009 Ig Nobel awards, an annual event in Cambridge, Massachusetts, organized to "first make people laugh and then make them think." I didn't laugh, though, because the question of naming cows had come up in conversations with dairy farmers I had sought out in Vermont, Switzerland, and England. A note of condescension might have crept into one or two responses, but managers of milk production know what's at stake in the question. Cows respond to sensitive, personalized handling; they're acutely aware of environmental factors, particularly so in the incongruous realm of modern milking machinery. Dairy farmers report their cows expressing joy, jealousy, sulkiness, and fear. When handlers interacted with cows with respect and affection, annual milk output averaged slightly more than sixty-eight gallons higher than in less friendly settings. Not for nothing did Wisconsin dairy farmers once use as their motto "Speak to a cow as you would a lady." And in 1997, a major milk corporation included a kindness clause in its contract with client farmers. If one locus in the contemporary world still finds economic value in chivalry and tenderness, it's the dairy barn.[1]

Dairies nowadays expect to host visitors yearning for an encounter with bovine charisma. This gives farmers an opportunity to educate the public and reaffirm popular support for a beleaguered industry, considered by most governments to be a revenue-draining public service.

The Ingénues, a female band and vaudeville act, serenading the cows in the University of Wisconsin-Madison Dairy Barn in 1930. The performance was part of a scientific test to see if cows would give more milk to the soothing strains of music. (Photo: Wisconsin Historical Society, WHS-2115)

Popular affection for cows has always helped the cause. Take, for example, the live cows bathed and milked on a merry-go-round (called a "rotolactor") exhibited at the New York World's Fair in 1939. Or the life-size "cow parade" figures that appeared in cities around the world in 1999, designed to elicit affection and then, it was hoped, hefty prices at charity auctions. Although not all cow paraphernalia relates to milk, most of it does in regions that depend on the industry. Departing from a Swiss airport requires passing through a gauntlet of cow backpacks, cream pitchers, and cow-print oven mitts. The Swiss nation is collectively in love with an animated "real" cow named Lovely, who dribbles soccer balls and ice skates her way through hilarious commercials for milk. I will confess to some advocacy of my own: our family car proudly displays a bumper sticker bearing a cow's face and the words "Got Vermont?" The humor, of course, comes from the cow itself adopting celebrity status in the manner of one of the most successful advertising campaigns of the twentieth century.[2]

Cow love is intimately tied to milk history and always has been. A quick look through the annals of world mythology turns up cows of plenty and visions of oceans of milk. Not all milk in history has come from cows, but then, our attachment to cows raises an interesting historical question: how and why have we become attached to cow's milk? This is the problem with which I began this book. As a historian of English dairying, I knew that the question was moot long ago: the eighteenth-century cow had already proven its worth as the most prolific producer of nourishment, at a time when quantity mattered more than even digestibility. As a food ingredient, cow's milk had more versatility and wider palatability (in other words, a blander taste) than others. Cows were quintessential docile bodies, suited to large-scale systematizing of production. And probably most important of all, cows were the favored domesticated farm animals of ambitious, commercial-minded western Europeans and Americans. Cows went where their masters went, which was nearly everywhere. It became apparent to me very quickly that the story of modern milk was one of conquest of space, energy, and dietary preferences. The commodity of milk today has triumphed as a universal icon of modern nutrition, despite all attempts to deny it supremacy.

But a narrative of conquest hardly does justice to what has turned out to be a story full of mystery, myth, and impassioned debate. Despite the compulsory feel of milk today, its history is just as often about suspicion cast on the opaque white liquid. For many centuries, milk was regarded as dangerous and even repulsive. (And, we might add, this remains true today for some people, for different reasons.) Anthropologists tell us that human consumption of animal milk and its products constitutes an aberration of animal nature that our ancient ancestors had to rework in their own minds. Denying young animals nourishment of their mother's milk puts progeny at risk; it requires inserting human agency where it does not belong. Transforming milk into butter and cheese represents another violation of taboo; note the telltale revulsion we feel toward, say, eating cheese made of breast milk. Non-milk-drinking cultures feel that milk is an unclean animal fluid, like urine. In eighteenth-century Buddhist Japan, milk was thought to be "white blood," which, if drunk, would bring down divine retribution. It is worth remembering that evolutionary geneticists see lactose tolerance, not intolerance, as the deviant trait that later spread across certain, often northern, populations. Over several millennia, religions endowed milk with added value, helping to convince consumers of its legitimacy as nourishment, apart from its identity as first food.[3]

Milk, a liquid associated with kindness and love, has never been free from conflict throughout history. Its most obvious purpose, that of feeding people, developed unevenly. The bestial origins of milk marked the liquid as barbaric, at least for southern Europeans with early pretensions to civility. Urban dwellers showed disdain for the liquid, or actively feared it because of Greek dietetic proscriptions. From the sixth century, milk and dairy products were regarded as forbidden foods on Christian fast days, following Saint Gregory's prohibition of all things that came from flesh. Few elites ever imagined drinking milk, except, perhaps, while on a recreational visit to the countryside. In small amounts, when consumed at the right times, milk products were considered safe by early medical experts. But because of its perishable nature, which could easily sicken careless consumers, milk ranked as a dangerous aliment well into the seventeenth century.

Fortunately for the ensuing history of human food, milk also conjured up a contradictory theme from the start. Its pallor and fragrance sang of bucolic purity and abundance. In answer to the sophistication of civility, milk struck notes of simplicity. Its bounty, commonly displayed in cheese and butter, was impossible to deny, as, for instance, in the record-breaking rounds of cheese coming from Italy in the fifteenth century. Charles VIII sent a giant sample as a gift to the queen of France in 1494, perhaps hoping to impress her. By this time, cows "of broken colors" were already identified as big milk producers in Lombardy and the Low Countries. In the Rhineland, Dutch milk earned comparison to wine and was regarded of greater value. The "Butter Tower" of Rouen Cathedral, constructed with donations enabling townspeople to eat butter during Lent (the Catholic Church granted "lacticinia" dispensations for those who offered charity), stands as a bold tribute to the French devotion to the dairy. And butter flowed from places like Bruges, where residents ate it with every meal. "Bring a knife," a resident instructed a French friend in a letter.[4]

Examples such as these convinced me of the special nature of milk and its products as subjects of history. It may be that every author sees the subject of her book as special, but in this case, I felt I had an arguable case. My hunch was buoyed up repeatedly by evidence from many fields of knowledge, besides history, that regarded milk as imbued with unique characteristics. From this, I made my primary discovery, which at first seemed simple: it was the formative — one might say definitive — role of context in shaping the path of milk through history. Each appearance of milk seemed deeply situated in its setting, or weighed down with "cultural baggage." And in establishing a relationship with the product, societies seem to have generated a surplus of imaginative thoughts about milk, at times so thoroughly enmeshing it within beliefs that the actual nature of milk — if we can use that term — was eclipsed by everything else.

There can be no milk without the contexts woven into its past: this absorptive relationship happens with all food products to a greater or lesser extent, but with milk, an added virtue lay in how its cultural history had something bigger to say about the history of food. Situated in culture, milk acted as a mirror of its host society, reflecting attitudes toward nature, the human body, and technology. Moreover, its larger presence as a liquid in diets since the beginning of the twentieth century became a litmus test of wealth and attitudes toward modern food production. In the historical record, milk repeatedly called attention to the larger forces of change. Milk, then, becomes a marker of the emergence of a peculiarly Western food culture and its path into the modern age.

Culture in its most general sense means shared beliefs, information, and technologies of a given group. Anthropologists and sociologists differ over the specifics of the term, but my use agrees with the broad definition used by evolutionary geneticists. As Peter J. Richerson, Robert Boyd, and Joseph Henrich have shown, looking at the long term of human evolution, "culture is fundamentally a kind of inheritance system" enabling people to inhabit and adapt to their environments. As a food, milk was very often crucial to survival. Populations in the past seem to have known this and were willing to make accommodations to keep supplies in production. As Richerson and Boyd have pointed out, culture in this case has actually altered the path of biological evolution: humans are still in the process of acquiring lactose tolerance, and those who have benefited from animal milk have enjoyed dietary advantages. Milk history claims a role in a much larger process than one might at first suspect.[5]

Milk figures prominently in a second big picture of history. The past hundred years or so, according to scholars of world food supplies, can be seen as a series of dominant global "food regimes": international systems of production in which agriculture, science, and industry combine in force, using a shared set of technologies to produce a shared set of priorities, namely, particular foods for a profit-driven market.[6] Milk is a leading component of the modern food regime we now inhabit, and its origins are apparent in the history traced in this book. My focus on western Europe and North America is deliberate: particularly in northwestern Europe, as Alfred W. Crosby puts it, one finds "champion milk-digesters," who worked out ways to maximize production for the market and eventually carried their values and technologies to other parts of the globe.[7] In a final brief look at milk in India, I show how the European pattern of industrial development encountered resistance in the late twentieth century.

This book makes no claim to being the first work on the subject of milk and dairying. But for years, much of what was written on the subject approached milk within narrowly defined realms. G. E. Fussell's early book on English dairying, for example, treated milk as a product of agriculture, a subfield of economic history which, until recently, existed in its own plot of intellectual ground. Scattered scholarly works have not exactly been lacking: the place of milk in religion, the role of milk as a carrier of disease, milk and infant feeding, and milk and modern nutrition represent just a fraction of the many subjects intersecting with mine, though often in the form of article-length monographs. Valuable exceptions include E. Melanie DuPuis's *Nature's Perfect Food: How Milk Became America's Drink,* a fascinating investigation of milk as a commercial enterprise in the United States. And Peter J. Atkins's *Liquid Materialities: A History of Milk, Science, and the Law* explores the many measurable ways in which milk created problems for British science and the state as the product evolved into its modern form. Stuart Patton's *Milk: Its Remarkable Contribution to Human Health and Well-Being* is an agricultural scientist's helpful empirical account of the substance. Outside the academy, culinary historians have devoted serious attention to milk. Most recently, in *Milk: The Surprising Story of Milk Through the Ages,* Anne Mendelson examines the many uses of a wide variety of milks across many cultures, focusing on its flavors, textures, and quintessentially milky characteristics. All four books celebrate different aspects of milk's power—as a part of economic life, urban modernity, biological enhancement, and as a food full of zestful and even whimsical characteristics.[8] Is there anything missing from the picture, given this array of books?

My own approach to milk history aims to tell a bigger, yet selective, story for the interested reader. The following chapters reveal the cultural malleability of milk alongside the emergence of a modern Western way of thinking about food. To my mind, none of these developments were wholly predictable. Milk really has enjoyed a life of its own, teasing and sometimes poisoning humanity with its organic wealth, a mixture of protein, sugar, and water, which enables it to support microbes like few other food substances. It has also rewarded consumers through its nutritional power, making it the central subject of passion-driven studies, particularly at both ends of the twentieth century. And the milk of humans challenged Western science to one of its most arduous competitions, namely, the attempt to replicate breast milk for infants. Infant feeding, especially in cities, posed a major challenge to nineteenth-century medicine and food purveyors. Women reformers, wealthy philanthropists, socialist cooperators, laboratory scientists, population advocates, and pediatricians had something to say about this, and their voices are part of a broad range of discussion in milk history.

Basic food, good health, human growth—such were issues that endowed milk with its modern job of sustaining an aura of goodness and purity in Western society. Milk's history shows how governments across Europe and North America, carried along by the trauma of World War I, recognized the political need for an abundant, affordable supply of milk. How else could populations thrive, if not through its newly discovered (or rediscovered) essential elements? A strengthened branch of science, nutrition, made a convincing argument for its daily consumption by everyone, not just the young. Celebrated by school advocates and health workers, the aura of milk was a powerful force in raising the commercial product to the status of universal commodity.

Business interests, along with technology, feature as important players in milk history. The market had served and exploited milk's aura for centuries. The fact that cows were kept in lactation during winter months as early as the fifteenth century suggests that dairy producers were responsive to the demands of the market. (The price of winter milk, five times that of seasonal milk, suggests how indispensable some consumers believed the product to be.) Entrepreneurs invaded and shaped activities in the dairy when they devised ways of preserving and transporting milk. And consumers responded by developing a dependency on milk that became habitual as soon as dairy farms surrounded big cities. Ice cream and candy added to the demand already made by cheese and butter on huge quantities of liquid milk produced in Europe and America. As technology developed, distributors and marketers seized upon standardized glass bottles and refrigeration as necessary aids in bringing milk to the people, literally to their doorstep. By the middle of the twentieth century, farmers were forced into economies of scale that produced a superabundance of milk. Advanced cattle feed, new chemical fertilizers, artificial insemination, and growth hormones would carry the logic of high output into contemporary times.

And yet the aura remains. When I asked a class of graduate students of nutrition what associations they had with milk, one answered "bowl of cereal," and another, "mothers." No matter how much milk has been subjected to familiar modernizing forces, from the laboratory to the factory, it still commands a powerful aura that drives its story. So the discovery that cows with names are actually better producers of milk rests on a longstanding tension in history. The most modern dairy farms now rely on sophisticated technology that Karl Marx would recognize as ruthless agents of alienation. At a small dairy farm outside Bern, Switzerland, I watched a robotic milking apparatus attach itself to a cow that had, of its own accord, walked into a closet-like station inside the barn. Cows don't like to loiter with full udders, so Flora, Bea, and forty other cows could be expected to queue up patiently when their time for unburdening approached. Certain others also came around in anticipation of the tasty soymeal treat offered at the end of the milking session. The early birds lumbered off when the farmer in charge recognized the usual culprits and shooed them away.

Along with their names and personalities, the cows bore serial numbers associated with something akin to modern-day medical charts. On each collar hung an expensive battery-powered identification sensor, known as "radio frequency identification," or RFID, which alerted the machinery when a cow stepped into the milking closet. Then other things went into play, once milk began to flow: automatic measuring apparatus recorded output per teat, detected the presence of pus and infection, and sensed when the udder was empty. Such tags are becoming mandatory so that inoculation programs combating BSE (bovine spongiform encephalopathy, or "mad cow disease") and other livestock diseases can be closely monitored. For a farmer struggling with labor costs and government regulations, new tagging technology can mean daunting expense, along with added administrative tasks. The farmer in charge of the Swiss dairy, Martin Tanner, reported that he spent time each night examining the day's computer printouts, another sobering reminder that machines for dairying, like those for housework, can produce more work for the people in charge.

Yet Tanner and his wife found time to construct a virtual domestic haven around the raw milk they offered for sale outside their suburban farmhouse. Alongside the dispenser, which they replenished with each morning's milk, stood a tall refrigerated vending machine not unlike the kind found in Automats of the 1960s. Jars of homemade preserves perched on revolving shelves, above coffeecake and jellyroll redolent of vintage kitchens and gingham-checked tablecloths. When it was time to sit down for discussion of the business of dairying, Martin invited us inside, where his wife brought out plates of the cake, served with adorable cow napkins embossed with Swiss flags and edelweiss along their margins. Her role as business partner immediately became evident from our conversation and extinguished any suspicion we might have had that she was somehow confined to the kitchen. Despite the high-tech robot in the barn nearby, the mental labor of managing a dairy was still the shared enterprise that it had been in earlier renditions of milk history.

Modern dairying owes much to the past, but what is so surprising about the story of milk is its power to reinvent itself. Every chapter of its history has involved a kind of cosmic drama, in which the magic or potency associated with the liquid food becomes urgently at issue. Milk has never meant quite the same thing in any two phases of its history; yet its value has been felt acutely enough to impel journalists to write inflamed articles, or to inspire advocates to organize collectively and even break the law on its behalf. Much can be learned from the humble, homely liquid; this, I hope, becomes apparent in the following pages.

PART 1

The Culture of Milk

CHAPTER ONE

Great Mothers and Cows of Plenty

According to the great Hindu story of the Churning of the Ocean, milk assumes a pure and simple guise as a limitless source of bounty. The tale begins with the quest for an elixir of immortality, when Hindu gods took charge of a still chaotic world and decided to stir things up, literally. Using a snake as a rope and a mountaintop as a churning stick, they pulled and writhed as the sap from plants from the mountain mixed with water from the sea. As the swirling progressed, the ocean water turned to milk and then—following laws of an ordinary dairy—butter. From a rich, congealed mass emerged the sun, moon, and stars, along with Surabhi, the Cow of Plenty. Her offspring have assumed sacred status as four-legged carriers of perfection and reminders of this extravagant genesis.

In the beginning, there must have been milk. As first sustenance of all mammals, not just humans, we know that milk has been around for a very long time. There is no question that it was valued as a building block of civilizations. Ancient reliefs offer immediate proof in depictions of cows being milked and maternal figures nursing infants. Milk was considered so fundamental in ancient Egypt that its hieroglyph was similar to the verb "to make."[1] Yet the history of milk is one of profound complexities rather than simple truths. Though ubiquitous, it was also scarce. Though utterly familiar, milk possessed mysterious powers. Given these two paradoxes, we should not be surprised by a history that generated contradictions and even conflict that qualify as both mundane and cosmic in their significance.

Scene from Hindu mythology depicting the Churning of the Primordial Ocean of Milk in order to obtain Amrita or Amrit, the nectar of immortality, in a relief from the eighth century. In areas of southern Asia, churning was performed by wrapping a rope around a stick placed in the center of the churn; laborers then alternately pulled the ends of the rope from opposite sides. (Virupaksha Temple, Pattadakal, Karnataka, India; photo © Luca Tettoni/The Bridgeman Art Library International)

Milk was far from plentiful at the dawn of time, according to the human imagination. In most myths of creation, other elements — primarily water — precede its appearance and remain in abundance on earth. In the case of the Fulani myth of western Africa, the world emerges from a single drop. In Norse mythology, a single cow is responsible for feeding human ancestors. A principal exception to this rule of scarcity is the Hindu legend of the Churning of the Ocean; it is the only account that arrives at a relatively harmonious outcome, a point that will later seem mythical indeed as we learn more about the global distribution of milk throughout history.[2]

Ancient religions communicated the universal desirability of milk through their stories and symbols. Milk meant survival, replenishment, and fecundity. Mentioned in Dionysian sacred rites of the ancient Mediterranean, the very idea of milk — along with the idiomatic expression of "milk and honey" — belonged to an imagined abundance of a messianic age, when a "universal mother shall give to mortals her best fruit" and "cause sweet fountains of white milk to burst forth."[3] On a more daily basis, the most powerful deity of the ancient Near Eastern world, Isis, served as the primary vessel of this universal form of sustenance. A stately seated figure, her breasts exposed, Isis was commonly presented as the "giver of life" in the act of nursing an infant pharaoh. Though her husband, Osiris, also possessed a vital organ, Isis outlived him, making copies of his lost phallus (a casualty of his final battle), which she then distributed for use in worship. Her primary job as mother, maid, and matron, however, required that she protect young women and cheer on mothers and midwives in their jobs sustaining human life. The demand for her presence spawned a virtual cottage industry of reproductions in the form of figurines, amulets, lamps, and funerary monuments bearing her image. The Great Mother in her various guises, her two prominent breasts prompting people to hope for plenty, ranked as the most popular of all the ancient goddesses.[4]

As a symbol of salvation, milk occupied an important place in popular religious festivals. Jars of milk surrounded the tomb of Osiris, husband of Isis, on the island of Philae. Priests gathered there to sing and pray, solemnly filling a libation bowl, one for each day of the year, with milk. For the common Egyptian, visiting this site constituted a sacred pilgrimage that could illuminate a lifetime. Another important ritual took place on the banks of the Nile, when Isis herself made a pilgrimage: priests and celebrants would transport her image to a ship in order to surround her with votaries before launching her, pilotless, into the sea. Herodotus noted that the Greeks borrowed festivals and processions from the Egyptians, providing another path of milky rituals across space and time. An account of a lengthy ceremony appears in Apuleius's celebrated story, *The Golden Ass,* which historians trust for its detailed description of elaborately attired people, priests, and animals who turned out for the event. In a long procession, priests sprinkled milk from a golden pitcher, and, just before shoving the ship from shore, poured milk upon the waves. It is not simply coincidence that this tale of transformation of Lucius into an ass and back again took place in the spring, a season replete with rose blossoms and an abundance of new milk.[5]

Isis, Mother Goddess of the World, sporting her signature cow headdress and nursing her son Horus, in an Egyptian bronze sculpture, possibly Late Period (ca. 712–332 B.C.E.) (Musée Municipal Antoine Vivenel, Compiègne, France/Giraudon/ The Bridgeman Art Library International)

As the source of the milk of life, Isis stirred deep emotions and a strong sense of identification in followers. It may be difficult for the modern-day Western reader to imagine a heartfelt attachment to a deity styled as a cow goddess, replete with horns on her head, yet it was exactly this vivid materiality that made Isis earthily available to her supplicants. How else can we explain the gigantic expression of a Mediterranean-wide adoration, which must account for the eighteen-ton granite statue (recognizable by its breasts) pulled out of the Bay of Alexandria in the 1960s?[6] Images of Isis, defying containment, acquired attributes from other protectress goddesses and their animals — bulls, cows, griffons, sphinxes, and snakes, to name just a few. Popular demand must have driven the design of icons, evident from the "peasant woman" statue, an Isis who sits on the ground, one leg bent casually, while nursing a baby, an endearingly human form who must have been reenacting an ordinary pose for her followers. At the other end of the spectrum stood the astonishing Artemis of Ephesus, a Greek adaptation of the goddess, who bested the original by offering several dozen breasts. Some art historians have contended that these globular appendages were meant to be bull's testicles, which ancient people nailed to totem-like figures as tokens of fertility. But ever since Isis acquired her first heavily endowed chest, followers have voted in favor of hyperfemininity. Comments by contemporaries referred to her breasts, and subsequent copies of the statue added nipples to make the point even more aggressively. Fountains built in her image centuries later realized the fantasy that lay at the heart of "Beautiful Artemis" by assertively spouting endless streams of liquid into the air.[7]

The spread of Isis worship across the Mediterranean spawned ardent rivalries and even physical conflict. Young women often donated cloaks to her Greek counterpart, Artemis—it was a common practice in Greek religions to provide all goddesses with new clothes, particularly at festival time—as gratitude for her help with conception and childbirth. The practice went too far, for sumptuary laws eventually strove to curb the competition and limit the yardage laid at her feet. In the Greek city of Sardis, where a cult formed around a temple dedicated to Artemis, local pride welled up in the fourth century B.C.E., when followers from Ephesus arrived with a shipload of garments. A group of defenders physically attacked the intruders; we are not told if the participants were male or female, though it seems likely that women were involved. The scuffle ended with death sentences for forty-five Sardis, suggesting that central authorities in Ephesus might have intervened. In view of this adaptable goddess's universality, such conflicts serve as forceful reminders of local attachments, proof of the investment adherents had made in making the religious site their own.[8]

Nevertheless, the milk goddess migrated across the Near East while aspects of her identity surfaced repeatedly in other religions: as Ishtar, the "benevolent cow" of Mesopotamia, who suckled princes; as Hathor, the cow goddess so prevalent in Egyptian iconography; and as the Virgin Mary of Christianity, who displayed more than one attribute borrowed from the ancient Great Mother archetype. But as Christian traditions eclipsed paganism, the notion of an all-powerful mother fell on hard times. Religious authorities found such popular materiality, with its earthy female presence, intolerable and even repugnant. Images of women "giving suck" were struck down by Jewish prohibitions.[9] The fathers of the early Christian Church also moved decisively against what was decried as pagan: divine cow images were zoomorphic rather than suitably philosophical and abstract. Later on, medieval monks would inveigh against vulgar identification with the animal world. Their account of "Woman Devoured by Serpents," a deeply misogynist tale, turned the friendly earthbound creatures associated with Isaic ceremonies into hideous attackers. Women who gave suck to all animals would meet the same end.[10]

But what about the milk of salvation? As Gail Corrington has shown, the liquid of promise proved to be a durable commodity over centuries of religious history. The subject will resurface in a later chapter, but it is worth tracing its immediate fate in the ancient world. We know for a fact that the concept survived the reaction against polytheism, but under new auspices: it was transferred into the realm of male jurisdiction. Christ and the Church, not Mary, would become the official purveyors of milk. The breasts of Isis had signified nourishment; Mary's breasts now were said to represent the institution of the Catholic Church. Early Church Fathers wished to elevate the Virgin Mary to the level of an abstraction, along with the act of nursing and even her milk. As a metaphor, the heavenly liquid represented "logos" — the word of God — rather than food for all.[11]

But popular belief and iconography seldom took orders from central authority. If we examine Christian art in the early medieval period, we can see signs of Isis's survival in the infant Christ's apparent enjoyment of his delectable drink. Sitting on Mary's lap, the baby raises his finger to his lips, as though taste-testing the milk of salvation emanating from his mother's breasts. In fact, an infant with a finger in his mouth was the ancient sign of Horus, son of Isis, who remained a celebrity among pagans during the time of Christ.[12] Were artists deploying images of an old favorite as a way of promoting popular identification with Christianity? Or did this pose indicate that, no matter how rigid a stance Church Fathers might take, the consumer of milk might have the final say?

The more familiar profile of milk handed down by Greco-Roman mythology is one of absolute scarcity. Milk belonged to the gods; it was the elixir of immortality. One sip from the breast of the chief female goddess,

The Origin of the Milky Way, *painted around 1575 by Jacopo Tintoretto, captures the moment at which Juno awakens as Jupiter holds his illegitimate infant son, Hercules, to her breast in order to obtain the gift of immortality from her milk. The tiny droplets sprayed heavenward created the wide band of stars known as the Milky Way; those that fell to earth sowed lilies. (Photo © National Gallery, London/Art Resource, New York)*

Juno, could confer divinity and an endless life. Mortals might remind themselves of this fact by gazing at the heavens. According to legend, the stars of the Milky Way — plentiful but unobtainable — represented droplets of Juno's milk, scattered when Jupiter stealthily installed his illegitimate and mortal offspring, Hercules, at the breast of his sleeping wife. The infant's energetic sucking woke the goddess and she pulled away, startled. But it was too late to avoid a divine mess: her let-down reflex sent milk spraying into the heavens, where the fluid congealed into stars, and down to earth, where it sprouted lilies. Thus, one of the earliest interventions into maternal milk production, despite so little cooperation, results in spectacular universal distribution.

In truth, milk remained scarce in the ancient world, bounded as it had to be by gestational cycles of mammals (including human mothers), the suitability of various climates and geography, and the simple fact of its perishability. Ancient peoples did not drink milk, at least not ordinarily, but they were intent on collecting it for purposes of adding it to grains, separating out butterfat and heating it to make ghee, or turning it into cheese, the only way of preserving the nutritious liquid. Its use in religious ceremonies suggests a value above the ordinary. In terms of convenience, daily nutrition, and caloric content, the "first nurture of humans" in Classical Greece, according to Phyllis Pray Bober, were figs and acorns, "one sacred to Zeus, the other to Demeter." Homer's *Odyssey* referred to barley as "the marrow of men." And as far as a universal beverage existed, the ancients drank beer, not milk, which historians have identified as "the first-born invention of culture" in Mesopotamia.[13] Milk remained a seasonally available commodity that was associated with pastoral life and a very particular understanding of nature.

For philosophers of the ancient world, milk *was* nature; that is, milk and its properties provided a laboratory for witnessing the very mystery of life. When Aristotle set forth his explanation of how a foetus came into being inside the uterus, he turned to the analogy of curdling milk to make cheese, likening semen to rennet as an activating agent:

> When the material secreted by the female in the uterus has been fixed by the semen of the male (this acts in the same way as rennet acts upon milk, for rennet is a kind of milk containing vital heat, which brings into one mass and fixes the similar material . . .) — when, I say, the more solid part comes together, the liquid is separated off from it, and as the earthy parts solidify membranes form all round it.

Aristotle's cheese-making analogy was not unique to the ancient world. In the Book of Job, which experts believe grew through the accretions acquired over time, the same folksy image comes to the rescue when Job himself asks the Lord, "Hath thou not poured me out as milk, and curdled me like cheese?" The emergence of cheese from milk provided a dramatic demonstration of creation, different from fermentation and rotting, two other processes that would have been wondered at for their embodiment of mystery and potential.[14]

Human breast milk also inspired wonder, though Aristotle once again did much to dispel the mystery by invoking a homely culinary metaphor.

According to his theory of concoction, milk was blood in a different form, "cooked twice." When cooked yet again by the superior heat forces of men, blood changed into semen. But more than just a stovetop ingredient, breast milk was understood to be alive and active. Through its intimate relationship to blood, it was capable of conveying the very traits of its mother, including disposition and intelligence, to the ingesting infant (hence, the considerable concern of Romans in choosing appropriate wetnurses). Its curative powers were near miraculous and many: infertility, eye ailments, the fragility of old age, among others. Not only babies had a claim to this highly potent fluid, which was dispensed by the dropper-full and must have brought a fair price on the market.[15]

Yet despite its wondrous impact on human well-being, the female breast and its milk helped to situate women a short step away from the beasts. In Aristotle's *Historia animalium*, female humans took their place in an eclectic list that included cows, sheep, horses, and whales as viviparous animals who nursed their young. Greek and Roman legends told of originating human characters who obtained milk from the wild: Zeus, who, with the help of some nymphs, nursed from a goat named Althea; and Romulus and Remus, founders of Rome, who nursed from a strangely cooperative she-wolf. The ancients did not draw a clear-cut boundary between beasts and human beings; having sex with animals, for example, would not have shocked their sensibilities, and neither would the exchange of animals and people at the breast of mothers. The human connection to milk-giving drew women into a liminal universe, where their natures became especially vulnerable to the siren call of wild.[16]

The effort to grasp the nature of milk was thus inextricably tied to a similar effort to comprehend the nature of women. The prevailing view of the female gender in the ancient world depended on a hierarchy distinctive to the age. Differences between the sexes were not just a matter of differing degrees of "hotness" of men, mentioned above in relation to the superiority of semen to milk. (Though in a later century, Galen would construct a rigid hierarchy out of the gender difference of "hot" and "cold" and add an interesting feature — horns — to his description of the uterus. An indirect tribute to Isis?) For the early Greeks, whose views on the female sex were constructed around the story of the first woman, Pandora, another explanatory model ruled the day. Like Pandora, women were said to be unruly and potentially destructive. In biological terms, their condition was

According to legend, Romulus and Remus, founders of the city of Rome, were suckled by a she-wolf, as seen in this bronze sculpture from the Capitoline Picture Gallery, Rome. Nursing directly from animals was well established in ancient legend, and knowledge of myths may have helped to condone the practice in later ages. Ancient authorities believed that in matters of feeding the young, the distinction between women and other viviparous animals was not especially great. (Palazzo Conservatori, Rome/Alinari/The Bridgeman Art Library)

based on Hippocratic theories of bodily fluids characterizing them as "wetter" than men and thus closer to the state of nature. Their repeated fluxes of blood only served to underscore their lower state, reminding Hippocratic writers of the carnality of animal sacrifice. Their soft, porous bodies — their breasts being an outstanding example of "sponginess" — indicated that they were more "raw," less cooked or finished, and more subject to emotional excess and sexual appetite than their male counterparts. Described as a kind of "jar" (the word doubled for "womb" in the Greek), a woman was imagined to contain a tube-like passageway that stretched from the opening of the vagina to the nostrils. (In order to test a woman's unobstructed fertility, a healer could place garlic at one end of the "tube" to see if it could be smelled at the other.) From top to bottom, the female body suggested a thinly disguised reproductive organ with voracious orifices at both ends. The only reasonable response among men, the less volatile sex, was a kind of watchful mastery of Pandora's daughters.

Not surprisingly, then, we find that volumes of Hippocratic writings focus on female maladies related to bodily fluids. To be fair, the ministrations of medical men must have demonstrated compassion as well as a certain amount of anxiety. At a time when worry about fertility was legitimate (a woman would need to give birth to six children in order to have the hope of being survived by at least two), consultations with healers served a larger social purpose. Historians have even suggested that "cures" were shared, communal events, requiring substances that were hard to come by and sometimes administered before public audiences. Nevertheless, the Hippocratic corpus astonishes by its attention to detail and its pursuit of female healthfulness. Ancient authorities displayed mighty valor and hubris in equal parts as they demonstrated a certain confidence in knowing what this state of being might be.

What constituted female health? Ideally, blood flowed through the central female passageway and enabled other processes, such as the production of milk, to proceed as they should. But all too frequently, Hippocratic healers heard complaints that sounded to them like an overabundance of blood, perhaps the result of incorrect flow. A direct connection between womb and breasts gave rise to a kind of hydraulic approach to solving problems. In some cases, milk itself could come to the rescue. For women experiencing difficulty conceiving, the best remedy consisted of pouring milk down (or up) the vagina. To counteract an unusually heavy menstrual flow, authorities recommended placing a cupping-glass to the breasts in order to draw the blood out of the alternative organ.[17] For women plagued by irregular menstrual cycles or violent nosebleeds (seen as identical in origin), a more complex series of fluid applications might help, particularly for those who hadn't given birth. First, sweet wine was poured into the mouth, foul odors applied to the nose, and sweet odors applied to the womb; then a drink of cooked asses' milk would induce purging (along with beetle pessaries), while the vaginal tract was infused with aromatics. The womb, viewed as a living organism with a will of its own (and a sense of smell), in this way might be coaxed into returning to its rightful place.[18]

Women, milk, and nature on one side of a divide, learned texts and the effort to contain them on the other: this time-worn dichotomy, still with us today, played a significant role in the history of milk. For the ancients, the polarity would find embodiment in two different modes of producing sustenance. On the side of nature lay pastoral life linked to herding goats and sheep, which was seen, sometimes inaccurately, as nomadic or root-less. On the side of learned containment were the achievements of seden-tary cultivation of grain crops — hence, our notion of "culture." The op-position carried with it the stigma associated with barbarians, who were ignorant and uncouth, compared with the denizens of civilized society. According to this version of human history, milk was the sustenance of not only babies, who were incapable of consuming anything else, but also unredeemed savages, who lacked a civilized sense of taste and distinction.

The opposition pitting nature against culture and milk against civiliza-tion was not invented by the Greeks. As early as the second millennium B.C.E., the Gilgamesh epic inscribed upon Sumerian cuneiform tablets im-plicated milk in the struggle between "natural man" and "civilized man." Enkidu, a shaggy inhabitant of the steppe, embodied all the features of prelapsarian existence, including a dependence on milk. Discovered by hunters, his appearance initially terrified and confused them: his head was so hairy that he was said to resemble a woman. Yet a female he decidedly was not: Gilgamesh's cagey men employed a harlot to lure him into civi-lized life. After seducing and exhausting him, she persuaded the beast-like creature to settle in the land of Uruk. His arrival demanded a transforma-tion from savage to sociable: he donned clothes and, more significantly for our purposes, changed his diet. His harlot friend, now the vehicle of his emerging civilized manhood, urged Enkidu to give up drinking the milk of wild creatures. Solid food (meaning meat) must now satisfy his ap-petite. As for beverage, Enkidu adopted the manly practice of quaffing "strong drink," which made him "cheerful" and led him "to rub the hair from his body and anoint his skin." This considerable makeover enabled Enkidu to advance to the privilege of joining Gilgamesh (albeit as a subor-dinate companion) in a series of heroic adventures. For those who prefer to read the subtext, it is abundantly clear that Enkidu was made to do Gilgamesh's bidding through regrettable slash-and-burn victories, includ-ing one against his former acquaintances in the wild. On his deathbed, realizing how he was exploited by the epic hero, Enkidu voiced Rousseau-like remorse for the loss of his Edenic innocence. Yet the reader is led to accept the verdict of the "winners": a vaunting civilization, vanquishing Enkidu's former ways, proved that roaming with animals and imbibing milk was a lesser form of existence, even in the third millennium B.C.E.[19]

Homer deployed the same way of categorizing the difference between barbaric nomads and civilized cultivators in *The Odyssey,* his epic poem originating in the eighth century B.C.E. The contrast is particularly evident in his account of the Cyclopes: these savage, one-eyed herdsmen lived on a diet of dairy products and wild flesh. Polyphemos, the star of the Cyclopes tribe and another famed milk-drinker, was a dedicated pastoralist whose cave was crowded with dairy tools and a surplus of ripe cheeses. Though the men wanted to make off with a load of luscious cheeses and lambs before the monster returned, Odysseus ordered them to watch and wait, hoping for greater largesse from their host. He and his men lay hidden in the dark interior of the lair while the giant, having just returned from his work in the pasture, set about his chores:

> Then down he squatted to milk his sheep and bleating goats,
> Each in order, and put a suckling underneath each dam.
> And half of the fresh white milk he curdled quickly,
> Set it aside in wicker racks to press for cheese,
> The other half let stand in pails and buckets,
> Ready at hand to wash his supper down.[20]

Such a systematic laborer just might offer hospitality, but Odysseus's hopes were soon dashed. Once the Cyclops discovered the men in his lair, he promptly showed his mettle by crushing and consuming two of them like so many morsels of snack food. Herein lay justification for the fate of Polyphemos, who, up until then, might have passed as a gentle vegetarian in peaceful cooperation with his environment. Thence forward, he forfeited any chance of winning credit for his skill in dairying or his wholesome diet, suffering condemnation as a crude cannibal instead.

The fact that the giant captor drank his milk "neat" and had no tolerance for the ambrosial wine fed to him by Odysseus marked him as an alien being, a brutish monstrosity of nature. But more than that, Cyclopian pastoralism, far from being a bucolic utopia, also appeared as fatally flawed by passive inattention to the productive potential of the island. Where were the cultivators, Odysseus noted rhetorically, prefiguring explorers who also sailed westward and ran into human-eating peoples. The Cyclopes tribe failed the crucial test of civilization through not only their cannibalism but their laziness. More than a literary trope of ancient writers, the construct of milk-drinking pastoralists as idle and ignorant was a necessary foil for the elevation of the cultivators as masters of the world.[21]

One final example of the connection between milk and unredeemed nature can be found in the ancients' account of the Scythians, the wild race of peoples inhabiting ancient Ukraine. Our contemporary image of Scythians is usually linked to their skill as horsemen (according to legend, they were among the first humans to master fully the art of riding horses) and their uncanny knack of coaxing milk from their mares. But the ancients had far less respect for these northern peoples, whom they ushered into common parlance as a synonym for barbarians. (In fact, when the German Huns invaded southern Europe, they were first mistaken for Scythians.) Herodotus identified several groups of wanderers north of the Black Sea, some of whom did, in fact, cultivate grain; nevertheless, he classified them all as nomadic, which, in other words, meant barbaric. He labeled the Scythian an "eater of meat" and a "drinker of milk" — definitive signs of a rootless, ad hoc existence.

Like Enkidu, Scythians inhabited barren steppes and, like the Cyclopes, they followed no laws, eating human flesh when necessary. According to legend, they mixed the blood of their victims in combat with the milk of their horses, a trademark celebratory cocktail unrivaled by the brews of squeamish flatlanders. Their craving for large quantities of mare's milk required the assistance of large numbers of slaves, whom they were said to blind before enlisting them in dairy labor. Horses have been known to be notoriously difficult to milk, which may be the reason behind a rather bizarre feature of the Scythian approach. In order to coax mares to let down their milk, according to Herodotus, the Scythians

> thrust tubes made of bone, not unlike our musical pipes, up the vulva of the mare, and then . . . blow into the tubes with their mouths, some milking while the others blow. They say that they do this because when the veins of the animal are full of air, the udder is forced down. The milk thus obtained is poured into deep wooden casks, about which the blind slaves are placed, and then the milk is stirred round. That which rises to the top is drawn off, and considered the best part; the under portion is of less account.[22]

Was this a demonstration of arcane bestiality or technological ingenuity on the part of Scythians? Or was it simply the imagination of the Greeks, working overtime to come up with a picture of the Scythians that clinched their identity as rude and barbaric? We can only wonder how the canny steppe-dwellers performed their job in the wild, no less if they were blind! And whether or not the account was true, we can enjoy the irony of Herodotus the historian, forced by his subject matter into the sub-disciplines of animal gynecology and labor management, the alpha and omega of dairying—a rude preoccupation for a learned man.

Hippocratic writers shared their contemporaries' fascination with Scythians: here, to their delight, they identified a test case of the ill effects of excessively cold humors. They described these peculiar barbarians in terms of their flaccid bodies and slothfulness, manifested in their sexual impotence. Though the medical men never treated the famed horsemen to physical examinations or interviews, their knowledge concerning milk helped them diagnose the Scythians as suffering from "excessive liquidity of their bodies," caused by their diet of dairy foods and worsened by "their continual horse-riding, which renders them even more sexually impotent." Applying humoral theory to both sexes, they determined that a voluminous intake of mare's milk and horse's cheese led to fat and sterile women and despondent and lazy men. So enervated by their condition, Scythian males sometimes castrated themselves and "felt an irresistible urge to carry out female tasks."[23] Through a strange twist of conceptual associations, uncivilized nature led to effeminateness, rather than to brute force. And so the pastoral race in the north became locked in the closed circle of custom, wedded to benighted animal husbandry, equine dairying, and matriarchal tendencies.

A Sumerian relief carved nearly five thousand years ago reveals much about the production of milk in ancient times. The narrow stone tablet, known to archeologists as "the dairy freeze [sic]," shows dignified cows in profile, each limb and feature carefully indicated. Hunched at their hind quarters are men at work, their heads tucked covertly beneath elegantly tassled bovine tails. Key information offered by this intimate milking scene appears at the front end of each cow, where ancient *techne* is on display. Young calves nuzzle their mothers' necks, prompting the physical reflex releasing the flow of fluid from the cow's udder. Employing stagecraft worthy of Shakespeare, Sumerians thus seduced their animals into a seasonal surrender of milk.[24]

So accustomed to thinking of milk as a natural product donated happily by domesticated animals, we often fail to realize how each step in the history of dairying represents a triumph of ingenuity. Milk, according to nature, belongs to a tightly defined relationship between female parent and offspring, and the success of any attempt to tap into the flow depends on considerable determination and know-how. Both of these factors — intentionality and technical skill — are highly contingent: details of time, place, and circumstance can alter the likelihood that milking animals and consuming milk will ever take place at all.

It just so happens that the question of what humans were doing with their goats, sheep, and cattle matters a great deal to the way we construct explanations of the first appearance of agriculture. Until the past few decades, archeologists mostly adhered to a "big bang" theory of settled agriculture, believing that some ten thousand years ago, a revolution of profound significance took place during the Neolithic era in Mesopotamia. In the Fertile Crescent, the "cradle of civilization" familiar to schoolchildren, human beings abandoned their method of haphazard hunting and gathering of food to follow a more purposeful existence: they began to plant grains, such as rye and barley, from seed, and they tamed livestock — oxen, goats, sheep, cattle — from the wild and made systematic use of their strength, meat, wool, and milk. This determined way of life took place in villages and "urban centers" of the Fertile Crescent and later spread outward to southern and northern Europe. The revolution brought civilization as we know it: settled life linked to steady improvements in the production of staple goods and, eventually, the birth of trade and outward expansion. Changes in agriculture thus provided history with a fundamental turning point, one that could be used as a touchstone for understanding the emergence of civilization around the world.

When this neat theory was elaborated by Gordon Childe, the titan of British scholarship on the subject, archeology was in its infancy. Judgments depended heavily on the cultural outlook of western Europeans, inclined to see light where there was progressive agriculture and darkness where populations had not evolved enough to move from the haphazard to the systematic. Educated in the values of classical culture, Childe and his contemporaries mobilized a perspective not unlike that of Homer and Hippocrates: signs of settled agriculture ranked above pastoral nomads, and so cereal cultivation, along with its products, grains, bread and beer, were of greatest interest. In the words of one recent archeologist, the whole enterprise was driven by the search for "farmers like us." (We might add "men like Gilgamesh.")[25]

The history of milk and dairy products posed a special problem for this set of expectations. Childe assumed that settled pasturage appeared alongside the first villages in Mesopotamia. But such civilized animal husbandry suggests an advanced set of practices, far beyond the extractive relationships developed by nomads and pastoral people, who depended on animals in an ad hoc way for meat, wool, and milk. In fact, archeologists now doubt that systematic animal husbandry developed all at once alongside settled agriculture. They have found that in agricultural habitats in the Near East and southeastern Europe, the production of "secondary products" such as yogurt, cheese, and butter occurred at least two thousand years after the appearance of agriculture. But the sequence did not apply to every location. In Britain, the two systems of production arrived simultaneously, and dairy products were of greater importance to diet than grains. Scrambling the puzzle even more was the discovery that in some places, "hunters and gatherers" were capable of cultivating fields of rye before ever settling down in villages. And in other settings, villages came into being before inhabitants began to cultivate crops. The old model of a transformation of agriculture failed to hold up against so many variations, and none of these findings explained how dairy production could be fit within a clear phase of the evolution of civilization.[26]

The "beginnings of dairying" in fact heads a current list of riddles in need of solving by archeologists. Thanks to innovative use of stable carbon isotope analysis, they now know from pottery sherds drawn from excavation sites that dairy fats were used regularly in cooking much earlier than previously assumed. One hypothesis is that early "civilized" peoples were not wholly settled and instead followed a mixed form of economy, relying on some nomadic pastoral activities while cultivating grains. Archeologists also admit that their assumptions may lack sufficient sensitivity to other worldviews, which may have regarded animals such as cows as kindred spirits, rather than exploitable milk machines. The ubiquitous presence of dairy fats in smashed pottery may indicate the use of milk in religious rituals rather than in cooking, for example. Given what we know of Isis and Osiris worship, this seems perfectly likely. Even the term "milk" deserves reconsideration. In Egyptian records, its use in the idiomatic expression "milk and honey" more accurately refers to "fat" derived from animal milk—bad news as far as rhetorical elegance is concerned, but a step toward greater accuracy in describing the material life of biblical Egypt. And in ancient Britain, the home of twentieth-century Oxbridge theories of civilization, archeologists have uncovered areas where cereal crops were devoted to religious sacrifice, while pastoral activities of tending animals provided the main means of sustenance. This is certainly the ancient world of production turned upside down.[27]

The mystery of milk in "prehistory" becomes more intriguing when we examine the earliest written records that give us the vocabulary, as well as the "administrative lists" (records of crop production), of ancient Near Easterners. These include a combination of cuneiform tablets from the Uruk III period (roughly 3100 to 3000 B.C.E.), combined with early Sumerian, Babylonian, and Hebrew sources. Fresh milk, first of all, constituted a rarity linked to the "suck" of babies and newborn animals, or as offerings to the gods. Common milk would have been soured—literally described as milk that "sits" for a while—and then turned into butter (always clarified, or boiled, to create ghee) and cheese. Another product, dried balls of buttermilk, or "milk cakes," was produced by placing the remnants of churned sour milk in high places, possibly on rooftops, to dry. (Their cuneiform symbols depict milk as a female cattle head spatially elevated in the texts.) These chalk-like lumps probably functioned as a form of instant milk, a way of storing protein for use throughout the year. Archeologists have unearthed petrified chunks, some of which have holes in the middle, suggesting that they were strung together and hung to dry. Along the Euphrates and in Kuwait today, such cheeses now go by the name of *bagel,* though across much of the Middle East, the modern name is *kisk.*[28] The earliest dairy workers, evidently adept at deriving knowledge from trial and error over lengthy periods of time, showed much ingenuity. Partly owing to their success (and that of their local game hunters), the protein consumption of ancient Mesopotamians reached a historical high.

The written record, as sparse as it is, nevertheless is haunted by the conceptual prejudices we have encountered in ancient epic poetry: the wilder side of milk is demeaned, or altogether left out of the picture. Cuneiform text writers gave no mention of rennet, for example. As a segment of lamb's stomach, the material would have been considered part of the domain of nomads and thus suffered from association with life in the hinterlands. (The authors of cuneiform tablets were Babylonian "urbanites" who, not unlike their modern counterparts, looked down on people from the hills.) The curdling process necessary to make true cheese would have required something like it, but perhaps Sumerians, like the Greeks after them, chose to use fig juice instead. Other usages support the idea that writers avoided picturing rennet because they perceived the shepherd, who sometimes wore a piece of rennet on his belt, as a repulsive outlier. In a surviving Sumerian-Babylonian dictionary, the word for "dung" (*kabu*) is used suggestively in ways that might mean "rennet." A word meaning smelly cheese — *nagahu* — is also used as a slur against an "uncivilized person."[29]

Current issues like climate change and dietary health have also made their mark on the early history of milk. Research on East Africa has shown that pastoral agriculture responds to every alteration of climate, and without sufficient rainfall, milk-producing animals do poorly. Until major climatic shifts in East Africa and the Near East, which correlated with a turn to more extensive dairying, inhabitants were wiser to concentrate on the meat, rather than the milk, of their animals. This deals one more blow to the evolutionary model that once ruled the field. Modern awareness of lactose intolerance has also provoked new questions about dairy production. The ability to digest milk has been shown to correspond positively to distance from the equator, so Near Eastern and Mediterranean populations would have found ingesting raw milk a decidedly unpleasant experience. This was less true, probably, for soured milk and secondary products, which contain lesser amounts of lactose. But the fresh liquid might well have gained its reputation as a purgative because of such intolerance. Northern populations, meanwhile, might have evolved as lactose-tolerant for a clear reason. In latitudes where sunlight is limited in winter months, the lack of vitamin D (manufactured by the skin when exposed to the sun's rays) would leave people vulnerable to rickets. By consuming dairy products, which are rich in calcium and offer a fair supply of vitamin D, those able to digest milk and cheese would have had an edge in survival. Fortunately, northern climates cooperated, so Scandinavia, the Low Countries, and the islands of Great Britain registered decisive progress in dairying early in the historical record.[30]

Given the obstacles — cultural, climatic, digestive — to dairying, we might ask what was working in the other direction to entice humans down a path of deeper involvement with milk. What early encounters with dairy foods encouraged them to develop a taste for — indeed, a passion for — milk and its products? Though some might argue that the predisposition developed at the breast would have spelled the inevitability of cheese making and yogurt consumption, we have seen how medical advice might have worked against such developments. Cheese, for example, was seen as an obstructive force in the digestive system, capable of clogging and slowing the human apparatus. Authorities advised consuming it at the end of the meal, as a kind of plug or stopper planted on an especially large intake of food. Yet we know from one of the earliest tracts on agriculture, Cato's *De agri cultura,* that cheese was a common item in banquet fare. Evidently, fears of constipation and dissipation were few, next to the temptations presented by special dishes. What exactly were people experiencing when they ingested milk products in the past?

Roman cheesecakes like the one Cato described in these pages were, in fact, served as desserts in the second century B.C.E. The elder statesman's magnum opus, recognized as one of the earliest works in classical Latin, offers a complete recipe for *placenta,* a cheesecake made of layers of pastry crust and sheep's cheese mixed with honey. (*Placenta* literally means "cake" — from the Greek adjective *plakoies,* meaning full of flakes, along the lines of the French *millefoie.* The uterine sack takes its name from the aspect of flatness implied by the Greek *plak.*) Cato indicated that cheesecakes stood as ideal offerings to the gods, but culinary historians, talented re-creators of ancient dishes, have poured some skepticism on this suggestion. It is more likely that Romans offered the simpler cake, *libum,* to the gods, or else they carted off the cold leftovers to their temples the next morning. Cato's recipe is simply too solicitous of tastebuds to be dismissed as a true final course of a fine human meal.

Contemporary culinary historians perfect their knowledge by devoting many hours to reproducing ancient recipes, and fortunately for the historian of the dairy, the placenta has received its due. Cato's instructions also provide a cornucopia of subtle information — indications of clean hands, clean bowls, and tools like a flour sifter suggest a certain refinement. The recommended quantity of sheep's cheese — fourteen pounds — suggests a certain affluence. The succession of pastry leaves (*tracta*) indicates the demand for considerable time and patience. As our cake mounts in elevation, each tracta separating one layer of cheese and honey mixture from the next, we are given the impression of a confection of architectural magnificence. The final step of baking must take place under a heated "crock," which historians have identified as either a *testum* or a *clibanus,* a baking cover that was buried among live coals remaining after the main meal had been cooked and taken away. (Attentive museum-goers may recognize the clay item by its profile.) Placenta probably took between a half hour and an hour to cook—Cato's command is "thoroughly and slowly." Once out of the oven, the cake was spread with honey. The resulting absorption of sweet within the *peccant* seems irresistible. Here is a hint of how the milk of animals made its way into the food history of humans.[31]

CHAPTER TWO

Virtuous White Liquor in the Middle Ages

What would an English gentlewoman summon up in the way of special fare for her New Year's Day celebration in 1413? The answer would not lie in the brewery. Alcoholic beverages, more reliable than water, were required but not special to such an occasion. On an ordinary day, a person might quaff an average of three and a half pints of ale simply to quench thirst. A wide assortment of meats and game might signal a certain expected extravagance in a holiday menu. But for Alice de Bryene, such offerings were not all that unusual. This prosperous widow of Suffolk regularly called for a procession of geese, rabbit, beef, and mutton (or fish on fast days) in order to meet the demands of a constant stream of dinner guests. Business on her estate brought bailiffs and workmen to her table, along with an average of twenty visiting friends at midday meals. So Alice understood and provided for lusty English appetites for meat on a daily basis. The one extraordinary item on her list of purchases for New Year's Day was milk: twelve gallons of it, much more than the small quantities that were sought after a mere handful of times during the winter months, when milk was in short supply because of its seasonal nature. Her holiday splurge is worth pondering. What became of the precious liquid? And what does it tell us about milk in the later Middle Ages?[1]

Medieval banqueting menus, for which we have actual recipes, help us puzzle through the mystery of the twelve gallons. Frumenty, a wheat porridge made into creamy, pudding-like mush by the addition of milk, stands as a likely suspect. The dish occupies a place of honor among folklorists and medieval culinary historians, probably because it fulfills all of their expectations of old-fashioned spicy reveling. Who could have passed up its seductive and expensive seasoning of saffron and cinnamon? (Never mind that these flavorings were combined with onions and sugar, a clear sign that we are in another gastronomic world.) Yet how easily did this mixture multiply into more than twenty servings? Did Alice keep a hoard of cinnamon in her cupboard? If so, then her trusty steward was unaware of the purchase and we see no record of it in her accounts.

Frumenty might have been the fate of at least some of Alice's twelve gallons, but once we investigate the guest list, doubt creeps in. Though Alice's steward recorded that thirty-odd friends, offspring, and their servants were at her table, the gathering was larger than that. One neighbor brought along "300 tenants and other strangers"—which must account for why 572 loaves of bread were on offer, along with a fair amount of ale. Perhaps the cook devoted the milk to a less special porridge, one known generally as a caudle. This mixture ranked one notch lower than frumenty, but one notch higher than the medieval cook's typical daily "stirabout" of milk and flour. Caudle wouldn't have called for costly spices and it wouldn't have warranted a recorded recipe: it was simply a mixture of eggs, butter, grain (oatmeal or wheat), and milk, according to what was available in the kitchen. This would satisfy a large gathering of neighbors without requiring much ado. Consider the main course on display in Pieter Bruegel's *The Wedding Banquet,* painted around 1568, in which the dutiful groom distributes simple bowls among the expectant guests. Alice, too, might offer simple fare, from what can be drawn from rather spartan menus on record for other days, so we have no reason to expect too much extravagance on New Year's Day. Plentiful porridge, yes, but multiple costly ingredients, no.

Winter holidays in northern Europe were, in effect, peace-making sessions with the forces of nature. Deprivation was inscribed into the calendar, steering eating habits into the realm of preserved or slow-cooked foods, such as stewed meat and dried fish. Animal cycles of reproduction dictated the use of fresh milk. Whether of cow, sheep, or goat, the calving season was in early spring, so milking ran from early May to late September. After that, farmers would take care not to tax their stock before the challenges of the winter months. Yet as early as the eleventh century,

English dairies worked hard to defy obstacles of nature. The demands of cities and towns, along with the market activity that sprang from them, ensured that at least some milk would be available year round. Dependent on the cash that the sale of milk, cheese, and butter would bring, farmers set about devising ways to manipulate milk-producing animals. Calves were coaxed from their mothers and then fed by hand, while the bulk of mother cows' milk was devoted to cheese making or brought to market. Even sheep, prized mainly for their wool, ended up in similar artificial arrangements so that their crafty owners could dictate their milk production.[2]

Here is an instance, then, of milk out of season acting as an indicator of an energetic, calculating commercial society in the Middle Ages. Contemporary cookbooks, a useful if not always accurate gauge, employed milk in 20 to 25 percent of their recipes. This suggests that well-to-do English households were reaping the benefit of prosperity and retailing acumen, two interdependent features of this particular phase of their history. Livestock, as yet, was not particularly productive: annual output per cow in fourteenth-century England (between 140 and 170 gallons) amounted to less than half of what it could have been (and would become in Holland by the 1570s) with better feed. But it was enough to satisfy current demand without too much monetary sacrifice on the part of the buyer. Alice's steward paid eighteen pence for twelve gallons, roughly three times the price of milk during summer months. Yet this was a permissible extravagance for holiday fare at a gentlewoman's household (costing roughly thirty-five pounds by today's standards, or around fifty-one dollars) and a tenth of the total outlay, apart from pantry stores, necessary that day. Our snapshot of communal celebration foretells a well-fed future for the island nation's elite.[3]

Medieval milk nevertheless remained a commodity apart from the ordinary, in a category unlike the one it occupies today. Suggestions of its separate realm went beyond its seasonal flow to other aspects of its identity that were widely known. Medical knowledge classified it as a bodily product related to blood. Religious proscriptions reminded medieval people of its organic nature: as the product of animals, milk, along with cheese and often butter, were forbidden on Christian fast days. These were numerous by the seventh century: three days of the week (Wednesdays, Fridays, and Saturdays), the eve of feast days, and the entire forty days before Easter throughout Catholic Europe. The assiduous cook might

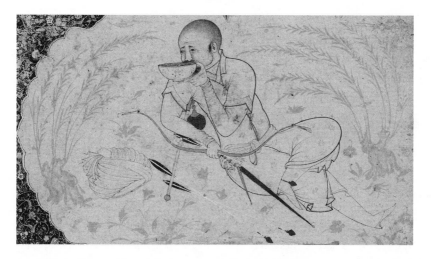

Il-Khan Hülegü, conquerer of Baghdad in the middle of the thirteenth century, perhaps resting from battle, drinks from a cup in a fanciful portrait from the sixteenth century. Though we cannot know whether he is enjoying wine or comos at this particular moment, it was said that his brother's castle boasted fountains spouting endless streams of fermented mare's milk and other intoxicants. (© The Trustees of the British Museum)

substitute almond milk, a rather labor-intensive alternative that involved pulverizing the nuts, adding hot water, and steeping the mixture until it thickened. Most dishes called for milk in small amounts, not more than a half cup at a time. Its role as a beverage was circumscribed in the Western world: few but the infirm or very young would ever actually drink milk from a cup. These customary inhibitions, from our vantage point, call attention to the more important aspect of medieval views of milk: its intimate connection to a liminal arena where religion and the natural world overlapped. Stories of its status in the East only underscored a sense that milk belonged to heroic leaders and the gods.

European powers of the thirteenth century were dimly aware of potential trouble in the East. News of the astonishing military campaigns of the Mongols had reached the emperor and the pope. The might of Genghis Khan, having united Mongolia and conquered China and Manchuria, now swept west, devastating Russia, Galicia, Silesia, and even nearby Hungary. Traditional Western enemies set aside their quarrels and contemplated joining forces. Though contemporaries couldn't have recognized its full significance, this was, in the words of one historian, "one of the greatest catastrophes in the history of the world."[4]

Prompted by troubling alerts, two solemn friars, barefoot and intrepid, set out to meet the enemy. The first among them — Giovanni da Pian del Carpine, known as Carpini — assumed the mantle of papal diplomat charged with the task of opening a channel of communication between East and West. He wrote down his recollections when he returned from his journey in 1247. The second — William of Rubruck — set out as a Christian missionary in 1253 and returned in 1255 as a chronicler: his promise to the king of France "to put in writing . . . everything I saw among the Tartars," without being "afraid of writing . . . at length," resulted in an account of the East of unrivaled significance. His name should be as well known to us as the more flamboyant and loquacious Marco Polo, who visited several of the same sites twenty years later.[5]

In the pages of the Franciscans' reports, the reader is struck by a common recurring fixation: both men report that the Mongols, unlike themselves, were avid drinkers of milk. The friars quickly grasped the significance of milk, given the clues on display at the threshold of every household: on either side hung felt replicas of udders, cows on one side (representing the female realm) and horses on the other (indicating the male), beneath simulacra of human forms made of felt. The Mongols' attachment to the drink called for regular recognition. A container of mare's milk stood at the doorway of every important household, guarded by a minstrel with a guitar. Downing a cupful, the master set off a chain of salutes: a steward on watch nearby shouted "Ha!" and the minstrel vigorously strummed his instrument. On special holidays, the doorway duo repeated their eruptions at the conclusion of festivities, providing a coda, as it were, to the sounds of celebration. Carpini, like Rubruck, noted how princes never drank without accompanying clamor: milk drinking made for noisy pleasure.[6]

In these mystifying moments of early culture contact, milk glittered with power and value. Everywhere Carpini and Rubruck went, they found Mongols busy sprinkling mare's milk. Forced into the role of anthropologists, they carefully recounted bewildering rituals. Carpini showed little patience with the seemingly endless interdictions punishable by death ("they have many things like this which it would be tedious to tell of"), yet he helpfully noted disrespect for milk on the laundry list of offenses.

Along with touching fire with a knife, touching arrows with a whip, and breaking a bone with another bone appeared the seemingly more practical law against pouring milk upon the ground. Lack of sympathy did not hinder careful observations. Viewing their rites through the lens of Christianity, Carpini reported that Mongols aimed no prayers or ceremonies at their "one god." But to their felt idols, they offered "the first milk of every cow and mare." This ensured a springtime bonanza for the stoic miniatures, whose job guaranteeing the prosperity of their makers was indeed a difficult one. The isolated tribes, understood by today's historians to have been abysmally poor, were nevertheless "rich in animals": "They have such a number of horses and mares," Carpini reported with fitting hyperbole, "that I do not believe there are so many in all the rest of the world."[7]

"The milk of three thousand mares daily"—this was the tribute of distinction given to Baatu, grandson of Genghis Khan and the thirsty leader carefully observed by Rubruck. Twenty years later, at the court of the great Kublai Khan, Marco Polo would offer the more extravagant (and even less believable) figure of ten thousand mares. The numbers, such as they were, were meaningless. The Eastern attitude toward their equine companions was either pragmatic or primitive, depending on one's point of view. Mares supplied the only food that the Mongols really cared for: *qumiz,* or *comos.* "In the summer," Rubruck recorded, "as long as their comos (that is, mare's milk) holds out, they care for no other food." Comos ranked higher than goat's or cow's milk, which Rubruck and his traveling partner were given in abundance; the Mongols knew better than to lavish too much comos on their guests. As for cow's milk, Rubruck hardly recognized the steppe variety, compared to what he had known back in France: it was "extremely sour," with all the butter extracted from it, closer to what he might have known as buttermilk, but stronger in taste. The Mongols often boiled it, then dried and stored the remains in bags for winter consumption, when the curds would be mixed with water to make a weak drink. But when mare's milk was in season, Mongols even passed up fresh meat. The steady summer supply of mare's milk took pride of place as the favorite food of these formidable people.[8]

And special drink it truly was: Rubruck provided detailed instructions on how mare's milk was obtained (by male milkers only), contained, and beaten into a fizzing, fermented beverage that made the portly monk break out "in a sweat all over from alarm and surprise" the first time he tried it. "While one is drinking it, it stings the tongue like *rapé* wine," he explained carefully, "but after one has finished drinking it leaves on the tongue a taste of milk of almonds. It produces a very agreeable sensation inside and even intoxicates those with no strong head; it also markedly brings on urination." A cut above ordinary comos was "black comos," a further distillation of the same product, made clear by the separation of solids from the milk. None but the "great lords" consumed this liquor, which Rubruck described as "certainly a really delightful drink and fairly potent."9

Given his missionary objectives in Mongolia, Rubruck felt obliged to attend many social and religious rituals, many of them punctuated by drink. One suspects that even the most bibulous friar would have been exhausted by the Mongol thirst for alcohol-laced socializing. Rubruck's diplomatic endurance on this score might have made possible the privileged audiences he eventually gained with Mangu Khan, where visitor and host exchanged views on the existence of God and other theologically oriented topics. At these events, Rubruck witnessed more milk ways, though by the second round, the friar was considerably less interested in the mystification surrounding comos than in telegraphing information about medieval Roman Catholicism to the benighted Mongols. Never mind, Rubruck seems to be telling us, that four silver lions at the foot of a large silver tree spewed four different kinds of drink, including refined mare's milk. And we needn't bother much with each time the great khan drank his comos, when those surrounding him quickly responded by sprinkling their felt idols with droplets of the same. Attendants must have been mightily active during the friar's final audience, when the Mongol leader laid out the tenets of his faith: during his introductory remarks alone, Mangu Khan drank at least four times. Rubruck's interpreter had warned him in advance that this would be his last conference, and so it was. From religious beliefs, the khan steered the conversation to plans for the future. "You have stayed here a long time," Mangu said to Rubruck; "it is my wish that you go back." When Rubruck finally took his leave, he recorded disheartening statistics: "We baptized there a total of six souls." He did not add that all but one had been offspring of western European captives.10

After the friar's dismissal, the account wears a different, more muted color; Rubruck's disenchantment is unmistakable in the text. This is unfortunate with regard to milk, for the tired author devotes a mere two sentences to the most subsequently studied event of all: the ceremony of the white mares, or *julay,* on the ninth of May. Modern-day anthropologists understand this event as a multifaith celebration of heaven, aimed also at a cast of deities that has multiplied through the ages. To one rather clinical twentieth-century eye, the event seemed "typical of horse and cattle breeding peoples of Central Asia."[11] But great interest lies in its ancient origins, which predate Buddhism and Lamaism. By the time Rubruck witnessed the ceremonies, they were already a laboratory of syncretism, demonstrating that seemingly incompatible beliefs and practices can meld into a single event. Rubruck saw priests of the Mongolian Christian sect called Nestorians gathering at the khan's palace; their job was to sprinkle the first comos on the ground to celebrate the new season. Rubruck likened the event to the dispersion of wine during the feast of Saint Bartholomew, although comparison with a holiday at the end of August seemed out of character with the springtime event. (Probably following the same association, Marco Polo recorded the event as occurring on August 28.) Did the friar by this time understand that mare's milk was the holy water of the Mongols? Or did his irritation with the "soothsayers" — Mangu Khan's special platoon of priests — overwhelm his ordinarily astute powers of observation?

Twenty years later, Marco Polo comprehended the meaning of the julay, garnishing his account with characteristic rhetoric:

> And the astrologers and the idolaters have told the great Kaan that he must sprinkle some of this milk of these white mares through the air and on the land on the twenty-eight [*sic*] day of the moon of August each year so that all the spirits which go by the air and by land may have some of it to drink as they please . . . so that for this charity done to the spirits they may save him all his things, and that all his things may prosper, both men and women, and beasts, and birds, and corn, and all other things which grow on the land.[12]

Comos represented survival, metaphorically and literally. If Rubruck had stopped to review his notes, he might have concluded the same from his own experience. More than once, Rubruck's traveling party failed to locate provisions (inhabitants offered fabrics, but not food for exchange), so they were forced to drink mare's milk for days on end. Wherever they went, it seemed, they encountered devotion to comos. In Russian territory, Rubruck nearly convinced a Saracen to convert to Christianity. But after consulting his wife, the man refused to receive baptism, "since this meant that he would not drink comos: the Christians in this country claimed that nobody who was truly a Christian should drink it, and he could not survive in that wilderness without the drink." Rubruck "was wholly unable to disabuse him of this idea," and he suspected that many potential converts were lost to this misapprehension. Mare's milk and western Christianity belonged on opposite sides of a surprisingly fixed barrier.[13]

What are we to make of this western European encounter with vigorous milk drinking? The impact of Carpini and Rubruck in terms of opinion and knowledge in the thirteenth century was nil. Very few people read these accounts before the sixteenth century, when travel literature and tales of natural wonders provided entertaining instruction for a growing readership. Initially, manuscript copies of Carpini's report were somewhat numerous, but they would have been carefully shelved in monasteries. As for Rubruck's remarkable testimony, the author delivered his treatise to King Louis IX of France, but the monarch did nothing to promote or preserve the text and possibly never read it himself. Thanks to the learned eye of Roger Bacon, the renowned English Franciscan, four copies ended up in England, where Bacon mined them for geographical information while writing his *Opus maius*. Centuries later, Richard Hakluyt, the famous travel compiler, recognized a potential best-seller in Rubruck's plangent tale and published it in Latin and English in 1598.

The texts nevertheless serve as sensitive compasses for the medieval history of milk; both irrefutably point upward to the heavens. Precious, sustaining, and elusive, milk was an accompaniment to conversation with divine powers. Wherever milk was found, we can expect to see a mix of bodily and spiritual concerns in the Middle Ages. Consuming milk in this era constituted a noteworthy occasion, calling for celebration and conviviality. Its presence suggested bounty; its absence marked a period of dearth and endurance. The ways of milk mirrored the seasonal cycle of living creatures before the subversion of natural rhythms by science and technology.

Not surprisingly, western Europeans honored their own variety of sacred milk. In the twelfth century, milk featured as one of the foremost attributes of Christianity, and for the next several centuries, discussion of the virtuous white liquid set the tone for a great deal of popular piety. As it turns out, Carpini and Rubruck, as Franciscan friars, were only exchanging one milky milieu for another when they lifted their cups among the comos drinkers of the East.

According to legend, when Bernard of Clairvaux (1090–1153) sought help from the Virgin Mary, the holy mother responded with maternal sustenance. Appearing as a vision to the monk as he recited the *Ave maris stella* before her statue, she exposed her breast and squeezed three drops of beatific milk into his mouth. This heavenly gift so greatly empowered Bernard that he subsequently defined medieval spirituality for his age, offering hundreds of sermons and treatises on the right way to holiness. Certain altar paintings and prayer book woodcuts from the fourteenth and fifteenth centuries depict Bernard's milky reward *in media res*. (According to recent count, 119 similar lactation images remain scattered across Europe, mostly in the Iberian peninsula and regions of the northwest.) The humble saint, kneeling with head upturned, imbibes a miraculous arc of milk. This startling literalness evokes amusement from modern viewers, separated as we are by seven hundred years from the unmediated physicality of the Middle Ages. Yet the legend and its image speak volumes about the powerful hold of sacred milk on the western European imagination.

Like most legends, this one is born of numerous conflicting sources. A telltale sign of its ambiguous pedigree is its mysterious origins. Bernard never actually narrated such an experience, though he preached and wrote endlessly of spiritual sustenance taking the form of milk—from Christ's breasts as well as Mary's. Historians of religion dispute exactly who claims the honor of being first, after Christ, in a long line of drinkers of Mary's milk. Most stories originate in the twelfth century, when Bernard lived, beginning with the miraculous cure of the theologian Fulbert of Chartres (d. 1028). Yet sick clerics in fact figure as numerous recipients of "the most holy milk," so the question of how Bernard claims pride of place remains. Art historians provide further clues: they have shown that the image of Bernard's arc preceded textual support for the story. Thus, when the church later printed up an authorized version of the story, oral tradition must have had some role in it. Two altarpieces in the neighborhood of a Catalan abbey of Poblet, one dated 1290 and the other 1348, depict the virgin squirting milk at a kneeling Bernard. This scene was to become

Alonso Cano (1601–1667), St. Bernard and the Virgin, *one of many depictions of the moment when Mary endowed the saint with pastoral wisdom in the form of a stream of breast milk, perfectly aimed at his mouth. (Prado, Madrid/ The Bridgeman Art Library International)*

widespread as a devotional image, but not right away and not exactly because of Bernard: its heyday occurred at the end of the fifteenth century, when Mary enjoyed more popularity than Jesus as an object of veneration among common Christians.[14]

One particular donation of milk may explain how the *lactatio* came to be attributed to Bernard. And it also tells us how sacred milk was understood as a special form of sustenance in medieval Christianity. According to a collection of miracle stories about the Virgin that dates from the thirteenth century, Mary came to the rescue of a nervous "unlettered abbot" who was facing a tough assignment given to him by the pope. Henry of Clairvaux, succeeding Bernard as leader of the Cistercian order from 1176 to 1179, evidently lacked the "powers of . . . mind" to take up the challenge of preaching a crusade. Asking Mary for help, "he saw the most beautiful virgin standing before the altar. She called him softly and gently held forth her most sacred breasts to him to suck from them. He with the greatest devotion went to her and through that most holy fluid, which he sucked, he obtained so great knowledge of letters that he received the office of cardinal at Rome."[15] Mary assisted the humble and powerless monk, lending power to his speech and, we might add, his cerebral capabilities. If Christ represented humility, Mary trumped that role by being the helpmeet of the humble. Her milk played an important part in discussions of just how democratic Christianity could be. For proof, we need only examine the fifteenth-century painting in the Palazzo Comunale in Chieti, which shows Mary's breast milk cascading into the mouths of a large population in Purgatory.[16]

In terms of this story, it makes sense that her miraculous lactation should become attached, over time, to Bernard. He, not the diminutive Henry, was responsible for aggressively preaching the dismally disastrous Second Crusade; better to make this man the recipient of such an irresistible offering from on high to ensure a spotless image for posterity. Other medieval images show less renowned friars literally sucking from the breast of the Virgin, but Bernard's signified a special kind of Marian fortification. The arc (sometimes a jet stream) of milk flew directly into his mouth, indicating that the Cistercian would use this sustenance to preach the word. And so Bernard did. He repeatedly spoke about milk as a source of knowledge of the holy spirit: it represented spiritual intimacy, wisdom, and "science divine." This evocative scenario satisfied two audiences at once: the popular taste for earthily immediate imagery, alongside a scholastic propensity for rich metaphor.

The ultimate message of the lactatio, then, was this: if medieval Christians were seeking a channel to knowingness, then let them get milk. Clerics invoked the fluid in its metaphorical sense. Food imagery, along with bodies and breasts, figure prominently in many medieval religious texts and, we may assume, in sermons from the pulpit. The celebration of breast milk should not mislead us into thinking this constituted an upsurge of feminist power in the Middle Ages: both male and female intercessors "lactated," and Jesus was understood to nurse his followers with his milk. Medieval historian Caroline Walker Bynum has shown how twelfth-century clergy regularly mobilized "consistent stereotypes of femaleness as compassionate and soft . . . and that they saw the bond of child and mother as a symbol of closeness, union, or even the incorporation of one self into another." The human breast and milk were vehicles of intense spiritual awakening, promising intimate union with Christ, the saints, and of course the Virgin Mary.[17]

Bernard stands out as a voluble authority on these matters, particularly in his many sermons on *The Song of Songs*. The articulate cleric (nicknamed "the mellifluous doctor") rendered this biblical book, renowned for its blatant sensuality, as verbal lovemaking between the spirit and the lord. Seeking the meaning of the words "For your breasts are better than wine, smelling sweet of the best ointments," he explained that the "bride" (representing the Church) who spoke this passage meant to say this: "If I seem to be high-minded, O my Bridegroom, you are responsible; for you have honored me so greatly with the nurturing sweetness of your breasts, that by your love and not by my own temerity I have put aside all fear."

Breasts were ubiquitous sites of hearts and mammary glands; this intentional blurring of gender distinctions made possible a powerful circulation of sacred and nurturing fluid. Granting the bride her wish, that is, a kiss, Christ the bridegroom endowed her with fecundity, so that she immediately conceived her faith, growing in goodness and love. The bridegroom explained the difference between this spiritual consciousness and a merely cerebral understanding of the word, depicted as wine: "Henceforth you will know that you have received the kiss because you will be conscious of having conceived. That explains the expansion of your breasts, filled with a milky richness far surpassing the wine of the worldly knowledge." Dwelling on the nourishing powers of authentic spirituality, Bernard inserted a more practical explanation of how it all worked:

> Men with an urge to frequent prayer will have experience of what I say. Often enough when we approach the altar to pray our hearts are dry and lukewarm. But if we persevere, there comes an unexpected infusion of grace, our breast expands as it were, and our interior is filled with an overflowing love; and if somebody should press upon it then, this milk of sweet fecundity would gush forth in streaming richness.

The signal property of milk was its potency, its power to make something grow and prosper. Bernard made the most of this ever-rich metaphor, a product too good to be confined to mothers and babies alone.[18]

Sacred metaphors of the body had unmistakable sexual overtones, which point to another facet of the medieval worldview. Physical life was understood to mirror things spiritual and celestial. Medieval listeners wouldn't have judged Bernard as prudish for translating sex into spirit via milk; they simply would have grasped the immediacy of what he was saying in concrete as well as theoretical terms. (For the record, not all of Bernard's milk references employed metaphors related to human breasts: in another sermon, he invoked a rustic flavor when he spoke of salvation having the sweetness of milk and added, "if pressed stronger the richness of the butter may come forth.")[19] Did listeners or viewers feel squeamish about stories of grown men and women drinking from celestial breasts? Classical antiquity had established this trope as the ultimate act of charity, enacted by the virtuous daughter Pero, who nursed her father when he faced starvation in prison.[20] The only dispute attached to the lactatio remaining in the historical record had to do with deciding how much of an ordinary human this made Mary. If she oozed milk like every other woman after giving birth, did she also experience ordinary menstrual cycles? Then her immaculate conception and freedom from original sin, ever debatable, became difficult to defend. (This is why, art historians tell us, the lactating Mary was sometimes painted unrealistically, with a breast emerging from her collarbone area.) Lactation was infinite in its meaningfulness and also very problematic.

When religious authorities praised sacred milk as the essence of spiritual life, they were actually fashioning a theory parallel with contemporary scientific opinion: following the ancients on this subject, medieval natural philosophers argued that milk was another form of blood, and blood was the essence of life. After childbirth, blood that had nourished the infant in utero traveled upward to the breasts and emerged as perfect white nourishment. Like blood, breast milk was personalized in its makeup: it conveyed not only nutrients, but the character traits of its producer. Thus, a child acquired temperament and morals from whatever breast milk he or she imbibed.[21] Bernard's account of his infancy showed that he had a proper grasp of this lesson in biology. He credited his mother, renowned locally for her exceptional charity, as the source of his own zeal. "She was careful to nurse [her children] herself, contrary to the custom of the time among those of her rank, believing that with the mother's milk something of the mother's spirit might be infused." The bond between clergy and mothers, crucial in sustaining the mission of the church, clearly owed something to milk and the breast.[22]

Bernard deserves credit for his role in increasing the volume of sacred milk in the Middle Ages. The ardent cleric was at the helm of a new trend that amounted to a revolution in devotional practices, which blossomed in the twelfth century: the proliferation of stories from the lives of saints. By hearing and reading accounts of saints' experiences, believers engaged their emotions as powerful vehicles to sanctity. These impulses prompted a veritable tide of milk and its attendant fluid, blood, and not only in the metaphorical sense. From the sermons of Guerric, we learn that the Apostle Peter was "filled with an abundance of milk" by the Holy Spirit. When Saint Paul the Apostle was beheaded for his beliefs, he emitted milk, not blood. Saint Victor bled both milk and blood together.[23] In the next century, many such images would be immortalized in *The Golden Legend,* compiled from oral sources in the latter half of the thirteenth century. These popular accounts of miracles often depicted milk as the ultimate litmus test of sainthood. Saint Catherine of Alexandria, one of the most popular female saints of the Middle Ages, replicated the experience of Saint Paul: her neck spewed a fountain of milk at her beheading. Milk appeared as miracle glue in the legend of Saint Cuthbert: it was used to make the plaster for repairing his leg, severed when the young man chopped it off in remorse after speaking angrily to his parents. Such vivid depictions of medieval wonders, concrete rather than metaphorical, received a boost from the advent of print culture in the fifteenth century. *The*

Golden Legend became one of the most frequently reprinted books of the era, supplying formidable competition for Bernard's lactatio.

Women responded powerfully to these invocations. The wish for sustenance and gushing abundance in the form of milk resonated with the daily challenges facing most medieval women. For mothers engaged in "the oldest vocation," according to Clarissa Atkinson, Christian images provided the material from which they scripted their biological and social identities and their passage through childbirth and child rearing.[24] Milk became not only the product of their own breasts, but the ultimate symbol of well-being. As food, milk assumed an important role in religious experience. Women commonly engaged in the practices of fasting and feeding in order to demonstrate their piety. Such rituals also enabled them to seize control of their situations, catapulting them out of the realm of the ordinary into the extraordinary. Countless tyrannical parents and undesired suitors were thus spurned as women focused on the powers of their own bodies. They had no trouble finding role models for their fantasies about milk in the stories of saints.

The theme of so many of these accounts was milky abundance and miraculous survival. Thomas of Canterbury told of the divine salvation of Christine the Astonishing: fleeing through the desert without sustenance, "God made her virgin breasts swell with milk, from which she fed herself for nine weeks."[25] When Lukardis decided to fast to prove her faith, she was said to feed from the breast of Mary, which sustained her for days without recourse to earthly food. Lidwina of Schiedam (d. 1433) became a font of transformative holy fluids, including milk, after falling on the ice and becoming paralyzed. As her body decayed and literally fell to pieces, Lidwina became the focus of local popular piety. Her body parts exuded sweet smells that inspired visions, cures, and other miracles. On the night of the Nativity, Lidwina "saw a vision of Mary surrounded by a host of female virgins; and the breasts of Mary and of all the company filled with milk that poured out from their open tunics, filling the sky." After this vision, "Lidwina rubbed her own breast and the milk came out," enabling her caretaker to drink to satisfaction. Though not quite as abundant as Mary, the lactating Lidwina offered living proof of the connection between milk and supernatural power.[26]

The same could be said of the Irish saint of the dairy and cattle, Brigid, who emerged from the early days of persecution of Christians. (She ranks as second only to Saint Patrick in popularity in Ireland.) From the auspicious moment of her birth in the mid-fifth century, Brigid bore a connection to milk. Her mother, a slave who regularly tended the household's cows at dawn, went into labor as she carried a pail of milk across the threshold. The baby Brigid, represented with "flames playing round her head," called forth one portent after another, including that of special appetites: no food would satisfy the infant until her mother's master set aside a "beautiful white cow" for her sustenance, milked by no one but a Christian woman. Brigid's earliest miracle concerned the simple but sumptuous commodity of butter: after giving away most of the household supply to beggars, she produced endless quantities for her mother's master (who happened to be a magician) and his wife. When it became clear what Brigid had accomplished, the magician offered her a dozen cows, but Brigid demonstrated her trademark pluck by responding, "Keep your cows, and give me my mother's freedom." Her wish was granted and the two women were happily reunited. Legends from Brigid's adulthood continued the theme of charity and often included cows and dairy products as vehicles of her kindness. One of the more common images of Brigid features her with a large bowl, the receptacle of a miraculous production of milk from a single cow. Enthusiasm for the saint was so great in 1220 that the Archbishop of Dublin, fearing charges of heathenism, ordered the flame burning at her gravesite extinguished. Followers succeeded in rekindling and tending it until Henry VIII's reforms led to wholesale desecration of the site and its church.[27]

Milk achieved its highest visibility, we might say, in the latter part of the Middle Ages, when the Catholic Church harnessed the groundswell of devotion to the Virgin Mary. Countless churches and shrines gave tribute to her powers of lactation. The Church of Sant Esteve de Canapost, called "Our Lady of the Milk," displayed an elaborate retable (positioned at the back of the altar in order to present didactic images of saints' lives) featuring Bernard's lactation, dating from the second half of the fifteenth century.[28] Tributes to Mary, public and private, could be found across Europe. When Calvin decried the popular mania for spurious sacred relics in the mid-sixteenth century, he singled out Mary's milk for special mockery. "There is no town so small, nor convent . . . so mean that it does not display some of the Virgin's milk," he complained. "There is so much that if the holy Virgin had been a cow, or a wet nurse all her life she would have been hard put to it to yield such a great quantity."[29]

Lactating saints generated cults of their own and kept milk in the public eye. The Abbey of the Holy Trinity and Saint Catherine in Normandy, claiming ownership of a collection of relics, was at the center of Catherine's following. Her popularity spread to England, where she was one of several saints proclaimed patroness of nursing mothers. Another favorite was Saint Giles, saved by a charitable doe who nursed him while he was living as a hermit in the wild. Pilgrimage trails to the shrine dedicated to the Virgin at Walsingham in England provided followers with an array of milk-related saints. Called "the Milky Way" by a recent historian, the churches en route presented images of Saint Catherine, Withburga (like Giles, nursed by a doe), and Etheldreda, who provided the village of Ely with milk (supplied by two does) during a famine, on rood screens in successive local churches.[30] On the eve of the English Reformation, no less a figure than Henry VIII signed the visitors' book at Walsingham, perhaps hoping for divine intercession in his own reproductive life.

Dreams of abundance have supplied a constant trope of literature, hardly unique to the Middle Ages. But this period was particularly soothed by visions of milky abundance, perhaps because the age was also haunted by dangers associated with its dearth. Especially with regard to breast milk, supplies were always precarious, seemingly at the mercy of some higher power. Yet ingenuity in the dairy promised to satisfy at least part of the yearning for milk and its products. Pastoral life would conjure up its own response to such yearnings, even as early as the Middle Ages.

A story about Charlemagne tells us of an early French triumph, allegedly originating in a dairy attached to a monastery. The celebrated emperor might have been traveling between Paris and Aachen (Aix-la-Chapelle) when he stopped at a bishop's estate for a meal. On this particular Saturday, the bishop discovered that the kitchen had no fish at hand. Given the requirements of fasting, which in this case would not be followed to the letter, the bishop set before his guest a decent cheese, "white with cream." In order to get at the inner portion of the cheese, Charlemagne cut away the rind. "Why do you do that?" exclaimed the bishop. "You are throwing away the best part." With this encouragement, the emperor bit into the snowy crust, to great satisfaction. He pronounced it "like butter," this in itself providing a revealing comparison. The imperious visitor — who was reputed never to like fasting of any kind — then ordered two cartloads delivered to his court at Aachen each year. And such is how a passion for (most likely) Brie cheese was born among the powerful.[31]

The anecdote alerts us to the fact that cheeses, like people, were local in character. And in an age of poor roads and transport, they usually didn't travel very far from home. Like people, their identities existed along a spectrum ranging from common to superior in status. And even superior guests might not know what to do with an unfamiliar type when they encountered it. High-quality products, made by artisanal techniques passed through generations, would be found in particular regions where dairying supplied a primary form of income. They would require a fair number of animals, a sufficient workforce, the necessary equipment, and considerable know-how. Large estates in the Middle Ages were only just beginning to command such products. Yet the snow-white cheese served by Charlemagne's host was not just any cheese: it was well aged and, given its buttery texture, perfectly ripe. In both recipe and technique, the cheese represented something of a culinary achievement. The estates of monasteries, at the forefront of sophisticated cheese making and viniculture, housed virtual laboratories devoted to experimentation and recording of natural phenomena. Cheese was an indispensable part of their dietary regimen. Four times a week, in addition to vegetables, monks were allowed "general" plates of either eggs or cooked cheese.[32] The more elite monasteries such as those run by Benedictines demonstrated a proclivity toward the finer enjoyments of the table, and cheese became one of their specialties. They developed trademark varieties, proven by cheeses today that still carry their names: Champaneac, Chambarand, Citeax, Cluny, Igny, Laval, Munster, Saint-Maur.[33] The supplier of Charlemagne's cheese could have been a monastery, given that his host, the bishop, may have had ready access to the products of nearby abbeys. Much about the story, alas, must remain pure speculation.[34]

Cheese at the humble end of the spectrum was an entirely different thing. Most likely made from the milk of sheep or goats, it might stand in for meat as the main protein, alongside bread, in the midday meal of laborers. Its reputation as a foodstuff was decidedly low. Nearly every peasant household in the medieval period seems to have produced its own cheese: we know this from the financial records of landlords, which show ubiquitous payments of cheese as partial rent. Perhaps landowning house-holds passed the ordinary stuff along, serving some of it to their day laborers and servants. Undistinguished but wholesome dairy products were sometimes wedded to staple dishes in remote rural regions. The Welsh national dish described by William Caxton in the late fifteenth century was a hearty soup called cawl, consisting of grain, leeks, butter, milk, and cheese: a sure sign that the Welsh were ignoring dietary pro-scriptions coming from Rome. Various kinds of cheese reached medieval markets, as peasants entirely dependent on livestock were the first to spe-cialize in production for sale: they had to translate their products — animal hides, meat, wool, butter, and cheese — into credit to purchase the staple crops they were unable to grow on their land. Fortunately for them, mar-kets increased in number and size during the later Middle Ages, though not without significant periods of disruption caused by famine and dis-ease. Trade routes across the Mediterranean, the Baltic region, and around the coast of England meant that some farm produce, including cheeses, traveled great distances even as early as the 1300s.[35]

If we were to measure the success of medieval cheese by distance trav-eled, then a singular star stands out in the firmament: parmesan cheese, hailing from the area south of the river Po in northern Italy, in what are today the provinces of Parma, Reggio Emilia, Modena, Bologna, and Mantova. Historians speculate that Tuscan merchants absorbed this ac-claimed product into their extensive trade networks extending from Pisa to Sicily, northern Africa, and coastal towns in France and Spain. The dispersal of parmesan must have owed, in part, to its relative imperishabil-ity. Highly salted and very hard, wheels of parmesan could have withstood even summertime temperatures across hilly terrain and aboard medieval ships. Here was a product with great commercial potential.[36]

Recorded early in the annals of cheese history, parmesan quickly estab-lished a prestigious pedigree. A ledger book of the Commune of Florence indicates that the Priors' Refectory (that is, the governors' cafeteria) pur-chased some of the cheese in 1344. The coup de grâce, however, arrived during the following decade, when another Florentine, Giovanni Boccac-cio, published a reference to the cheese in his *Decameron*. Wound into a story told to a gullible character named Calandrino, we hear a description of a peasant utopia: "The district is called Bengodi, and there they bind the vines with sausages, and a *denier* will buy a goose and a gosling into the bargain; and on a mountain, all of grated Parmesan cheese, dwell folk that do aught else but make macaroni and raviuoli."[37]

A geography lesson accompanies this tall tale. The narrator, a clever Florentine wittily dubbed Maso del Saggio (roughly, "a mess of wisdom"), tells his listeners that the wondrous land lies somewhere in "the land of the Basques." Poor Calandrino asks in bewilderment if this is farther away than the Abruzzi, cuing us to enjoy the joke at the expense of the wide-eyed peasant. We are thus treated to one of Boccaccio's trademark inventions: humor that rests on an encounter between country bumpkin and city slicker. In tracing the history of parmesan, we can appreciate yet another cue regarding geography: this whimsical cameo appearance of parmesan proves without doubt that by the latter half of the fourteenth century, it had secured a place in a "gastronomic culture . . . beyond the limited frontiers of its production area." Boccaccio would have assumed that his readership, which reached far beyond his native Florence, would have had to know parmesan in order to appreciate the image.[38]

But what exactly did this reference to a mountain of cheese mean? Gastronomic culture drew its elements not only from foods eaten, but also from foods dreamed about. We need to ask: who was eating parmesan cheese at this date? Much of what we know about the cheese comes from inference, but we can safely speculate that only the wealthy could have afforded the specialty item. The price for parmesan was relatively high, for good reason: it was made from the milk of cows, not sheep or goats, and a large number, at that — at least fifty, possibly eighty to ninety head of cattle in order to produce a single wheel of cheese weighing forty-four pounds. Moreover, its production process was labor intensive. The milk was cooked twice — could medieval monks have tried to mimic the body's transformation of blood into milk? — and the curds required copious amounts of salt for a soaking process in brine. These were techniques and supplies not likely in the repertoire of the typical Italian peasant. Furthermore, the cheese was not always an "eating cheese," but one that was evidently grated and added to dishes such as soup or macaroni. Like other items sprinkled on food, such as salt, parmesan was meant to enhance, not overwhelm. These indicators signaled a commodity to be savored by the discriminating palate.[39]

Select cheeses, along with cream and butter, eventually won a place for milk on the luxurious side of the dietary divide running through European foodways. An early notice of this triumph occurred in a thirteenth-century poem, "The Fight of Lent with the Meat-Eater." Two major powers, Meat and Fish, call their citizens to war: ordinary and feast foods thus join battle against the foods of Lent. Dairy products technically belonged in the category of meats, owing to their derivation from creatures of the earth.[40] As the ranks of soldiers convene, fresh cheese, butter, and cream assemble alongside large and small sausages. Their adversaries are Lenten fish and vegetables. The foods of abstinence are no match for the hearty, blood-strengthening foods on the side of celebration. The fable happily reenacted a theme of oral culture that stretched back to the banqueting of ancient times: the good kingdom is one that provides ample sustenance.[41]

The conflict between abstinence and full indulgence continued through the heady years of the Renaissance into the era of religious reform, becoming especially pronounced in northern Europe. Pieter Bruegel's *The Battle Between Lent and Carnival,* painted in 1559, pits the energetic exuberance of the days before Lent against the prescriptive duties of organized religion. The contrast may also be a representation of the tensions accompanying the Reformation, which insisted on greater attention to discipline and restraint. In a slightly later painting, *The Land of Cockaigne,* Bruegel inserted a moral lesson. Transporting the viewer to a plebeian Elysium, he laid out the consequences of satiety. Three sleeping figures, legs splayed and mouths open as they snore, force us to consider the ultimate price of total satisfaction. Above them, a table displays an array of meaty delights, their abundance hardly depleted by the revelers. Complete gratification exhausts human creatures, Bruegel suggests, prone as they are to gluttony and excess. As the prostrate figures indicate, the Land of Cockaigne is as likely to extinguish as to renew life.

But Boccaccio's delectable mountain of parmesan cheese gestures us toward a different legacy, one in which milk takes on a stronger, more zestful role in the human diet. Abundance in the Renaissance assumed a more solid form, quite literally, as a groaning board of delicacies enjoyed by European nobility, permissible if consumed wisely and in moderation. Elites might associate Boccaccio's mountain with the tradition of classical banquets, enjoying a certain element of parody. And in an odd but typical parallel, popular culture celebrated its own version, as an imaginative Edenic place, a land of plenty. Print culture eventually picked up and circulated the popular tradition, enabling us to "hear" the place described

Pieter Bruegel the Elder (ca. 1525–1569), The Land of Cockaigne, *1567 (oil on panel). Bruegel's mocking portrayal of the Land of Cockaigne shows the popular fantasyland of edible delights as a comically stocked alternative to the more conventional Garden of Eden. Three social types (soldier, peasant, merchant) lay asleep, overcome with satiety. The empty eggshell in the foreground signals the painting's message about the spiritual vacuity of gluttony and laziness. Viewers are nevertheless treated to the vision of countless cakes, a fence made entirely of sausages, and a mountain of pudding in the background on the right. (Alte Pinakothek, Munich/The Bridgeman Art Library International)*

in later centuries. A sixteenth-century Italian version of Cockaigne displayed the typical geographical features, not unlike those described in Boccaccio's *Decameron:*

A mountain of grated cheese
Is seen standing alone in the middle of the plain,
A kettle has been brought to its summit . . .
A river of milk gushes forth from a cave
And goes flowing through the town
Its embankments are made of ricotta . . .[42]

Accompanying this overwhelming delivery of food was the liberation of social constraints: in Cockaigne, everything was free and easy, no one worked, and all property was in common. The divisions of medieval society disappeared along with limitations on luscious dairy products.

Though moralists would harness the trope of utopia to regimens of austerity, the provocative François Rabelais put the theme to more irreverent use. In *Gargantua and Pantagruel,* published in the 1530s, his Gastrolaters ("Belly-worshipers") offer baskets of non-Lenten foods to their gods, including cheeses and clotted cream among the final course of "reinforcements": "eggs in snow," "curds in cream," and puff pastry undoubtedly made with rich butter. These cowl-necked gluttons enacted a double-sided tale of a land of plenty, where milk products provided a surfeit of satisfaction from the table of privilege. In fact, Rabelais and his creations straddled two eras: one age was grounded in medieval abstinence, when milk indicated holy divine presence and monastic exhortations. The other looked forward to a more secular banquet, where the pleasures of fresh cheeses and puff pastry might be spread beyond elite guests to those who did not know the sins of idleness. Like fire, water, and earth, there was something elemental about milk that suggested a more democratic realm of food.

CHAPTER THREE

The Renaissance of Milk

Sometime around 1465, a feisty scholar named Bartolomeo Sacchi (known to posterity as Platina) penned one of the earliest self-help books to appear in print. *De honesta voluptate et valetudine* (On Right Pleasure and Good Health) offered much more in the way of enjoyment than advice: it consisted of a mere dozen sections on how to maintain good health, followed by a lengthy catalogue of foods and recipes, some of them apparently well tested. In draft form, the manuscript had circulated so much that it became "filthy and besmeared," looking as though it had "crawled through ointment shops and taverns." At least this is how the author, lacking delicacy, described the packet dispatched to a noteworthy cardinal before its publication. As a work in progress, its pages offered up a kind of communal space for deliberation. "Rub out whatever you want, change, turn around, move, and if anything seems imperfect and too little, add whatever and as much as you want," Platina instructed the cardinal. Even when entrusted to the finality of print, the book never lost its quarrelsome quality. Debatable opinions ran rampant as dozens of the author's friends came and went in illustrative anecdotes. The project ultimately produced a remarkable window onto a rich Renaissance food culture, made possible by Italians who apparently enjoyed fretting over their plates as early as the fifteenth century.[1]

The Italian Renaissance is important in the history of milk. Powerful aristocrats and learned humanists, elites we associate with the celebrated inner circle of the Renaissance, ruminated endlessly over matters of the table, and not without devoting considerable attention to milk and its products. More than most other comestibles, milk was hedged in by a thick barrier of cultural assumptions that contemporaries, even the illiterate, would have known about. *Caveat emptor* would have described a typical approach to the liquid. The medicinal benefits of milk were understood, but only according to strict specifications. For those who were physically hardy, milk was thought to "fill the veins" with nourishment. But for the majority of privileged urban elites, who apparently harbored one ailment or another, its reputed ill effects stirred up a great deal of anxiety: to consume or not to consume? The answer was never very clear. The pathway from fearful restrictions broadcast by the ancients to a fulsome appreciation of milk and its products was beset with contradictions.

Thanks to Platina's communally shared pages, we can see how rules and practices parted company in many areas of diet during this time, enabling milk to enjoy its own kind of renaissance. Early in the annals of printed books, *On Right Pleasure and Good Health* was in fact only the latest entry in a near millennium-long conversation on how to eat sensibly. The ever blunt Platina said as much in his dedicatory preface: "Let those who pass judgment on everything . . . stop carping because I have written about health and the theory of food, which the Greeks call diet."[2] From the sixth century B.C.E., when Pythagoras inaugurated a program of vegetarianism, philosophers devoted careful attention to the relationship between diet and health. Many fundamental assumptions remained the same in the Renaissance. Western European physicians inherited two important features of Greek medicine: a willingness to set aside supernatural origins of disease and a holistic approach to the health of mind and body. "Food was considered among the crucial regulating factors in maintaining health," Ken Albala points out in his history of eating practices in the Renaissance, "and diet or a change in regimen was among the most common therapies for illness." Hippocrates, Aristotle, and especially the towering figure of Galen (129–200 or 216 C.E.) established the ground rules still in place as late as the seventeenth century.[3]

The ancients believed milk to be blood in a "twice-cooked" state, and blood constituted one of the primary humors in charge of regulating the body. Standard views drew from the dietary rules of Galen, who had developed the humoral calculus that informed European medicine for

Giovanni Bellini and Titian, Feast of the Gods, *1514/1529. This masterpiece of Renaissance painting was commissioned by Alfonso de'Este, Duke of Ferrara, and depicts a familiar scene of Bacchanalian amusement in ancient mythology. The pleasures shown here indicate a new era of gratification, which included not only intoxicating drink but edible delights of a pastoral kind. The female figures in the painting were originally clothed in modest peasant garb, but the duke seems to have requested that Bellini render their appearance more faithful to Renaissance ideas of ancient nymphs. (National Gallery of Art, Widener Collection; image courtesy of the Board of Trustees, National Gallery of Art, Washington, D.C.)*

centuries. The four humors that made up the body—blood (hot and moist), phlegm (cold and moist), choler (hot and dry), and melancholy (cold and dry)—appeared in different proportions in each person. All foods belonged to categories embodying the same four qualities: vegetables were cold and watery; meats were warm and beneficially dry, though they varied according to the animal source. To maintain health or to correct illness, one needed to add or subtract elements of these qualities in one's diet.

Milk presented a conundrum to this relatively straightforward system. As a form of blood, we might expect it to be hot and moist, but according to Galen, its heat had been extracted in the process of production, making it dangerously cold. (Another school of thought categorized milk as warm, but bracketed certain elements within it as cold, once they were separated from the original liquid.) To what extent these cold and moist properties yielded nutrition depended on the consumer. For the very young and the very old, who were by nature colder than people in the prime of life, milk proved especially nutritious. It harmonized with their already cool systems and contributed to the building of flesh and blood. Lacking teeth, infants and the elderly would likely consume milk on an empty stomach, which ensured its proper digestion. But for consumers in mid-life, milk presented a slippery slope of nutritive seduction and betrayal. The viscous liquid most likely would spoil in the process of digestion, sending putrid fumes upward to the brain and chalky deposits downward to the kidneys, where it would create blockages. Milk was most dangerous in cases where the consumer was ill: the weak and sickly, those who were melancholy, and those with headache or pain were warned away from it. Only the sturdy (and, in the case of hard cheeses, the vulgar) could tolerate the digestive challenges of milk.[4]

If we take Platina's opinion of milk as a litmus test of current attitudes toward the regimen of old, we see a startling reversal: caution appears to be pushed aside in favor of hoping for the best. "Some think the force of milk is mild because it inclines a little toward coolness and moisture, and because of this, it is agreed among all doctors that milk nourishes well, generates much blood (since blood may be drawn from udders and breasts), and warms the brain, is good for the stomach and lungs, and increases fertility." Platina had excised the usual tedious discussion of possible dangers and spoke mainly of nourishment. If consumed under the right conditions, preferably at the beginning of a meal and still warm from the animal, raw milk delivered big benefits: it helped to build blood and enhance every aspect of vitality. He listed a familiar ancient hierarchy: goat's milk was "best because it helps the stomach, removes obstructions of the liver, and loosens the bowels." Next came sheep's milk, followed by that of cows. Holding on to a bit of ancient prejudice, he advised milk drinkers to make sure to sit still after having their fill and to abstain from other food so that the milk would settle well and not turn sour. The stomach, viewed as a container, or, in Ken Albala's words, a kettle on a fire, could cook its milky contents safely this way. "Drunk with either sugar or honey," Platina advised, the milk could be "kept from spoiling." (This was pure and simple Galenic chemistry; sugar was understood to add heat to foods that were classified as cool.) Adhering to good humanist philosophy, Platina advised moderation in milk drinking, as in all other alimentary activity. The new world of milk was going to be a positive and refreshing alternative to the former age of anxiety.[5]

Advice books usually deceive with their emphasis on ideal behavior, but Platina's offered a mirror image of the virtuous society: here was damning evidence of predictable human weakness. Italians appear to have developed a passionate attraction to the pleasures of the dairy, regardless of the apprehensions of yore. Fifteenth-century Italian gourmands were happily tucking into *blancmange,* capons cooked in milk, and rich ricotta pies. His panoramic scope of the Renaissance table, a boon to readers of a later era, exposed the sins of self-indulgent eaters. "Let Hirtius not lap this up because he takes away much of our sleep with his cry because of the violent pain in his sides," we are told at the bottom of a recipe for polenta. This recipe required grinding a pound of "absolutely fresh cheese until it seems converted into milk," which was then added to "eight or ten eggs" and "half a pound of sugar." "Let Philenus Archigallus beware of this," we are told after a description of herb pie, "for it digests slowly, dulls the eyes, makes obstructions, and generates stone." No wonder: herb pie contained a pound and a half of fresh cheese, a half pound of butter, and fifteen or sixteen beaten egg whites.[6] Admittedly, the clever humanist was fawning to potential patrons by publicizing dishes devised by a celebrated chef named Martino, employed by a cardinal and several Milanese dukes. Platina himself must have enjoyed many happy dining occasions at the country homes of his patrons, which included that of Martino's employer. As the first librarian of the Vatican collection, our guide to the age was giving us the inside story: Italian elites relished inventive dishes and were willing to set aside worry about the ill effects of certain ingredients.[7]

Not all responsibility for this change lay with consumers: cooks and chefs seem to have continually raised the bar by producing prodigious feasts for elite tables. Perhaps, historians at one time speculated (incorrectly), this had to do with coaxing the deadened appetites of Romans who were suffering unknowingly from poison in their water supply, leached from lead pipes.[8] Platina's commentary also hints that fads for delicacies like birds and game had squeezed out plates of simple whole-some foods from the Renaissance table. So while experts on the stomach recommended eating cheese as the third course to ensure proper diges-tion, cooks apparently slipped it into every kind of dish; herb pie was one such constructed temptation. The dutiful Platina injected medical judg-ments throughout — "This is even more unhealthy than the one above," "This mixture has more bad than good in it" — yet behind such statements lurked an invitation to cast inhibition to the winds. "And who would want what is most abundant to be restricted?" asked Platina rhetorically in an introductory to fried foods. "And who would be pleased to be reduced to extreme leanness?"[9]

Milk, cheese, and butter populate the pages of *On Right Pleasure* in the manner of kitchen staples. Many dishes called for milk by the potful and cheese by the pound. Vermicelli and other pastas, gourd soup, and all kinds of pies required the liquid, sometimes specifying goat's milk as preferable. Butter became the ingredient *du jour* in chef Martino's innova-tive shortcrust pastries, which could be eaten with pleasure along with a pie's contents.[10] Cheese, often understood to be in a fresh state, not hard, enriched recipes for soups and meatless pies like Bolognese, made with chard, herbs, and lots of butter. Cheese omelettes, however, called for a half pound of the dangerous hard variety. The good life among Renais-sance elites included tasty "snacks" (*ofella*) made from parmesan and fresh cheese mixed with raisins, cinnamon, ginger, and saffron. Fried tidbits — *frictellas,* or fritters — conjured up associations of gluttony, according to the honest author, yet here were twenty delectable examples, including the more dairy-laden (fresh cheese fritters, sour milk fritters, rice fritters) and the fruit-stuffed (elderberry, apple, fig), which depended on milk and egg whites as binding ingredients. Delectable buns used dough to "enclose fresh rich cheese" (or small birds). We also hear of the popularity of ricotta cheese — "cooks mix it into many vegetable ragouts" — and are given simple instructions on how to make it ourselves.[11]

Milk products were deeply implicated in a new regime of taste evi-dent at the tables of the elite. Platina lets us in on the winners of a veri-table competition among cheeses, indicating that consumers had culti-vated their knowledge of the different types available. "Today there are two kinds of cheese in Italy which vie for first place," he reported: the "rotten" kind from Tuscany, made in March, and "the Parmesan which is made on this side of the Alps and can be called *maialis* from the month of May." Fast days required abstention from milk and cheese, but judging from Platina's evidence, skilled cooks had already drawn humanity down the road of refusing to do without. Recipes for almond gelatine promised to create a product "like milk" and concoctions of artificial ricotta and butter (*Recocta ficta* and *Butyrum fictum*) proved the lengths that cooks might go to satisfy their clientele — a sure sign of dairy products entering the ranks of customary table items.[12]

Accompanying this panoply of rich dishes, Platina provided the obligatory harangue against luxurious habits of contemporary times. No doubt, the author of this lengthy catalogue of earthly delights had to defend himself against charges of gluttony if he wished to preserve his credentials as a bona fide student of the ancients. He also needed to qualify with care his notion of pleasure in order to avoid charges of Epicureanism, a philosophical stance related to paganism and seen as a menace to society. Not surprisingly, then, Platina repeatedly urged a return to lessons of moderation and frugality, which he insisted were lost on the present age of indulgence. "We are all so given over to gullet and stomach that what was then clear seems obscure," he lamented. The "pampered tastes of our contemporaries want meat pies" and shun those made with vegetables. If old ways could be recovered, he argued, "we would not see today so many so-called cooks in the city, so many gluttons, so many dandies, so many parasites, so many most diligent cultivators of hidden lusts and recruiting officers for gluttony and greed."[13] The author's colorful language clues the reader in to his equally colorful and at times unhinged personality: known for hotheaded speechifying and violent outbursts, he was imprisoned twice. His was an odd profile for a librarian. Perhaps these character traits led the fulminating Platina to ponder the possibility of tempering his humoral makeup through a reformed diet.[14]

Platina was not alone in his celebration of the delights of milk products. As chief librarian at the Vatican, he must have been gratified to receive the beautifully embellished manuscript of a doctor from Piedmont, Pantaleone da Confienza, known as "Pantaleo," titled *Summa lacticiniorum*. The work was first published in 1477, just a few years after Platina's *On Right Pleasure*. Historians have translated his title as *A Treatise on Milk Products*, but Pantaleo's testament loses something in such a clinical reduction. We might ask, after all, who ever reduces Augustine's *Summa* to "treatise"? Better to enjoy the irony that informed this court doctor's love song to cheese. As part of the household staff of the Duke of Savoy, Pantaleo gained firsthand experience of delicacies of the dairy while traveling through France with his employer. His description of the genesis of cheese from milk endowed the substance with life: it grew a "skin," it "breathed," it "dried up" in the same way an organism might become desiccated. In a paean to the general subject of milk, Pantaleo devoted his first chapter to human breast milk, recognizing the nourishing properties of the substance.[15]

Platina might have wielded more influence as a source of advice on diet if he had sported medical credentials. Even before the advent of the printing press, an information-craving public in Italy leaned heavily on the opinions of medical men in order to adjust their sails in the high winds of the Renaissance. As successive waves of plague swept through the peninsula (Platina died at sixty owing to the disease), opinion vacillated over approaches to life and death. For the generations following the second outbreak of that horrific disease, medical authorities came to the fore as trusted advisers. Of course, plague was only one scourge of many in an era of chronic illnesses such as dysentery and influenza, sporadic food shortages, and endemic poverty. Yet it was both a profound reminder of the transience of life and a spur to meet circumstances with the best available knowledge. And in pursuing these objectives, too, milk could be of use.

In the aftermath of catastrophic disease, psychiatric advice was born. How else can we interpret the approach of a respected doctor of Padua, Michele Savonarola, to the peril of plague? The greatest enemy of health, he argued, was "heavy thoughts" (*gran pensieri*), especially "the thought of death." For a long life, Savonarola and his medical contemporaries in the first half of the fifteenth century adopted a new strategy. Instead of concentrating on its environmental transmission, they argued that "states of mind" determined susceptibility to the plague. It was now imperative to work against "melancholy, anger, struggle, sadness, trouble, torment, envy, fury, and confusion." Pleasurable pastimes must distract people from their worries. Singing lovely songs, playing games, spinning stories, reading history: such delights would ward off danger better than any other strategy. Moreover, to achieve *mens sana in corpore sano,* a sound mind in a sound body, survivors of the second wave of plague should cast aside fasts and seek out fortifying food.[16]

Melancholy, the most unfortunate — and now, dangerous — of the four basic temperaments, had long been stigmatized in European culture. The saturnine complexion, marked by black hair and dark skin hues, occupied the unfortunate end of a spectrum of personality types. According to medieval theories, such people were suited for little besides counting and measuring, a sad fate, given the medieval prejudice against association with money. The signature juice dominating the melancholic — black bile — suggested fears associated with darkness and a realm of unseen demons and terrors. More commonly, melancholics lived out their destiny in introspective or pensive inactivity, which was how students and scholars came to be identified with the temperament. Not all authorities viewed the melancholic as disabled. According to one ancient text believed to be authored by Aristotle (but now referred to as Pseudo-Aristotle), melancholy constituted the stuff of heroes and geniuses — Hercules and Plato were cited as good examples — who were able to translate their despondency into a "furor" of creativity. Plato's works also supported this theory. But exactly how creative frenzy came about remained something of a mystery. Actual management of the condition lay beyond medical (and psychological) expertise.

In an era with a heightened appreciation for the production of knowledge, melancholy posed a threat to a key resource: the Renaissance intellectual. The foot soldiers of the age of humanism found a staunch protector in Marsilio Ficino (1433–1499). Standard histories of the Renaissance often omit him; given his synthesis of magic and astrology, he is difficult to describe as a "renaissance man." (Though a modern reader might wonder about the compatibility of astrology and medicine, their interdependency was assumed in the fifteenth century.) Ficino obviously read Pseudo-Aristotle on melancholy and was prepared to connect the theory of creative genius to the medical knowledge of a later age. As an ardent Neoplatonist, Ficino would also enlist that philosopher's ideas about genius. Here was a bold synthesizer, the author of a best-selling *Advice Against Plague,* who had been ordained as a priest and educated in philosophy and medicine. Thus well equipped, Ficino made the case for melan-

Albrecht Dürer (1471–1528), Melancolia *(engraving). Dürer's famous representation of melancholy links the mental state to the contemplative life of the scholar and artist. Though the condition was the least desirable of temperaments, the Renaissance revitalized its reputation as a source of creative genius. According to Ficino, milky foods could counteract the black bile that dominated the body of the sufferer. (Guildhall Library, City of London/The Bridgeman Art Library International)*

choly as the wellspring of genius and constructed an all-encompassing cosmological system around his theory.

A winningly candid voice speaks from the pages of Ficino's *Three Books on Life*. Completed around 1480, the treatise remains wonderfully readable. A wedding of dietary medicine, psychology, and the occult, each section assumes a confidential, chatty tone. "Hail, intellectual guest!" one of the books begins. "Hail to you, too, whosoever you are who approaches our threshold desiring health!" In a preface, Ficino begged both patrons and readers to pay attention to the welfare of intellectuals, whom he presented as a precious natural resource. These were "the priests of the Muses," "the hunters after the highest good and truth." They worked hard, but they neglected the condition of their bodies and minds. "As much care as runners habitually take of their legs, athletes of their arms, musicians of their voice, even so it behooves literary scholars to have at least as much concern for their brain and heart, their liver and stomach," Ficino argued. Along with dietary advice, Book One offered information on body care and hygiene, optimum times to work, and strategies to enhance concentration. ("Comb your head carefully and moderately with an ivory comb, drawing the comb back from the forehead to neck about forty times. Then rub your neck with a rather rough cloth.") Book Two, on how to prolong old age, and Book Three, a user-friendly guide to astrology for the benefit of health, only expanded the appeal of this eccentric trilogy. *On Life* became Ficino's second most popular work, next to his book on plague, transmitting his opinions across Europe in nearly thirty editions over the next 170 years.[17]

Ficino's support for milk sprang from the crucial importance he placed on the blood supply as the source of creative and vital juices. Above and beyond blood stood the all-important spirit, the "vapor of blood" that moved around in the body in erratic fits and starts. In different states, spirits informed the workings of the heart, brain, the senses, and the soul. John Donne later captured its significance more poetically in "The Exstasie":

> As our blood labours to beget
> Spirits as like soules as it can,
> Because such fingers need to knit
> The subtile knot which makes us man.[18]

Together, these substances enabled the intellectual to function. But often-times, the scholar suffered from a kind of compression caused by an aberrant flow of blood. As thoughts dispersed spirits throughout the body, eating up the best of the vital fluids, the blood was "rendered dense, dry and black." The conditions under which the intellectual labored made things worse: "with too little physical exercise, superfluities are not carried off and the thick, dense, clinging, dusky vapors do not exhale. All these things make the spirit melancholy and the soul sad and fearful."[19]

In the midst of this malfunctioning, black bile built up. Enter Ficino, with a revisionist's clear-sighted insistence: this fluid, unjustly loathed, was precisely what made the intellectual a wise and intelligent person. Like responsive inner strings of the soul, black bile vibrated with stimulation from outer forces such as music, art, and philosophy. He described its appearance with a chemist's admiration: "it burns and shines much like red-hot gold tinged with purple; and it takes on in the burning heart various colors like a rainbow."[20] To communicate a sense of its power, Ficino compared it to *aqua vitae;* it had the quality of pure essence, extracted after a lengthy process of fermentation. Though too much black bile rendered the intellectual immobilized by sadness, a generous store of the stuff transported him or her to a state of mental ecstasy. Here was Ficino's original contribution to the theory of poetic genius. "Let black bile abound, but very rarefied."[21]

Fortunately for its subsequent reputation, milk appeared friendly to black bile. According to Galenic principles, milk was moist and, when in a fresh state, warm, so it could replenish a dry, cold body like that of the sedentary intellectual. In offering a highly technical set of instructions on what to ingest in order to coddle the precious humor, Ficino gave milk high marks. "Make sure [black bile] has the moisture of the more subtle phlegm surrounding it," he explained.[22] "Of advantage are all milky foods: milk, fresh cheese, sweet almonds." Most melancholics and scholars suffered from sleeplessness, both a cause and an effect of a "dried out" brain. Here, too, milk came to the rescue: "It is especially good to drink milk mixed with sugar on an empty stomach, provided that the stomach can tolerate it. These moist things are of great advantage to melancholics, even if they get enough sleep."[23] Comfort in the form of a cup of warm milk was thus launched on its own career.

In the second book of *On Life,* dedicated to the objective of achieving a long life, Ficino presented a memorable treatment for old age: human breast milk, drunk from the source. He pronounced his opinion with a certain and measured authority:

> Immediately after the age of seventy and sometimes after sixty-three, since the moisture has gradually dried up, the tree of the human body often decays. Then for the first time this human tree must be moistened by a human, youthful liquid in order that it may revive. Therefore choose a young girl who is healthy, beautiful, cheerful, and temperate, and when you are hungry and the moon is waxing, suck her milk; immediately eat a little powder of sweet fennel properly mixed with sugar. The sugar will prevent the milk from curdling and putrefying in the stomach; and the fennel, since it is fine and a friend of the milk, will spread the milk to the bodily parts.[24]

Ficino's specificity may seem uncomfortably related to his astrological practices, but he was neither the first nor the last of the dietetic authorities to recommend fresh human breast milk for its curative potency. Platina had identified it as a cure for "chronic cough and sores of the lungs," among other ailments, and ancient practice utilized breast milk for sore eyes.[25] Ficino was simply repeating contemporary wisdom on the point about character traits of the lactating female: these were believed to be communicated through breast milk, as qualities present in the blood from which the milk derived. Nevertheless, his advice flew in the face of the ancient proverb, *Lac senum existit vinum,* "Wine is the milk of the aged." In a perfect world, according to Ficino, infants and the aged might share a lactating woman's breast.

Ficino himself suffered from melancholy; his letters reveal both his unfortunate horoscope (he was born directly under Saturn, the planet ruling over the melancholic temperament) and his determination to make the best of it. In many ways, *On Life* was a testament to his personal efforts to turn melancholy into genius. Whether he purposefully drank milk, we cannot know. Scholars usually point to Ficino's use of music as a solace; yet *On Life* also suggested a therapeutic use of nature as an antidote to melancholy. Bathing the senses in "pleasant smells" and "frequent viewing of shining water and of green or red color, the haunting of gardens and groves and pleasant walks along rivers and through lovely meadows" promised the restoration of a favorable mental state, we are told.[26] In this,

Ficino was not alone. Renaissance literary figures had already found a resource in bucolic pleasures that completed the circle with respect to milk. For it was through nature, newly glorified by the Renaissance, that milk could escape its associations with barbarism and beasts.

"The day we spent at the property boundary was so happy," Laura Cereta recalled in the only surviving letter addressed to her mother, "that it ought to be commemorated with a special token." The sixteen-year-old thus composed an epistle demonstrating her literary accomplishments, one of eighty-three letters written between 1485 and 1488. In her account of the innocent pleasures of that outing, the reader senses that the delights of nature were fortifying the young writer while ornamenting her pages with poetic expression:

> We gazed at meadows blooming with flowers and glistening with small stones and winding streams, and we felt full of contentment. Noisy birds were singing in the morning sun and as soon as we got out of the wagon, some rural folk came to meet us, and, garlanded with chains of wicker, they sang songs and made music with oaten pipes and flutes of reed. Some were herding flocks while others pressed milk from the swollen udders of sheep, and now and again the fields resounded with the lowing of approaching cattle.

Remindful of the pastoral strain of ancient poetry, these references to milk-bearing animals signaled the goodness and abundance of nature's gifts. Their restorative powers carried special meaning for the well-to-do mother and daughter: "with charm and freshness, the sensations aroused by the land filled our city-bred minds," Cereta wrote.[27]

It may not be simply coincidence that this spirited young woman also offered a redeeming perspective on milk. Though her writings focus on many of the standard themes of humanist scholarship — the human quest for pleasure, the potential of all human beings to learn and excel, the sources of human (and female) folly — Cereta showed unusual boldness in weaving material from everyday life into her texts. As a fledgling humanist, Cereta was nothing short of extraordinary. While most other women scholars remained as close as possible to the classical models they imitated, Cereta repeatedly rebelled, turning each example into a startling surprise for the reader. Particularly in her engagement with rural life, observed

LAVRA CERETA BRIXIEN*SIS*,
LITERIS ORNATI*SS*IMA.

Portrait of Laura Cereta, shown as a contemplative and somewhat solemn figure. Her modest dress and concealed hair appear in striking contrast to the inscription, which denotes her as the most accomplished and distinguished of (female) writers. (Image from Laurae Ceretae Brixiensis Feminae Clarissimae Epistolae iam primum e MS in lucem productae, *Jacopo Filippo Tomasini, ed. [Padua, 1640]; courtesy of Mannheim University Library)*

around her native Brescia in northern Italy, Cereta celebrated a wholesome simplicity that would rescue milk from condescension.[28]

Cereta's literary career, though brief, proved intense and remarkable. She appears in many respects motivated by a gritty self-reliance and resolve. "For an education is neither bequeathed to us as a legacy, nor does some fate or other give it to us as a gift. Virtue is something that we ourselves acquire," she wrote in a letter decrying the gossip-mongering of women who wondered at her learning. "For those women who believe that study, hard work, and vigilance will bring them sure praise, the road to attaining knowledge is broad." Cereta learned such independence early in life. As her letter to her mother implied, Laura lived apart from her and received visits infrequently, perhaps because her parents were formally separated. Only this, it seems, could explain why the young Laura would have assumed charge of her father's household and the care of her younger siblings at the age of eleven or twelve. "Thus it was my lot to grow old when I was not far from childhood," she confided in one of her early letters.[29]

Laura's father saw to it that his daughter received a somewhat serious education, at least for a girl, beginning when she was seven. His sense of the dangers of ignorance and idleness for a young woman, according to Laura, led to a succession of arrangements, first at a monastery and later at home. But from age twelve or so, most of her study took place at night, after her household duties had been performed. The same would be true after her marriage at fifteen. Her husband, a Venetian merchant, did not seem opposed to her pursuit of learning, nor to her continuing responsibilities for her father's household. Their marriage seemed amicable, if distant, for a brief eighteen months before he died, possibly of plague.

By what means did this young writer come to the subject of milk? Italian literary conventions of the Renaissance were turning to rustic themes just as their ancient models had done. In imitation of Vergil's *Eclogues,* for example, writers would have used rural settings as vehicles for ideal happiness or beauty, or an evocation of the untamed power of nature. Cereta's writings reflect just this trend, especially in her celebration of the withdrawal from the civilized world that such a perspective offered. Yet Cereta's treatment of milk may reveal how much the subject could expand the powers of a female literati. Cereta seized upon milk en route to one of her favorite causes: the virtue of rural simplicity over the worldly arrogance built into life in the city. Given the preoccupations of her era— Platina's tell-all account of the elite table is an indication—this was a means of assuming a bold stance for the "city-bred" woman. Cereta has been called one of the most feminist of women humanists for her insistent discussion of the subjection of women, but she did not limit her campaign to that instance of injustice alone. In rustic life and its simple virtues, she found the perfect vehicle for her animus against the hypocrisy of the powerful against the weak.

Cereta set to work on the theme in her earliest piece of writing, "Dialogue on the Death of an Ass," an homage to the popular ancient novel by Apuleius. The novice writer flaunted her intellectual agility by greatly complicating the tale, a murder plot in which dishonesty and greed victimize an ass prized by a humble peasant. Three voices debate the weakness of human nature; along the way, the story examines the details of rural life. Drawing her evidence from Pliny, Cereta introduced a description of the many beneficial uses of asses' milk, one element providing a defense of the virtue of the lowly beast:

Poppaea didn't blush to walk in the company of a flock of asses. The great utility of a bath of asses' milk for smoothness and softness of the face is zealously prized. But nature (which lets nothing remain a secret) opines that nothing is more efficacious or healing for the sudden fevers of infants and the aching breasts of a woman newly delivered of a child than asses' milk.

Not for the first time, milk acted as a bearer of goodness and virtue. In the same spirit, Cereta inserted a brazen female character named Laura — no surprise — whose funeral oration made sense of the obscure circumstances of the ass's death. Making a case against human chicanery, the speech seemed much like a winning courtroom defense of rural innocence.[30]

More important to the elevation of milk, however, was the theme of pleasure. Cereta, like her contemporaries Platina and Ficino, wrestled with the philosophical and moral problems associated with this concept. All three authors were schooled in the debates related to Epicurus, the ancient Greek philosopher who defended the rational enjoyment of moderate pleasure (*voluptas*). Fifteenth-century Italy was not entirely hospitable to the recovery of Epicurean thought; though scholars had returned to the philosopher's work, he was still stigmatized by centuries of vilification by Christians, who condemned him because of his implicit denial of immortality and his suggestions of hedonism. Platina distanced his *De honesta voluptate et valetudine* from Epicurus by his use of the word "right" (*honesta*), aiming to correct any implication of decadence. Ironically, three years after the publication of his book, Epicureanism constituted one of the allegations made against him and other members of the Roman Academy, leading to their imprisonment by the pope.[31] Ficino's *On Life* devoted several pages to the careful modulation of pleasure: his target, predictably, was mostly "that secret and too constant pleasure of the contemplative mind." Pleasures could vary, however, even for the intellectual: the philosopher added dire warnings to "shun deceitful Venus in her blandishments of touching and tasting."[32]

Cereta's approach to pleasure appears innocent in contrast: she met the experience head-on in nature, relishing the countryside through her own bold synthesis of philosophical and Christian inspiration. Though launched on paper in the privacy of her study rather than among bibulous academicians, Cereta's arguments were careful and elegant in their scholarly examinations. The lone young woman at her desk elbowed her way to the table of her philosophical ancestors. In "A Topography and a Defense of Epicurus," a description of a day's outing at Mount Isola, she wove together strands of Plato, Pliny, and Petrarch. Then, in a virtuoso ending, she added her own rendition of Lorenzo Vallo's two dialogues *On Pleasure* and *On Free Will*. Especially from the latter, she developed an image of the intellect's instinctual yearning for pure insight into God's goodness. In each of her borrowings, a startling inventiveness informed her argument.[33]

Through no coincidence, the subject of milk was hovering in the wings during these discussions of pleasure. In the assemblage of opposing sides of the debate — virtue against moral corruption, purity opposed to luxury, moderation against satiety and excess — milk came down on the side of simple, humble sustenance. Drawn from a countryside free of contrived pleasures and embellishments, milk constituted one of the elements of nature. It stood as a metaphor for the essence of nourishment. Its human sources were virtuous viragos: humanists like Cereta turned to the many examples in ancient legend, rather than to the central Christian image of the Virgin Mary. In an essay devoted to the wondrous deeds of women, Cereta recounted the story of the young Roman woman who nursed her mother, imprisoned and starving, through the cell window. (Christine de Pisan also cited this legend, obviously preferring its mother-daughter relationship to the less comfortable tale of Pero nursing her father Cimon through the bars of a prison window.)[34] Animal sources of milk appear throughout Cereta's letters; as innocent philanthropists, they, too, donated their products to the maintenance of human life. A celebration of nature's goodness and divine charity extended to the female breast, lauded for its selfless, nurturing strength. Through her use of metaphor, Cereta established a firm connection between the nurturing breast and a concept of enduring, as opposed to ephemeral, pleasure.

The connection between milk and the vitality of nature figured prominently in Cereta's description of her climb of Mount Isola.[35] This lyrical letter constituted one of her most audacious (and some would say feminist) moves, in which she took Petrarch's account of his ascent of Mount Ventoux and bested it. Petrarch's essay focused on the subject of virtue, but unlike his intrepid female admirer, he chose to make human nature, not rural nature, the vehicle for his moral lesson. Petrarch presented himself as a lazy hiker: his difficulty in climbing a steep slope became a trope for his avoidance of the hard effort necessary to reach the summit. His insights came from sitting in a valley, momentarily defeated, where he reproached himself for trying to find an easy way up the mountain. Cereta presented the opposite image: no shrinking violet, she plunged into an outing of several days with the zest of a famished city-dweller. Harmony between her group of climbers and the environment pervaded the occasion; whereas natives warned Petrarch and friends against their ascent of the mountain, the villagers who greeted Cereta's party urged them on in a celebratory uproar. Nothing in Cereta's experience displaced the centrality of her inspirational encounter with nature *qua* nature. Our gaze is directed outward, in a brilliant paradoxical stroke of humanist humility, to the sensual pleasures of rural experience.

Cereta described every sound and sight available on Mount Isola, sometimes intertwining them with encounters with sustenance. At the foot of the first peak, she recorded her wonderment: "At this level, the deep vales echoed with the resonant lowings of cattle; and from the highest crags, the pipes of shepherds could be heard, so that I might have thought Diana and Sylvanus inhabited these woods." The climbers literally consumed the mountainside:

> We sat, caught our breath, and pushed on, peacefully picking strawberries and flowers in the lush meadow. We were led to other small gardens where climbing vines with knotty tendrils and trees heavy-laden with apples provided shade. Here freedom from care — and this is what helped us seize grapes by the handful and gobble them down, since hunger was definitely the architect of our tour — gave us the strength to continue, and laughingly we put aside all thought of turning back.

At the summit, the climbers "launched a cascade of rocks" and watched rabbits as they "flew down the slopes ahead of us." The author seemed too absorbed in the fun to interrupt her account by philosophizing. Instead, she depicted the group giving vent to playfulness as it reveled in its triumph. The summit was hardly contemplated; almost immediately, the group turned to go home.[36]

It was at this point, when Cereta and friends became aware of how long their return journey would be, that the group stumbled upon "great flagons of milk," which they "thirstily drank down." Any reader of this text, then and now, is beset with questions: what kind of naive hospitality afforded such generosity? Without knowing the customs of this fifteenth-century locality, we cannot be sure, but clearly Cereta wanted us to view the countryside as abounding with sustenance. Did she insert milk into her narrative as a sign of divine goodness? The discovery of the flagons occurred after the summit was reached, perhaps as a reward. And where was citified apprehension of this rustic beverage? Cereta and her friends simply trusted in the freshness of the liquid and helped themselves, almost greedily. Presumably, the milk came from the goats they were seeing along the way, and the hikers found it agreeable and restorative. After this refreshment, the party continued the descent, "sliding down the breast of the mountain." With their stomachs still full and their feet hot and tired, they bathed in a pristine river and lay down for a nap as a "skylark perched above us gently filled the air with song."[37] Once again, milk was associated with peaceful slumber.

"The sage is content with plain foods," Cereta wrote in the passage that followed her description of the climb of Mount Isola. "He asks for that which nourishes rather than that which delights." The banquet awaiting the adventurers upon their return consisted of chestnuts, turnips, and polenta, along with barley bread, hazelnuts, and fruit — all products of the field and woods. For beverage, they "passed around" "one small jug of wine." The consumption of milk thus remained confined to the "breast" of the mountain, enveloped in nature and surrounded by animal life, freely given by a mysteriously absent donor.[38]

Cereta's celebration of milk, in light of what followed, took on a valedictory quality. Within the next year, she composed her final letters, and for the remainder of her life she retreated into silence. Perhaps the persistent taunts and rumors generated by both male and female detractors finally undermined her determination. (Skeptics demanding proof of her authorship sent a female acquaintance to observe Cereta in the act of writing such polished Latin prose.) It is also possible that she ultimately turned her energies to teaching, some have argued, in order to pursue a more conventional life of female piety. Her brother, a poet, hinted that she had tried poetry as a vehicle of expression, but if so, no evidence of her output has been found. Another theory holds that her father's death removed a crucial support for her intellectual life. Without his presence, she may have simply given up writing. She died, eleven years after her last composition, at the age of thirty.

Literary praise of milk remained at the sidelines in the next century in Europe. The extravagance of Platina and chef Martino also fell into the shadows, eclipsed by a more anxious approach to the consumption of food and drink. Yet the intellectual legacy of the Renaissance discussion of milk lived on in dietary literature that multiplied, unabated, in the north. Particularly in England, where Italian humanists provided a staple of university learning, the familiar chestnuts of advice appeared in renewed form.

Anxiety at the table was so great by the end of the sixteenth century that, according to one account, every host invited his physician to dinner in order to have his advice at the ready. Doctors complained that questions flew at them so relentlessly that they were unable to finish their meals.[39] Judging from the surfeit of advice books on food in this period, dinner guests were probably behaving badly in spite of warnings. "What a reproach is it, for man whom God hath . . . endowed with reason," intoned an English author, "to be yet beastlike, to be moved by sense to serve his belly, to follow his appetite contrary to reason?" He struck a chord typical of the age: "all appetites are to be bridled and subdued."[40]

Many familiar teachings survived the voyage to England: Galen, Avicenna, and the School of Salerno were all there, in recognizable phrases. Though the permissiveness of Platina had been banished, Ficino's ghost appeared at the table, recommending music to soothe the troubled spirit. Among the many facts assembled for common use were several relating to milk. First, "it is especially good for them which be oppressed with melancholy, which is a common calamity of students. And for this purpose it should be drunke in the morning fasting abundantly, new milked warme from the udder," with a little sugar or honey. Cambridge and Oxford students were evidently the latest catchment group lined up for the cure. (This is perfectly in keeping with what historians have found out about their generous intake of two other commodities, sugar and coffee.)

More remarkable was the way in which the miraculous curative properties of female breast milk appeared in testimonies from England and

America. "Common experience proveth that womens milke sucked from the breast is without comparison best of all in a consumption," Thomas Cogan's *Haven of Health* reported. The author bolstered his case with local evidence:

> Whereof a notable example was shewed of late yeares in the old Earle of Cumberland, who being brought to utter weakenesse by a consuming Fever, by meanes of a Womans sucke together with the good counselle of learn'd Physicians, so recovered his strength, that before being destitute of heires male of his owne body, he [be] gate that most worthy gentleman that now is inheritour both of his fathers vertues and honour.[41]

More remarkable still was the case of John Caius of Caius College, Cambridge, who lived on breast milk for the final years of his life. According to local legend, Caius acquired the temperaments of his various female sources during the course of his regimen.[42]

By the seventeenth century, dietary advice and learned opinion decried overindulgence in food. The line between enjoyment and gorging oneself proved increasingly difficult to draw: even a well-known French dictionary listed "glutton" as a synonym for "gourmet." Across Europe, the "gourmandise" was seen as the "consequence of luxury, which leads to decadence." Once again, the pendulum swung against "the sophisticated gourmets of antiquity," who were "denounced as a warning" to the later generation. French cuisine of the late seventeenth century eventually adopted refinements that prohibited "the prodigious overflowing of dishes, the abundance of *ragouts* and *gallimaufries*." The age of classicism in cooking was born from the reaction against undiscriminating excess.[43]

Out of this fire of condemnation, the humble country commodity of milk emerged stain-free. Particularly in the north, among the Swiss, Scots, Irish, Dutch, and English, milk was considered a delicacy.[44] The courtly style of Italian cuisine never transplanted successfully in northern soil, where taste did not recoil from rustic commodities. What we now consider "traditional English country cooking" was once the cuisine of the well-to-do and the aristocracy.[45] According to the season, milk drinking was well established as a custom and even a staple of ordinary consumption. As for the secondary product of butter, the Dutch were renowned for eating it all day, not as a luxury commodity, but rather, as a spread slathered all over their bread. Butter was "the Flemmins Triacle" (treacle, or sugar syrup), according to the English, who believed it should be used only for cooking. Only the laboring classes ate it in great amounts.[46]

Then, as now, scientific knowledge sometimes surrenders when faced with longstanding custom. Medical opinion in this northern setting thus placed milk on firmer ground than one might have expected. Thomas Moffett's *Healths Improvement, or, Rules Comprizing and Discovering the Nature, Method, and Manner of Preparing All Sorts of Food Used in This Nation,* probably written at the end of the sixteenth century, remained unpublished for a half century, until his progeny apparently saw its value (and no doubt its potential for financial return) years after the celebrated doctor's death. Moffett's promotion of cow's milk stands out as note-worthy, given that in previous works on health, other types had surpassed it on the list of "best milks."

> Now and then taken of sound men (not subject nor distempered with hot diseases) it nourisheth plentifully, encreaseth the brain, fatneth the body, restoreth flesh, asswageth sharpness of urine, giveth the face a lively and good colour, encreaseth lust, keepeth the body soluble, ceaseth extream coughing, and openeth the brest; as for children and old men they may use it dayly without offence, yea rather for their good and great benefit.[47]

The list of milk's benefits had grown noticeably lengthier. Perhaps, too, the general public was deemed sturdier, or at least sturdy enough to toler-ate a fair intake of milk and its products. Of particular goodness were curds and whey, from which children stood to benefit. And so milk found its general defender in the shape of a doctor universally known today not for his medical advice, nor for his groundbreaking work on insects, but for his homely rhyme about his daughter Patience and her taste for fruits of the dairy.

PART 2

Feeding People

CHAPTER FOUR

Cash Cows and Dutch Diligence

The achievements of the Dutch during their Golden Age were fair game for an elevated form of gossip in the seventeenth century. "The Old Hollanders were formerly despised by their neighbors," an English statesman remarked on a visit to the Low Countries, "for the grossness of their temper, and the simplicity of their life. They were us'd to be call'd *Blockheads, and eaters of Cheese and Milk:* but [just] as they formerly had the reputation of silly, so now they are esteemed as subtil and understanding a Nation as any in Europe." In this insult aimed at dairy products, it's clear that the ancient measure of barbarism was still at work, at least rhetorically. Yet signs of Dutch prosperity and civility were everywhere: in the appearance of Dutch painters studying in Italy, in the renown of ingenious Dutch inventors across Europe, and in the obvious material comforts enjoyed in Dutch homes. "Good pictures are very common here," William Aglionby reported back to his English readers, "there being scarce an ordinary Tradesman, whose House is not adorn'd with them."[1] As invaluable cultural capital, their artwork no doubt helped the Dutch assume a higher station as tasteful consumers in the 1600s. Launched by their stupendous economic success, the Hollanders managed to shed the stigma once imposed by milk.

It was no accident that many Dutch paintings depicted robust cows and diligent milkmaids, along with pyramids of butter and cheese. A delicious irony lay in the fact that at least part of the explanation for Dutch

Vincenzo Campi, The Cheese-Eaters, *ca. 1585. Fresh dairy products, in this case ricotta cheese, constituted rustic fare, regarded in the seventeenth century as best suited to the coarse physical nature of the lower orders of society. Campi produced many paintings of food and peasants. According to scholars of his art, he presented his humans as close to nature, not always in flattering ways. (Musée des Beaux-Arts de Lyon, Lyon; photo Alain Basset, © Musée des Beaux-Arts)*

triumph as a "civilized" global power could be found in the microcosm of dairying, that bastion of gastronomic barbarism. Dutch cheeses, butter, and milk became the focal point of excited discussion about agricultural improvement. In the dairy, knowledge of nature, commercial acumen, and hard work came together with almost combustible power. Here was a worthy proving ground for what historians now call "the Dutch miracle," the emergence of the United Provinces as a leading commercial power buoyed by a flourishing, well-fed population.[2]

One could find no better source of information about the Dutch than the English, who wavered between mild envy and fierce rivalry when confronted with their doppelgänger across the Channel. The two nations shared many traits: an orientation toward trade, a strong work ethic, a progressive Protestantism that valued scientific and rational inquiry, and a championing of unpretentious domestic virtue. Yet as commercial competitors, Holland and England went to war three times in the seventeenth century, making relations more suspicious than affectionate. Dryden summed up English disdain in *Amboyna,* his drama about Anglo-Dutch rivalry in the Far East: the Dutch were upstarts from "seven little rascally provinces, no bigger in all than a shire in England." Most English people believed that their avaricious neighbors were plotting to take over the known world. Why, a visiting Dutchman asked his hosts in a English country tavern, were his countrymen called "butter-boxes" by the English? Because, he was told, they "find you are so apt to spread everywhere, and for your sauciness, must be melted down."[3] Slurs against the Dutch by way of their devotion to dairy products abounded in English propaganda. "A Dutchman is a lusty, Fat, Two-Legged Cheese-Worm," ran one pamphlet. "A Creature that is so addicted to eating Butter, Drinking Fat Drink and Sliding that all the world knows him for a Slippery Fellow." Responding to the Dutch claim to being the new Canaan, an English writer added "a Canaan, but seated in a Bog and overflowing with milk and water instead of honey."[4]

The Dutch approached their milk and their cows with a high regard that the English nevertheless would strive to emulate. Visitors like Aglionby lavished praise when describing the magnificent beasts producing the liquid beneficence; they were "bigger and taller than in any other place." The meadows "feed a world of Cattle," he reported, "particularly a large sort of Cows, which give great store of Milk, of which is made excellent Butter, and rare Cheeses, which are sent all the world over. In some places there are Cows that yeeld three great Pales full of Milk a day." Compared to the output of livestock from elsewhere, including England, this was remarkable indeed. The lushness of the meadows, modified by human design, deserved some of the credit. Nearby regions like Holstein and not so nearby Denmark would send their cows and oxen to fatten up on "these excellent pastures" in a near-miraculous three weeks' time. Aglionby became increasingly smitten with the cornucopia that was the Dutch countryside, much of it reclaimed from the sea, when he reached Edam in the north: "Here is made the best Holland Cheese with red Rinde, so much sought after by all Nations," he trumpeted; "and indeed," he added, sounding dangerously close to having crossed over enemy lines, "it yeelds not to the Parmesan."[5]

Where dairying reigned, nearby resided simple virtue. This maxim was most apparent in Holland, where, among all the provinces of the Low Countries, energetic labor and its fruits were on perpetual display. Aglionby gazed upon a terrain "full of navigable Lakes, Rivers, Channels, all of which are night and day loaded with Boats and Passengers." Dairymaids ferried from house to pasture twice a day to milk the cows, while cows themselves ambled onto boats on their way to greener pastures or town butchers. Country people shipped their milk, butter, and whey to town, where the middling sort depended on it for daily sustenance. The movement of goods for sale, foremost among them butter and cheese, never ceased. "It is an ordinary saying in *Holland,* that *He that will work can never want,* and it is a very true one," Aglionby observed, "for there are so many Trades kept going by their great commerce, that no body can want work." He might have added a few English proverbs, perfectly suited to the Dutch economy: "God giveth the cow, but not by the horn." (The cow's bounty will come to nothing if you don't work at getting it.) "God gives the milk, but not the pail." By the 1660s, the Dutch were foremost masters of both horn and pail, along with cheese room and churn.[6]

The wellspring of general Dutch industriousness remained a favorite subject of speculation among the English. Some theorized that it was all due to habitat: forced into a front-row seat in the theater of water transport, according to this line of thinking, Dutch people simply made themselves the entrepôt for long-distance trade between points north and south. Aglionby endorsed this view: "The situation of this noble Province [of Holland] is such," he observed, "as if Nature intended it for the generall Mart of Europe." Yet it was more than just geography, given that the Dutch seemed ineluctably drawn to dealing in goods, trucking and bartering wherever they went. "They never complain of the pains they take, and go as merrily to the Indies, as if they went to the Countrey Houses." If not geography, then why not biology, such as it was understood in the seventeenth century? "They are rather given to Trade and getting," Aglionby wrote in a moment of reflection on Dutch manners and dispositions, "and they seem as if they had suck'd in with their milk the insatiable desire of acquiring." Here was supply-side economics in its infancy, the wellspring of the "industrious revolution" that history professors of today lay out for their students. Perhaps it is not surprising that the germ of so much lay hidden in milk.[7]

A paradox informed the birth of the commodity of milk: though dairy products remained rooted in small-scale, rural settings, their potential as capitalized commodities produced for the market lay in the urban nature of seventeenth-century Holland. The Dutch example provides a perfect case study of how, because of the scale of demand in cities, small, dispersed dairy enterprises transformed into intensified and extensive operations aimed at the market. After 1600, the history of milk would forever re-capitulate the experience of the Low Countries, where city and country quickly established a successful symbiosis in producing, selling, and consuming great quantities of milk.

Urban growth following the European Renaissance went hand-in-hand with a demographic revolution: population grew at a remarkably fast rate between the early sixteenth century and the mid-seventeenth century. For the Low Countries, this lasted longer than in nearly all other parts of Europe. The reasons remain a subject of debate, but the reality of the explosion is uncontested. Headcounts were unsystematic, but estimates show that the region of the Netherlands grew from roughly 450,000 people in the early 1500s to 1.1 million in the mid-1600s. Numbers concentrated in particular areas: by this time, the province of Holland "could claim to be the most highly urbanized and most densely populated province in western Europe." Of particular note was the *Randstad,* a blossoming circle of eight cities within a twenty-one-mile radius of one another: Amsterdam, Haarlem, Leiden, The Hague, Delft, Rotterdam, Utrecht, and Gouda. This heavy burden of human demand gave the Dutch an entirely novel set of problems to solve.[8]

A population expert once wrote that the city is like a minotaur that hungrily consumes labor from the surrounding countryside, but perhaps the image more aptly describes how urban areas have devoured food supplies throughout history. Prosperous Dutch burghers in Amsterdam and Utrecht, like all well-to-do city dwellers, required a full range of provisions in the seventeenth century. Growing numbers of laborers, however, along with tradesmen and shopkeepers, presented an entirely different kind of demand for food. Cheap staples like bread, ale, and beer were a must for the common people, making the price of grain the key to social harmony as well as survival, especially in an age of epidemic disease. As the continent's agents of transport, the Dutch were doubly fortunate: they were able to import whatever provisions they couldn't conveniently grow at low cost. By midcentury, they were relying heavily on grain imports from the Baltic region, the breadbasket of much of Europe. As a result, the Low Countries enjoyed plentiful and inexpensive food relative to other nations in the seventeenth century: most households spent only 30 percent of their budget on provisions. In an era before the cultivation of the potato, this was a remarkable feat indeed.[9]

But bread alone did not satisfy the appetites of Dutch commoners: their taste for dairy products, along with plentiful fish, enabled them to live and eat well. "Milk is the cheapest of all Belly-Provisions," observed one seventeenth-century Englishman while visiting the Low Countries. Milk was not simply a beverage like beer: it was hunger-satisfying nutrition that proved suitable for laborers needing to maximize their caloric intake. It was also cheaper than beer, probably because of its perishable nature.[10] Dairy products, the "white meats" of northern European diets, served as inexpensive substitutes for the real thing. The fact that poorhouses served milk and cheese indicated how these products fit into the common diet: they functioned as carefully doled out sources of protein and fat for diets scarce in nutrients.[11] Cheese was also relatively inexpensive, though sometimes still the victim of an earlier century's apprehension. According to one seventeenth-century doctor, cheese was fit only for "gravediggers and the poor." And a Dutch professor of science, determined to snuff out what remained of common prejudices, confided in 1658 that "many people" still believed that it was "deadly and that its consumption breeds diseases that may lead to death." His treatise even warranted a second edition (with the subtitle "A Diatribe on the Aversion to Cheese"), though we can be fairly sure that ordinary consumers wouldn't have been reading him. Perhaps elite consumers were paying attention, however, and that may be the relevant point here: by the mid-seventeenth century, everybody was eating cheese.[12]

For the better off, cheese was more than just cheese: it was a rose of many different varieties. The travel writer John Ray jotted down a careful catalogue of "the four or five sorts" that "they usually bring forth and set before you" as a visitor to their homes:

(1.) Those great round Cheeses, coloured red on the outside, commonly in England called Holland-Cheeses. (2.) Cummin-seed Cheese. (3.) Green Cheese, said to be so coloured with the juice of Sheeps Dung. This they scrape upon Bread buttered, and so eat. (4.) Sometimes *Angelots*. (5.) Cheese like . . . our common Countrey Cheese.[13]

Ray's mention of sheep dung proved the resilience of popular prejudice; the green color of this cheese would have been produced by the addition of herbs like parsley or sage. "Angelots" were small, round, creamy young cheeses, a sure sign that well-to-do consumers were enjoying Ray's menu of products, possibly as dessert. The final reference to "Countrey Cheese" implied something wholesome and mild-flavored. Such bounty occupied a constant place on Dutch buffets, as supplements to a diet rich in meat, fish, and vegetables — Ray mentioned "Boil'd Spinage, minc'd and buttered (sometimes with Currans added) is a great Dish all over these Countreys." The fact that the general diet of the Dutch could be described as bountiful and superior spelled hope for the lowly commodity of cheese. In the history of food, this was a case of a rising tide lifting all boats.[14]

Dairy cows were the favored providers of milk in the Low Countries, being better suited to the lowland pasture than goats. And in a region of small holdings, cows were more economical producers: one of them produced the same amount of butter and cheese as ten sheep.[15] Farms closest to the cities were in luck: their produce could be transported directly into urban locations, and as cities absorbed commodities, farms devoted more effort to the production of dairy products. Even before the big push of the mid-1600s, farmers complained of having to cart goods ten miles to town markets and so got their way: merchants agreed to intermediate locations, where dairy products were sold more frequently than before. Areas surrounding Amsterdam and Rotterdam found that the urban thirst for milk had skyrocketed. The use of butter became nearly ubiquitous and in steady supply in this industrious region. The Dutch were wealthy enough to afford the generous use of salt to preserve it, too; as a result, some supplies lasted as long as three years.[16] Waterways expanded and also multiplied in number to accommodate boats full of milk, butter, and cattle from surrounding areas. By the 1640s, the city of Gouda was marketing nearly 5 million pounds of cheese per year; by the 1670s, the figure had surpassed 6 million pounds. The spectacle of so much trade inspired foreign visitors to Hoorn, an important market city with many dairy merchants, to tally how many wagons passed in and out of a single gate of the city on a single market day. The count: just under one thousand. This was a society that was constantly on the move with goods to sell.[17]

Assured of urban markets, assiduous farmers in even the most un-promising low-lying areas set to work draining and ditching land. Whole lakes disappeared. Dikes, watermills, and windmills appeared across the landscape: the trademarks of Dutch identity had been born. Villages that had previously supported mixed economies turned entirely to food and sometimes only dairy production. In the town of Broek, outside Amster-dam, the prospect of supplying the city with fresh milk resulted in the drainage of three lakes, creating 750 morgen (about 460 acres) of new land, enough for forty dairy farmers. Investors with an eye to profitable development helped to transform other maritime villages. Urban thirst changed the Dutch landscape, increasing not just acreage but the literal size of the nation.[18]

If we stop to consider the size of these new farms, each on an average of about eleven acres, a question arises: how did such small enterprises sup-port the kind of output needed to serve the conurbations of the region? Even if the dairy farmer maximized his potential by owning as many cows as his land would hold (in the case of Broek, that would have been around seventeen), the cost of equipment and labor per farm would have elimi-nated hope of much gainful reward. Plenty of small farmers tried and failed to adapt to the new ground rules of rural life, draining and ditching land that they then could not afford to rent. Their paths merged with those of other migrants to the cities. The only way that such arrangements could succeed financially would be through intense efforts to maximize produc-tivity: each cow needed to produce an extraordinarily high yield of milk. And that is precisely what the Dutch and their cows succeeded in doing.

Before the development of modern chemistry (or pharmaceuticals), one method alone promised to boost a cow's output of milk: a rich diet obtained from dense pastures and grassland. And in order to enhance the land that would fatten and fuel Dutch cows, farmers harvested that un-ceasing by-product of urbanization, manure. The Dutch surpassed nearly all other Europeans as gatherers and carters of every variety of dung. Regions surrounding cities absorbed street sweepings, pigeon droppings, and manure from urban cows, along with industrial waste like ash from soapboilers. A thriving trade in night soil, heaped up in boats moving out of urban areas, lent a different coloration to scenic water traffic. The constant movement of manure gave a kind of literalness to the English epithet for the Low Countries as the bowels of the earth. Symbiosis between country and city was a literal reality. Urban visitors made note of the wonderful cleanliness of streets and doorsteps, compared with other cities in Europe, but they seldom connected this particular virtue to the commercial drive of Dutch farmers.[19]

The intensity of the effort to fertilize can hardly be imagined. In regions of arable land, where crops like wheat, hemp, and especially tobacco demanded high fertility, the project reached formidable heights. One scribe-like farmer in the 1570s recorded every cartload of manure that crossed his property: on average, he reported spreading about thirty-one and a half tons per acre every six years. In Groningen in the first half of the seventeenth century, edicts ordered settlers outside the city to plow in the public night soil collected by municipal authorities. The city offered bounties to those who would fill their boats on the outward voyage after delivering goods from the countryside. Even the most modest strip of farm land received its due. In dairying regions, where resident cows would offer an on-site supply, farmers felt the impetus to improve tiny plots of arable land for fodder crops. On narrow margins and ridges of property, they transformed neglected strips into productive hayfields. Cutting-edge techniques disseminated with a vengeance in this kind of environment, making the Low Countries one of the most progressive agricultural regions on the continent.[20]

According to every textbook on European modernization, Dutch "convertible husbandry" takes first place among methods of the new agriculture. This regimen of crop rotation and livestock management reaped more food from the same land, thus making it possible to feed urban people busy pursuing non-agricultural work. No new technology, apart from knowledge itself, was necessary. Convertible husbandry demanded a level of understanding of soil nutrients, as well as a basic minimum of land. The method consisted of a constant rotation of crops, including winter sowing of nitrogen-fixing plants like clover and legumes, in order to replenish the soil. While other countries burned turf or simply left fields fallow to prop up their depleted soil, the Low Countries found a way of reaping two benefits at once. Sowing the right crop at the right time, or stationing animals in a field according to a careful calendar of rotation, would result in enrichment of output. In any language, the term was understood in more ways than one.[21]

How successful were these efforts to wring more milk from the land and livestock? One Dutch historian of agriculture, B. H. Slicher van Bath, gathered impressive evidence of milk yields for the period before systematic record keeping of the nineteenth century. Our late-sixteenth-century recorder of manure spreading, Rienck Hemmema, not surprisingly, outstripped all others in output at nearly 357 gallons per cow; he was light years ahead of an assiduous Englishman by the name of Robert Loder, whose best cows each gave between 208 and 225 gallons annually in Oxfordshire in 1618. No ordinary farm would record an achievement like Hemmema's until the beginning of the nineteenth century.[22] Figures for cheese on Dutch farms point to the same conclusion: a rise in production "can only have resulted from an increase in the milk yield." True, herd sizes and barns show an upward trend during this period, but the overall output also indicates that each farm was turning out its maximum of commodities. What else could explain the steady increase in the capacity of cheese kettles found on Dutch farms? Between the 1550s and the 1650s, the average size of these vats increased by two and a half times. And what else could explain the exponential rise in the worth of a single Dutch dairy cow? In the one hundred years after 1565, its price increased more than four and a half times. No wonder that many households hung paintings of cows on their walls; these were tantamount to pictures of Dutch piggy banks, or, more literally, cash cows.[23]

Such industry and abundance belonged to a larger picture of Dutch culture. In the realm of food, the Dutch weighed in as hearty eaters and banqueters in the seventeenth century. Not for nothing was this period called the golden age, when mountains of food stood as bountiful evidence of God's providence, evident in Dutch paintings and also in real life. The accounts left by a Frisian innkeeper for refreshments provided at his own funeral in 1660 suggest astonishing appetites (and hospitality). In the small town of Sloten, the wake would have offered food for all, yet the list of meats (totaling 1,850 pounds, not counting the 28 breasts of veal, 12 sheep, and "18 great venison in white pastry") meant that the people in attendance must have eaten themselves into a stupor. Tellingly, cheese, bread, and butter, along with mustard and tobacco, were simply measured by the words "in full abundance." These were Dutch amenities that would have been lower down the ladder of prestige, but nevertheless counted upon. Butter and cheese graced every table.[24]

The sanctity of the domestic hearth, together with the groaning board, always included foods from the dairy. The best thing about such items was their double-sided identity: from one angle, milk, bread, and cheese were simple and wholesome foods, badges of humility in dietary habits. As the Dutch humanist and physician Heijman Jacobi advised, the healthiest life could be lived enjoyed on "sweet milk, fresh bread, good mutton and beef, fresh butter and cheese." (Note how the good doctor slipped in his mention of meat between so much rustic simplicity.)[25] On the other hand, cheese and its near cousin butter were also symbols of richness, "fatness," and abundance. The medieval reputation of butter as an aristocrat's spread lingered over the lavish use of the stuff in the Low Countries. The Dutch were known to lay down a carpet of butter on slices of bread before adding cuts of meat, a habit which, to the English, smacked of culinary entitlement. Such expectations appeared to be ubiquitous in the seventeenth century. Even poorhouse residents received their portion of butter with certain dishes, along with sweet milk and buttermilk. The fact that everyone in the Low Countries ate their bread with butter hinted at a startling element of dietary egalitarianism in Dutch culture.[26]

Successful dairy farming produced an abundance of material wealth in the Low Countries. Seventeenth-century farming families did well by the standards of the time. A typical modest means could be found in the estate of one Jan Cornelis Schenckerck of Oudshoorn, who died in 1700 owning seventeen milk cows, slightly over the median for Dutch dairy farms. His material goods included three books and eight paintings, suggesting that he and his wife were not only literate but had aspirations to an aesthetically pleasing domesticity. The rest of Schenckerck's inventory confirms this impression: twenty-nine bedsheets, seven tablecloths, and eighteen napkins, along with a valuable collection of clothing and woolen items. In fact, many Dutch farming families in the 1600s owned silver items in the form of spoons, beakers, and head decorations, in some areas rising as high as 67.5 percent of all households.[27] Clearly the rise of commercial agriculture promised great reward, if conducted in a way that attended to new opportunities in land development and marketing.

Dutch culture of the Golden Age has been called materialistic, yet if this was so, then a relatively down-to-earth perspective governed its taste.

Proof of this rested in aesthetic icons of Dutch prosperity, still-life paintings. In particular, the monochrome *banketje* (literally, "little banquet")
functioned as a celebration of ordinary commodities. These table scenes
conveyed a message and spirit captured perfectly by the poet Jan Luiken:

> a table daily laid with care,
> with nourishment and other ware,
> where in companionship and peace,
> all troubles that are elsewhere cease.

The banketje was a pastoral image in a domesticated setting: it invited the
viewer to reflect on the goodness and beauty of a bountiful earth, fetched
from nature and now on display as indoor personal property. Such paintings, like the objects that they depicted, were remarkable performances of
careful attention and skill. Each item was worthy of precise reproduction.
The term "still life" was of subsequent invention; at the time, these works
were regarded as pictures of particular things, and they were titled as such.
"Painting of a layed table with a pie" or "breakfast with a smoked herring"
would have appeared on sellers' lists. We can imagine prospective buyers,
perhaps influenced by what they had seen in the houses of others, leafing
through the catalogues: "What shall I have? The Van Vlecks have a lovely
picture of fruit. What would I like to see on the wall over the sideboard
table?"[28]

"Breakfast paintings" proved to be a favorite. First developed in the
1620s and 1630s, displays of cheese, bread, and butter multiplied in number through the 1650s. Such images advertised the successful marriage of
virtuous Dutch labor and providentially rich land. Ale and sometimes fish
joined the sacred trilogy, along with cake or some form of biscuits, usually
appearing simple but substantial alongside an attractive beaker or cheese
knife. Massive wheels of Edam and Gouda radiated an ochre ambiance
within otherwise modestly dull scenes; the coloration of the setting tele-
graphed a domestic aura of well-aged paneling and privacy. Signs of the
human hand were evident in the cuts apparent on the cheese's surface,
remindful of the true function of the commodity as sliced sustenance.
These basic foods constituted the morning menu of all Dutch people, high
and low, and as such, embodied the Dutch reverence for moderation and
simplicity. Yet cheeses, butter, grapes, and bread became much more than
the sum of their parts: they celebrated the abundance of real substance in
Dutch life.

How much symbolic weight should we attach to the cheese in breakfast paintings? Debates among art historians have belabored many suggestions hidden in the pictures, from maggots (indicating the dangers of excess and the inevitability of mortal decay) to heavenly milk (signaling the reverse process, immortality, via the miracles of Christ). Yet it is unlikely that Dutch artists would have known or wished to emphasize Catholic metaphors derived from Tertullian texts from 100 C.E.: this much, critics now agree upon. At least some paintings point to a deeper moral message, warning of the dangers of heedless prosperity. Especially because the menu was everybody's menu, plus or minus a slab of meat, breakfast paintings could speak powerfully about differences of social class and the implications of that for future reward or punishment. One pair of breakfast pendants by Antwerp artist Hieronymus Francken clearly headed in this direction: by showing rough containers of food and drink in one painting and a sumptuous display in another, the artist pointed to biblical lessons about the way to heaven. (A window opening out onto a landscape featuring a road helped this interpretation along.) But perhaps the ubiquity of cheese in Dutch discussions of material life has led art historians astray. Svetlana Alpers has argued that not all Dutch realist paintings should be analyzed as symbolic texts; she suggests that these pictures can be seen simply, as literal maps of the Dutch world.[29] Could it be that many paintings of cheese only celebrated the material reality of the subject in an exclamatory way? Here, one might say, lies a huge delectable heavyweight of provender. Just *look* at it!

But surrendering to common sense is not as much fun as pursuing a mystery, particularly one connected to food. A final foray into the chiaroscuro world of still life is worth making. As a subgenre of the subgenre of breakfast paintings, some works coupled cheese with butter, another ubiquitous necessity, albeit one with an aristocratic pedigree. One strikingly distinctive master of the cheese-and-butter display, Clara Peeters, was not afraid to produce an over-the-top version. Very little is known about Peeters, apart from an assumption that she was born in Antwerp and worked for part of her life in the northern Netherlands, where she painted a number of breakfast pieces. She follows several other contemporary artists in stacking her cheeses, one on top of another, in an oddly hyperbolic way. (This has helped art historians to pin down her dates to the first two decades of the seventeenth century.) Yet her "pièce de résistance" (these are words of the foremost historian of Dutch still life, Eric

This breakfast painting by Clara Peeters suggests moral as well as visual tension through its precarious placement of a dish of shaved butter on top of several cheeses. The "devilish feast" that combined butter with cheese represented material excess and even gluttony, according to Dutch practices. (Photo: Digital image © 2009 Museum Associates/Los Angeles County Museum of Art/Art Resource, New York)

de Jongh) turned out to be her willingness to balance a plate of butter on top of her cheeses. Surely this teetering pyramid foreshadowed ultimate calamity. Eating cheese with butter, in many quarters, was not done; either one or the other was coupled with bread, but not both. De Jongh pointed to a quotation from a satirical comedy of 1682 to explain the contemporary attitude. Encouraged by a devil, a servant buttered slices of bread before laying on thick slices of cheese, and this was described as "a devilish feast." It is very likely that Clara Peeters was pointing to the troubling effects of "too much," made possible by the overflowing riches of the dairy.[30]

As one observer put it, "Holland is a country where the demon gold is seated on a throne of cheese, and crowned with tobacco."[31] The success of Dutch commerce inspired many an envious comment. A seventeenth-century globetrotter could stumble across the celebrated Edam cheese in Venezuela, where Spanish colonists would supply hides and cocoa for the renowned red rinds. Dutch butter traveled well across the northern Baltic. In several varieties, it was the queen of its kind and fetched higher prices than the plentiful Irish commodity. This was no secret to the poorer inhabitants of the Low Countries, who provided a market for Irish butter in Dutch cities.[32] And not least of Dutch exports was the wholesome Dutch cow. From the early decades of the 1600s, settlers in New York carried along their cows, necessary as sources of dairy products and meat. The first such food ambassador stepped foot on Manhattan soil in 1629 and enjoyed the pastures there. When the Dutch governor demanded payment of rents in cattle, as many colonial administrations in North America did, settlers refused. Rather than giving up their cows, they paid in butter, a form of cash perfectly expressive of Dutch identity.[33]

Back in Europe, the golden age of the Dutch dairy became a more qualified matter as the century wore on. If prices for Dutch butter are the most reliable indication, then the first three quarters of the seventeenth century stand as peak years. During that time, farmers in Friesland and southern Holland concentrated their efforts on producing the commodity for the market, moving out of arable crops like wheat and rye, which they could afford to purchase from Baltic countries thanks to their capable fleets and trade connections. Northern and southeastern Holland, for similar reasons, concentrated on cheese production.[34] But by the last quarter of the century, agricultural prices declined significantly, finally dragging the Dutch Republic into the European-wide economic crisis that was well under way elsewhere. Cost-cutting and further specialization became the only way that farms could stay afloat. A certain sign of the effort to cut labor costs appeared on the horizon of technology: horse-powered butter churns come into scattered usage in the 1680s. Such inventions would have been feasible only on large farms, which were in the minority at this date. Though a logical outgrowth of a market orientation, they foreshadowed a distinctly different approach to the activity that had claimed near kinship with kitchen and barn.

Word of Dutch farming methods, meanwhile, spread to England via an anonymous manuscript describing a visit to farms in Flanders and Brabant in 1644. Such was the forward-looking atmosphere in England, where a tract on how to reap more from the land would end up being published by a second party and sell like hotcakes. Like Platina's *On Right Pleasure* sixty years earlier, *A Discours of Husbandrie Used in Brabant and Flanders* first circulated as a dog-eared manuscript. When the work fell into the hands of a progressive-minded publicist named Samuel Hartlib, such insiders' secrets were destined to become public knowledge. Hartlib had sponsored other projects aiming to make treasures out of earthbound substances; one of these, based on the current-day belief in alchemy, attempted to transform lead into gold as a way of increasing the money supply of England. Here was another means of enriching the nation: farming with a new eye to serving the market while making plentiful profit. "For no man with reason can deny," he asserted, "but that *Land is best, which will bring forth such Commodities as will yield most money to make one wealthy and rich.*" Hartlib designated himself the "conduit-pipe" of the latest wisdom on agriculture, "one of the noblest and most necessary parts of industry" of the nation, and published the *Discours* in 1650.[35]

"So *Regina Pecunia,* Money is the *Queen* that commands all." These were useful words, considering that by 1652, the English had decapitated their monarch and were trying out a republican form of government like that of their neighbors the Dutch. The author of the *Discours,* later revealed to Hartlib, was Sir Richard Weston, who had visited Flanders as a Catholic exile from England during the turmoil leading to the Civil War. As Catholic districts, Flanders and Brabant offered Weston a more hospitable environment for his inquiries and diary keeping than Holland. It took him a while to fathom the miracles being worked on farms there. Interviewing a merchant on the subject (for it seemed that knowledge of these methods extended beyond the farming community), the Englishman was befuddled by what he heard. "I must confess at first I thought his discourse to be some kind of riddle," wrote the English visitor. But Weston was readily convinced and, like scores of converts to various causes before and after, immediately set pen to paper to publicize what he had learned. His somewhat repetitious pamphlet included a monthly schedule of tasks for the faint at heart, along with plenty of accounting for costs of seed, lime, and labor. Weston's magic bullets turned out to be the homely turnip and re-mown clover, but along with these he promoted the miracle of masses of dung. Weston couldn't resist adding his own homespun advice on how to collect sheep dung in night stalls spread with sand, which should be piled up "until the quantity be grown so great, that the sheep cannot conveniently go in or out." It was difficult to convey just how much of the stuff would be necessary.[36]

By the time Hartlib published the improver's manuscript, Weston was back in England and already engrossed in another project: the enlargement of a river appropriately named Wey, which flowed into the Thames and thus to London. The ultimate success of Weston's efforts meant that dairy products from the district of Guildford in Surrey could make their way cheaply to urban markets. Weston thus boosted the entire economy of the district, making dairying "a very prosperous undertaking" in Surrey for years to come.[37] Weston's own fortunes were not so sanguine. The government of Cromwell had seized Weston's land owing to his Catholicism, so he was faced with rent payments that were never fully paid by the time of his death. His enthusiasm for the waterway to London had outstripped his savings: according to records that remain, he invested thousands of pounds worth of timber in the venture. Yet his fellow Englishmen, including non-Catholics, eventually lauded Weston for his contributions to commercial agriculture. His words proved prophetic: "When your Neighbours see your Labors thrive and prosper . . . when they once see your Crops, and somewhat understand that you do reap som[e] benefit by them, they will com[e] to you as to an Oracle to ask your Counsel."[38] Weston's name is now synonymous with the promotion of clover in English agriculture, even though a careful reading of the *Discours* tells us that he backed a less fragrant item as the secret to increases in output. His efforts inaugurated a new era of improvement, which students now dutifully call the "agricultural revolution." The next century would witness a phenomenal rise in the productive power of England, as the northern nation strove to outdo the Dutch miracle. Milk and the dairy, as food commodity and laboratory, were now permanently installed as participants in the race.

CHAPTER FIVE

A Taste for Milk and How It Grew

A melancholy Samuel Pepys took stock of his office in Seething Lane on the third day of the horrific Great Fire of London in 1666. Earlier, he had laid his papers in a hole in the garden, and that evening, with the help of Sir William Penn, he decided to make a second excavation. Here, Pepys solemnly recorded that night, they placed their wine stores "and I my parmazan cheese." In their haste, the servants must have viewed the commodities as inessential, but the two men obviously felt otherwise. An emblem of merry nights of good food and company, the cheese cried out for rescue in the face of the approaching blaze. "Now and then walking into the garden," Pepys wrote, "and saw how horridly the sky looks, all on a fire in the night, was enough to put us out of our wits." Promising the return of better times, the golden bulk sat in the ground until the conflagration subsided.[1]

Close readers of Pepys's *Diary* know that this seventeenth-century gourmand loved milk and every one of its by-products. His enthusiasm is palpable as he tells of a bright spring day outing to the Lodge in Hyde Park, where he and his wife, Elizabeth, "there in our coach eat a cheese cake and drank a tankard of milk." (Drinking milk on the spot was the best way to ensure a high-quality product.) Few of his excursions lacked a treat from the dairy. While visiting Islington, where the cows of the London milk market were pastured, he brought his party to "the great cheese-cake house" for midday refreshment. At the home of a friend who was "a great butter-woman," he and his friends "filled our bellies with cream." Pepys regularly downed afternoon "messes of cream" (sometimes with fruit) and evening meals of bread and cheese at alehouses. Before bed, he often snacked on bread, butter, and milk. Even the sight of dairymaids lifted his spirits; the assiduous diarist made note of the mirth and music of "country-maids milking their cows" in the meadows of Portholme, north of London. When he adopted a regimen of buttermilk and whey "to find great good by it," he could not have found it disagreeable. Only his twenty-first-century readers, aware of the effects of such a calcium-rich diet on a man predisposed to kidney stones, might now object to his penchant for milky pleasures.[2]

If we map dietary dependence on milk onto the past, the northern regions of Europe, along with its mountainous areas, would, of course, leap into view. Not surprisingly, the map would replicate areas of lactase persistence — an ability to digest raw milk — across the continent. This is the area that food writer Anne Mendelson has called "the Northwestern Cow Belt" in order to distinguish its culinary traditions from "diverse sources" of milk in Eurasia, the "Bovine and Buffalo Belt" of the Indian subcontinent, and the Northeastern Cow Belt of Europe and Russia. Mendelson points out that, "geographically, this used to be the smallest of the major Old World regions where strong traditions of fresh dairy foods developed. But after some five millennia it suddenly ended up as the largest."[3] The historical reasons for that spread are addressed in the later sections of this book. The point here is to explore some of the ways in which milk products that were part of customary diets among rural people became thought of as fashionable and desirable for everyone, including urban elites.

Histories of food consumption in these areas reveal remarkable varia-tions on the theme of milk. In Sweden, *drickasupa,* or "small beer sup," served as breakfast food, sometimes alongside (or mixed into) a plate of herring or bread. The mixture of low-alcohol beer with milk, possibly soured, displayed all the hallmarks of a regional dish: it was eaten regu-larly, sometimes from a common plate. (Twentieth-century literature and film have brought this regional specialty into a more general spotlight: Babette is taught to prepare a similar soup in *Babette's Feast,* and a charac-ter in John Steinbeck's *Cannery Row* orders up a "beer milk shake" at a small-town restaurant.)[4] In Pepys's time, hot milk and cold ale, or "poset ale," was regarded as a good remedy for fevers. More predictable were the many cuisines, such as those of Holland and Scotland, dependent on porridge (often oatmeal) mixed with milk or ale or both. The Irish lived on oatmeal and milk; the liquid part was the equivalent of a national currency, which was partly why the Established Church demanded that the tithe be paid in milk in the seventeenth century.[5] Whey, the watery liquid that remained after cheese curds formed in a vat of milk and cream, ranked as a favorite drink across the British Isles. In the Orkney Islands, Daniel Defoe reported that inhabitants drank a strong fermented variety, which they kept in barrels.[6]

Julius Caesar took note of the English predilection for dairy products when he visited the British Isles: "They live on milk and meat," he observed, probably with a note of Roman condescension toward this menu of barbarians.[7] Much had changed since 55 B.C.E., yet some things had stayed the same. The English in Pepys's time were renowned for a rich, carnivorous diet, and now they enjoyed their milk by the tankard, or in the form of syllabubs (a whipped-up mixture of warm milk and imported spices) and combined with chocolate or coffee, depending on their inclinations and neighborhood supplies. Like the Dutch, the English had leapfrogged over Renaissance apprehensions about ingesting milk. By the dawn of the modern age, they had assembled their own set of beliefs about the properties of milk, some of them borrowed from the past and some of them newly imposed on acquired tastes. A selective history of British milk in the eighteenth century can tell us a great deal about how this humble commodity very soon found its way to acclaim in the land of industry.

Anxiety marked the approach of the year 1700 in the British Isles. Despite the triumph of Newtonian science, scores of people, the learned along with the less literate, expected a rude arrival of the New Age. One theologian even used Newton's inverse square law to supply specific dates for a cataclysmic unveiling. Intense interest fastened on the weather — England was experiencing the first half-century of what historians now call "the little Ice Age" — and impending war abroad. Some people waited for the first sign of divine wrath to appear near Rome. Many braced themselves for a direct landing, given that their native land had plenty of sinfulness to account for: excesses of finery culled from Asian and American trade, bogus financial schemes in the City of London, and military ventures on the continent proved that the nation was doomed. Plague and the Great Fire of the 1660s had been a mere prelude to what was to come.[8]

Although the turn of the century passed without an apocalypse, doubts about the materialism of the age remained in full view. Voices from political, economic, and religious circles sounded strikingly similar notes. How was England to come to terms with what the age called "luxury" — the refinements of material wealth and bodily enjoyments so evident in recent years? All of Europe, according to one authority, was sinking "into an Abyss of Pleasures," "rendered the more expensive by a perpetual Change of the Fashions in Clothes, Equipage and Furniture of Houses." Trade was "a pernicious thing," carrying with it "fraud and avarice" and snuffing out "virtue and simplicity of manners." When religious prophets called the Camisards arrived in England in 1706, fresh from a social rebellion in the south of France, their preaching stirred up unrest in the streets of London. Authorities put one of the prophets in the pillory. The fact that he was an accomplished mathematician and European member of the Royal Society displayed to all just how broad the band of disquiet was.[9]

The destiny of milk converged with these events in a peculiar yet powerful way. Religious visionaries invoked milky metaphors reminiscent of medieval religious movements: milk stood for wisdom, the word of God, and spiritual enlightenment. London was awash with preachers and pamphlets. Odd echoes of Galenic medical knowledge were sounded in writings for the common people. For example, in John Pordage's *Theologia mystica,* doggerel verse reminded readers of basic physical chemistry:

Do not torment the Text; nor from it run.
With Words don't darken Knowledg [*sic*]: that's not good.
The pure milk of the Word don't turn to blood.
Don't wrong us of our food: for that's too much.

Female visionaries added their vivid descriptions of "the Heavenly Table," stocked with food for the "Heavenly Palate," and milk was there in all its glory. Jane Lead, a blind widow who became the designated leader of the Philadelphians, a vocal millenarian group, depended on maternal imagery through which she dispensed "the Milk of the Word" along with the "Bread of Life." "The milk of grace" promised to nourish believers, according to Jeanne Guyon, another French visionary. She urged followers to be like infants at the breast, "swallowing down what is given in." The

Francis Wheatley (1747–1801), Milk Seller, *ca. 1795. The sale of milk door-to-door rests on a very long history. The typical eighteenth-century milk seller in London was seldom so lightly laden with her product, but Wheatley's rendering is a charming representation of the ideal destination of the vendor's efforts. (London Metropolitan Archives Print Collection, City of London)*

white serum represented an alimentary blank slate, a starting point for fresh beginnings.[10]

Surrounded by so much apparent moral corruption, those who embraced virtuous eating found a personal path to physical enlightenment. Luxurious foods, including meat, wine, and spices, came under attack as signs of a more general social decline. The solution lay in a simple, wholesome diet that replaced meat with milk products and vegetables. Readers of Benjamin Franklin's autobiography know that experiments with diet contributed to his philosophy of self-help. As a "Tryonist," Franklin registered his support for the vegetarian principles of Thomas Tryon, a free-thinking reformer who rejected meat on spiritual and ethical grounds. (It was Tryon whom Franklin invoked during his famous inner debate about eating the fish in his frying pan.) Moralists and medical men ultimately agreed on many points. One need only consider gout, the disorder *du jour,* to see that "the *Rich,* the *Lazy,* the *Voluptuous*" were being punished for their indulgences. Through no coincidence, when prophets of the millennium wished to describe the torments of hell, they reached for analogies to the pain of this "patrician malady" to drive home their message. Dietary reform thus held out the possibility of salvation, whether secular or sacred. "No sooner is the Divine Eye opened," Tryon explained, "but immediately Man sees that Meats and Drinks are the very substance of our Bodies, Souls and Spirits, and that all the Dispositions spring from thence." The new era embraced the notion of "we are what we eat," and milk provided a sure means of seeking purification and the restoration of a good life.[11]

The career of one physician offers a vivid account of how several channels of thought converged to carry milk into public favor. Few readers today will know of George Cheyne (1671/2–1743), a Scottish-born doctor whose writings now seem eccentric, except for a work tantalizingly entitled *The English Malady*. Cheyne's reputation flowered at the height of the Enlightenment and displayed the perfect combination of old and new styles of thought. His advice spoke to the needs of an age that craved scientific solutions, in this case, for psychological as well as physical ailments. Yet, in many respects, his approach borrowed freely from old-fashioned dietary wisdom. His advocacy of a milk regimen was reminiscent of Galenic principles: milk was a cooling and often purgative substance, and this quality made it ideally suited to counteract the effects of a heavy, overheated contemporary diet. Cheyne's role as prophet at the spa town of Bath gave him ample opportunity to dispense such wisdom to the wealthy and powerful. His patients included the prime minister, Sir Robert Walpole, and his daughter, along with Samuel Richardson, the author of *Clarissa,* and the Countess of Huntingdon, who became the leader of a breakaway group of Methodists in 1783. Word of Cheyne's methods spread across Europe. When French restaurateurs of the late eighteenth century recommended a diet Anglaise, they were paying homage to this advocate of milky abstinence.

Cheyne's personal experience provided the basis for many of his medical insights: his was a tale of dashed ambitions, a search for religious truth, and a struggle against obesity and melancholy. As a specialist in mathematical medicine (iatromathematics, a marginal pursuit in the age of Newton), Cheyne migrated to London in 1701. Why he ventured south remains shrouded in mystery, though a suggestion of a habit of excessive drink appeared in a scurrilous pamphlet attack. He and his mentor, the eminent mathematician and physician Archibald Pitcairne, were known

Portrait of George Cheyne, M.D.
(1671–1743), mezzotint from 1732
(Wellcome Library, London)

for a hearty enjoyment of "the cup that cheers," though how many more cups than the average among Scottish intellectuals is hard to tell. Nevertheless, a bibulous Cheyne applied himself in earnest to scholarly life in London. With his teacher's sponsorship, he became a fellow of the Royal Society and attached himself to the circle surrounding Isaac Newton. But Cheyne obviously had difficulty fitting in. After publishing his mathematical ideas, he earned only the displeasure of Newton and estrangement from colleagues. Medical patients proved hard to come by. Cheyne frequented coffeehouses and taverns in an attempt to build a practice; he also waited upon the wealthy in the hope of their patronage. After the publication of a second book and more disappointment, Cheyne returned to Scotland, depressed and overweight. In his own words, he "grew excessively fat, short-breath'd, Lethargic, and Listless."[12]

Cheyne now wrestled with the pathologies of his mental and physical states, which were clearly inseparable. His enormity, by his own account, reached epic proportions: at one point, he weighed over 448 pounds. Plagued by "a constant violent *Head-ach, Giddiness, Lowness, Anxiety and Terror,*" he "went about like a Malefactor condemn'd, or one who expected every Moment to be crushed by a ponderous Instrument of Death, hanging over his Head." Medical historians have supposed that he was suffering from manic depression, coupled with cardiac arrest and asthma. It was clear, at least to Cheyne, that his ailment was both troublingly common-place and poorly comprehended. Suicides owing to melancholy, especially among intellectuals and clergymen, were numerous during these years.[13] (Richard Baxter, celebrated cleric and friend of Cheyne, believed that women were particularly susceptible to a religious form of melancholy, which he believed was caused by "a *weak Head* and *Reason,* joined with strong *Passion.*" Presumably, Cheyne did not confide overmuch in Baxter.)[14] Cheyne's sympathy with people afflicted by moral and religious doubt placed him in a potentially fatal category. If he was to survive this dismal period of his life, the doctor would have to cure himself.

We know exactly how Cheyne fared during this time because he catalogued his ailments and self-treatment in a piece called "The Author's Case," a favorite genre of medical writers at the time. He appended this medical memoir to one of his most successful publications, *The English Malady,* in 1733. Here was the Enlightenment's answer to Richard Burton's *The Anatomy of Melancholy,* the gold standard of the previous century's thoughts on the subject and no doubt one of the reasons why Europeans generally believed that the English had a special claim to "low spirits." (The plight of Hamlet probably contributed additional proof.) *The English Malady* captured a sharp image of the landscape surrounding this later age of melancholy, which historians have in fact linked to a peculiarly high rate of "self-murder." According to Cheyne, the advanced state of English civilization was causing unprecedented numbers of Britons to fall victim to lowness — he estimated that one-third of all complaints were about the same chronic ailment of melancholy. English air, fertile English soil, English trade, crowded English towns, and, of course, extravagant English luxury combined to bring down the English through this state of mind and body:

> Since our Wealth has increas'd, and our Navigation has been extended, we have ransack'd all the Parts of the Globe to bring together its whole Stock of Materials for Riot, Luxury, and to provoke Excess. The Tables of the Rich and Great (and indeed of all Ranks who can afford it) are furnish'd with Provisions of Delicacy, Number, and Plenty, sufficient to provoke, and even gorge, the most large and voluptuous Appetite. The whole Controversy among us, seems to lie in out-doing one another in such Kinds of Profusion.[15]

Luxury had delivered the final blow to the English environment, Cheyne argued. The nation was drowning in its effects and was destined for despondency and disease.

As proof, Cheyne pointed to the effects of manipulating livestock in order to produce meat for the market. Like a scientist citing laboratory experiments, he showed how Epicurean taste threatened the animal kingdom. Cheyne sounded very much like his hidden mentor, Thomas Tryon, when he railed against the self-interested treatment of animals. Here was a demonstration of how humans acquired their nervous complaints:

> Instead of the plain Simplicity of leaving Animals to range and feed in their proper Element, with their natural Nourishment, they are physick'd almost out of their Lives, and made as great Epicures, as those that feed on them; and by Stalling, Cramming, Bleeding, Lameing, Sweating, Purging, and Thrusting down such unnatural and high-season'd Foods into them, these Nervous Diseases are produced in the Animals themselves, even before they are admitted as Food to those who complain of such Disorders.[16]

Cheyne's moral message was as important as his dietary recommendation. While moderate eighteenth-century opinion would have recommended fresh air, pleasant surroundings, music, coffee, exercise, and religious contemplation as legitimate treatments for melancholy, Cheyne advocated a self-imposed moral and meal reformation. Both attitude and diet would have to change in order to get at the root causes of the malaise that plagued the highly advanced English nation.[17]

For Cheyne himself, the path to milk and enlightenment was paved with deprivation and despair. The doctor first attended to his weight: he gave up suppers and ate hardly any meat at his midday meal. Nighttime carousing was no longer permissible. His "bottle-companions" soon found that their old friend was no fun; they "dropt off like autumnal Leaves," he recalled, leaving him alone "to pass the melancholy Moments" while they sought out the "cheer-upping Cup" on their own. Isolated, his melancholy only grew worse. Cheyne next prescribed for himself a rest cure of country air, along with a regimen of "frequent Vomits, and gentle Purges," punctuated by herbal concoctions and mineral water. At least one aspect of his disability began to improve: "my body was, as it were, melting away like a Snow-ball in Summer," he recorded; but his misery continued.

Recovery was still a long way off and would ultimately demand spiritual and moral components as well as a more substantial dietary discipline.[18]

Religious enlightenment led Cheyne to consider a holistic cure. As one historian aptly put it, mystical millenarianism supported people like Cheyne as part of a "huge underbelly of their so-called Age of Reason."[19] The contemplative doctor had no trouble locating other Aberdonians seeking to reject not just luxury but an excess of cold-blooded reason. His circle of associates included supporters of two French-speaking mystics, Antoinette Bourignon and Jeanne Guyon. Both women urged followers to pursue an intense inner quest for spiritual simplicity. Both rejected the rationality of present-day philosophy; Bourignon's followers singled out mathematics, Cheyne's former yardstick of truth, as a particular enemy. Both women emphasized a mystical method of plain living, condemning the tyranny of the senses. "For we see by Experience, that many have become sick, and even died, by too much eating or drinking, or Gluttonously eating things contrary to their Health," remonstrated Bourignon in her *Admirable Treatise of Solid Vertue*. Religious belief of this era, particularly the ecstatic kind, depended on a distinct awareness of physicality. "If you do not carefully order Temporal things," the preface of her *Light of the World* warned, "your body will fall into a thousand sorts of Miseries, Anxieties, Maladies, Disquiets, Confusion, Poverty and want."[20] These were serious threats to a man searching for a path out of a personal crisis. His baptism by asceticism succeeded in making Cheyne "one of England's staunchest anti-luxury campaigners — perhaps *the* fiercest opponent of luxury anywhere."[21]

Thus prepared, Cheyne finally landed on the prescription that brought his fits under control: a regimen known as the "Milk Diet." According to Cheyne's account, a clergyman friend recommended that he consult a remarkable Dr. Taylor of Croydon, who had cured himself of epilepsy by ingesting nothing but milk. Cheyne knew of cases of gout, scurvy, and consumption that had responded "miraculously" to milk, but he wanted confirmation of this latest of applications. "In the Middle of Winter," he recounted, he rode to Croydon and found the good doctor Taylor sitting down to a "full Quart of Cow's Milk (which was all his Dinner)." Taylor gave Cheyne his own case history: all the best physicians of London had failed to cure him, he told his visitor, but now, after seventeen years on a strict diet of milk, he "enjoyed as good Health as human Nature was capable of." The cure could conquer other ailments, too, and Taylor boasted that his own satisfied patients included aristocrats frustrated by infertility, who had finally begotten heirs by drinking milk. Here, finally, was a substantial yet simple food, more satisfying than Bath mineral water and yet not considered rich or harmful. Milk, when taken alone, was capable of jump-starting stalled systems and restoring health to the seriously ill.

Cheyne returned to London, secure in the proven efficacy of milk. Whether Taylor was the actual source of Cheyne's ultimate cure is doubtful: Cheyne's diary of self-treatment followed recommendations of Thomas Tryon's handbooks on health almost to the letter. (Tryon even advised milk for treatment of "corpulency.") A good supply of milk would require special effort in the large city; Tryon and Moffett before him, as milk advocates, would have decried Cheyne's choice of residence. London milkmaids were notorious for adding water to their stores, though the fat content of any serving no doubt depended on how many deliveries had been made before one's own. But Cheyne, immediately upon his return to the Great Wen, wisely engaged "a Milk Woman, at a higher Price than ordinary, to bring me every Day as much pure and unmix'd, as might be sufficient for Dinner and Breakfast." For good measure, he supplemented the liquid diet with "seeds, Bread, mealy Roots, and Fruit." ("Seeds," in this instance, meant whole grains of wheat and barley.) In five or six months, he announced himself "considerably recovered." Though his weight and mood continued to bounce between high and low for another decade or so, Cheyne "always resolved, upon any great Change in my Health, to return to my *old Friends, Milk and Vegetables.*"[22]

Cheyne regarded the milk diet as a last resort for the chronically ill, yet it is surprising how many times he recommended the cure to others in his books and case records. As a physician in residence at Bath, where he moved permanently in 1718, he made the acquaintance of many who profited from his advice. By then married and with children, Cheyne's moment had arrived: with the publication of his treatise on gout, and later *An Essay of Health and Long Life,* he found himself at the head of a considerable following. It was in Bath that he encountered the young Selina Hastings, Countess of Huntingdon, still in her twenties, who was searching for relief from a number of physical ailments. The two struck up a solid friendship, exchanging views on religious matters while Cheyne attempted to alleviate the countess's sufferings. Many of her difficulties appear to have been gynecological, probably stemming from her almost constant state of pregnancy and seven births within a decade. Part of the correspondence between doctor and patient from this period survives and reveals much about the application of the milk diet to a woman with symptoms regarded as typical of the age.

"Drink your whey," advised Cheyne, "all the hot months, take your medicine as long and as constantly as your condition will admit, keep to your diet inviolably, let who will say to the contrary." This was in the spring of 1733, the first year in the previous five when the countess had not given birth. In the year before, Cheyne had advised her to mind a "breeding diet" of white meats and "milk meats, butter, new cheese, light pudding, rice, sago, and the like," along with Bristol water and milk. He expected her ladyship to encounter resistance from friends for following such a diet, but he told her not to be frightened by popular opinion. "For I know no mean [sic] to strengthen a child unborn than to feed the mother after the same manner the child ought to be fed," he assured her. Cheyne coached her through numerous purgings and "coolings" by means of milk products and waters, at one point recommending a "milk punch" as the "best cordial, made with oranges, milk, and arrack." In another instance, he advised her to bathe with milk and warm water, adding a little brandy, if she pleased, to relieve itching.[23]

Though her exact ailment remains unclear to us, a letter from the countess to her husband around this time suggests she was in fear of not being able to conceive again. Already the mother of four children, Selina hoped to restore her reproductive capacities to optimal condition. (She would eventually give birth successfully three more times.) They were devoted to each other, and the couple's correspondence is full of affectionate exuberance. In letters from her husband, Selina became "Goody" (short for "goodwife"); she referred to him as "My Jewel." Cheyne, on his part, appears to have felt strongly attached to Selina, chiding her more than once for her occasional lapses of correspondence. He also urged her husband to allow Selina to remain in Bath when her condition remained weak; he wrote letters to this end, and so did she. "Old Doctor Cheyne assures me he has made my case quite his study and he says he loves me so that he will do as much as art of Physick can do to send me well home to you," she wrote to Theophilus, earl of Huntingdon. "I believe I might even come sooner

Portrait of Selina, Countess of Huntingdon, circulated to advertise a Methodist mission to Georgia that she sponsored in 1773. A full thirty years after Cheyne's death, the countess was in her mid-sixties and capable of carrying on an active life as a sponsor of evangelicalism. (Published June 10, 1773, by Carington Bowles, after John Russell, mezzotint; © National Portrait Gallery, London)

than I mention, has he not absolutely resolved upon my bathing, for he says he thinks it is the only chance I have for going on with another child . . . and that by that means I may even want this place while I live again."[24]

These were difficult years for the countess, who from childhood had nurtured a religious disposition. In Cheyne, she found an eager partner in spiritual conversation, and the doctor shared some of his favorite religious works with his patient. Selina must have probed her inner depths while working on regaining her bodily strength. By the end of this period, she converted to Methodism, a move that shocked some of her acquaintances as much as her eccentric diet had unsettled them. The rest of her long life was marked by her passionate commitment to building her own evangelical brand of Methodism.

In the case of the countess, milk fulfilled a dual role: it revitalized reproductive organs while introducing a purified state of being that had decided ascetic overtones. Repeatedly, Cheyne hung back from placing the countess on nothing but milk and vegetables, though it is clear from his correspondence that he wanted to do so. "Milk is the only certain and infallible remedy for the scurvy in nature," he inserted into a lengthy letter. When she retired to the country to recover her health in 1733, he offered to introduce her to a few of his success stories: "You have three neighbours—vegetable, milk, and low livers—patients of mine in your neighborhood," he reported. One of them "lives entirely on milk and vegetables with seeds, and drinks milk and water. He is infinitely mended, and another summer finishes the cure." By the spring of 1734, Selina had succumbed: Cheyne's letters indicate that she had given up animal meats and was dependent solely on milk and vegetables. As a prize to his best pupil, the doctor sent her a copy of a letter he had recently received from a vegetarian of seventy-two years, "a considerable person in the House of Commons," "for your encouragement to persevere and to hope all."[25]

Cheyne often coupled his recommendations with recipes, which added a homely tone to his otherwise deferential promptings:

I bake all my bread with sweet cow milk and fine flour, without salt or yeast, in biscuits, with a quick and hot oven so as not to burn but brown them. Two quarts of milk with half a pound of these is food for a farmer. Dr. Taylor played six hours at cricket, with only 2 quarts a day. I have long dropped sugar and butter, as fouling the stomach.

Because Selina suffered from what Cheyne diagnosed as a "hot scurvy," this form of cool diet was designed to loosen her system and flush out impurities. His instructions included a step-by-step method of vomiting, recommending the use of tincture of ipecacohana, chamomile flower tea, and a finger or a feather to help along the process. The countess was never without the most detailed instruction.[26]

Cheyne remained attentive to the countess as long as he could, seeing her through the difficult period following the death of her next child in infancy and two more successful births. When hot flashes preyed upon her spirits, he assured her that she had "no fever, slow nor quick." His advice must have sounded repetitious by this time: "I beg you to believe me, have patience, and try whey, cow milk, asses' milk, bread meats, and seed meats, but never too much at a time," he advised. His later letters convey a deepened sense of the friendship that had evolved between doctor and patient over the years. On her part, a final mention of him, a year before he died in 1743, described his "setting with me" and "talking like an old apostle." He did not live to witness the meteoric rise of his dietary disciple as a leader of a dissenting sect of Methodists, which formally separated from the Church of England in 1781 and eventually comprised more than sixty chapels. As a widow, Selina exercised her formidable personality as a powerful organizer and administrator, earning the title of "Methodist archbishop" from her detractors. Though her health was never good, she lived to the age of eighty-four. By then, her biographers suspect, she was addicted to the use of opium, which Cheyne had sanctioned for occasional use in earlier years. Nevertheless, the "dietetical ghospel" of milk no doubt contributed to her long life.[27]

Launched into the Enlightenment, the milk diet lived on, winning more attention and praise from John Wesley in his own book on health, the immensely popular *Primitive Physick,* first published in 1747. The Methodist founder's handbook became a bible of practical advice for more than just his direct followers, enjoying many new editions in later years. (No wonder, given the startlingly inventive suggestions it offered. A head cold, for example, might be attacked through an early version of vitamin-C therapy: "Pare very thin the yellow Rind of an Orange. Roll it up inside out, and thrust a Roll into each Nostril.") Cheyne can be credited for much in Wesley's book and may also deserve credit for reviving an aphorism attributed to Plutarch: "Every man is a fool or his own physician at forty."[28]

Just as significant, Cheyne-ish wisdom traveled to that vortex of Epicureanism, France, where the milk diet earned the appellation of "English-style nouvelle cuisine" in at least one fashionable Parisian restaurant of the 1760s. Menu offerings at such places offered literal *restaurants:* simple preparations that promised to restore patrons to good health and strength. For "the weak-chested" and "the vaporous," hosts served up Breton porridge made with milk, delicately flavored puddings called rice creams, fresh butter, and cream cheeses. Rice creams were a particular favorite, recommended as baby food, a cure-all for ailments, and, not least, the perfect maintenance food for men of letters.[29] Plain milk remained de rigueur for those, like Diderot, who were afflicted with chest complaints. "I have been and still am pretty badly off in my own affairs," Diderot wrote to a friend in 1755 while turning out volumes of his famous *Encyclopédie.* "But things are going much better at the price of a drastic remedy: bread, water, and milk for my whole diet. Milk in the morning, milk at noon, milk at 'tea-time,' milk at supper. That's a lot of milk." The ordinary intellectuals could get by with more occasional servings of rice and milk. By the 1780s, cafés in Paris would make *rice au lait* a regular feature of their menus and thus cater to a growing clientele caught up in pamphlet wars and underground literary life before the Revolution.[30]

Medicine and intellectuals on the one hand, dietary encounters with nature on the other—the paths toward a cultural celebration of milk were in full view by the last third of the eighteenth century. Continental physicians adopted the milk diet with their own set of assumptions and aspirations. For them, the clash between civilization and nature could be mitigated by a program of fresh air, exercise, and simple food, preferably milk-based and vegetarian. The leading authority on nervous diseases, Swiss physician Samuel Auguste Tissot, devoted an entire book to a health regimen for men of letters, recommending this same program. Not coincidentally, the enthralled readers of *La nouvelle Héloïse,* Jean-Jacques Rousseau's best-selling romance of the 1760s, would encounter a puff for milk, too. In this story of frustrated love and tragic suicide, characters discover a dietary path to coveted physical purification through dairy products, along with fresh fruit and sweets. Rousseau was actually promoting the eating habits of peasants in his native Switzerland; these were hardy, hardworking people who dwelled close to nature. Emulating their ways was a sure way to counteract the excesses of upper-class life, which were

This charming goblet, intended for tasting milk, is a reproduction modeled on a unique set designed by Lagrenée at Sèvres for Marie Antoinette in 1787–88. Although Versailles featured its own pleasure dairy, Louis XVI built a second "temple dedicated to milk" at his newly acquired hunting lodge at Rambouillet as a surprise present for his wife. The goblet's pattern was made to appear Etruscan according to recently excavated Italian artifacts, the latest development in porcelain design. (Photo: Courtesy of the Lamplighter Shoppe, Williamsburg, Virginia)

making aristocrats febrile and faint. That peasant food became the rage in Parisian restaurants in the 1760s and 1770s is one of the delightful ironies of Enlightenment history.

Many forces worked together to bring the dairy and its pastoral setting into the limelight of fashionable taste by the end of the century. Literary and artistic interest in the pastoral, which stretched back to the Renaissance revival of such themes, grew in popularity during the Enlightenment: in fact, it was during the eighteenth century that "pastoral" as a term referring to an idealized rural setting came into being. Its persuasive influence radiated from Italy throughout Europe from the seventeenth century onward. The movement affected everyone from Dutch burghers in the market for bucolic paintings to English and French *salonnières* eager for aesthetic transport to Arcadian bowers. As art historian Alison McNeil Kettering has pointed out, the pastoral came to mean more than just a shepherd's song; it was "a point of view and an attitude towards life."[31] A pastoral work of art could have the same effect as milk: it offered a tonic for those who were weary of the demands of a rich and overrefined society.

Life also imitated art: the site of milk production, the dairy itself, became a place of fantasy in the hands of French aristocrats. Such idealized constructions, set in French gardens as "pleasure places" much like gazebos, dated back to the 1600s. Dairies were descendants of the Italian Renaissance grotto, another imaginative re-creation of a site of intense encounters with nature. The dairy promised to transport its visitors into a sensual world of pristine nature in what amounted to a temple dedicated to milk. (In one famous example at Chantilly, no longer in existence, the room was decorated with paintings of cows and dedicated to Isis.) In the mock dairy, no real milk need intrude. Though it might be situated, like an actual dairy, in a "cool, ground floor location," visitors would find walls covered with marble and counters featuring porcelain vases and bowls, each displaying a different pastoral image. Marie Antoinette's rustic retreat at Versailles included a dairy attached to Le Hameau, on the grounds of the Petit Trianon. Even more elaborate was the Laitterie at Rambouillet, a present to her from Louis XVI in 1786.[32]

The Rambouillet project became a veritable arts fair of the Neoclassical, inspired by Greek, Roman, and Etruscan precedents. Marble reliefs of Amalthea, the goat that nursed Jupiter, and Apollo in the dress of a shepherd decorated the walls, while niches displayed specially designed jugs intended for cooling milk. Multiple sets of Sèvres porcelain populated the counter spaces. (Wedgwood of England supplied less impressive pieces for the dairy at Le Raincy and for an English pleasure dairy of Lavinia, Countess Spencer, in 1786.) Such places provided a lovely excuse for a ritualized excursion to a dessert course after dinner; accounts from Chantilly describe outings that included musical accompaniment to gondola rides down streams leading to the make-believe rustic enclave. Marie Antoinette may not have used the dairy at Rambouillet much at all, but the goblets designed for the place became runaway favorites in the porcelain market of the 1790s; orders were still coming in during the years of the Revolution.[33] Milk and its accoutrements held a special charm for aristocratic and prosperous Europeans. The age of enlightenment had rescued the food of barbarians and deposited it in an idealized world of porcelain and purity. A great deal of distance stood between these fantasies of the dairy and its reality. It now remains to be seen how the Enlightenment invaded the rural realm of milk production, prompted by market demand and changing attitudes toward work in the dairy.

CHAPTER SIX

Milk Comes of Age as Cheese

A twentieth-century writer once quipped that cheese is "milk's leap toward immortality," but the idea has been around for much longer than a half century. The ancients also understood the point, but more important, so did the British Navy. Most of milk's perishable qualities derive from its fat content: remove the cream, and cheese made from milk becomes ever more resilient—and less expensive. So it is not surprising that the British Navy, during its glory days of the eighteenth century, made a practice of buying "three-times-skimmed sky-blue," a local nickname for Suffolk cheese. More commonly known as "bang," the cheese was universally eaten and reviled. "Bang" must be onomatopoeic in origin, given that another name for it was "Suffolk thump"—perhaps what one would do with a piece of the hard stuff when seated at table. John Ray's seventeenth-century book of proverbs notes that "Hunger will break through stone walls, or anything except Suffolk cheese." Among the many jokes about the cheese, a favorite went like this: "A parcel of Suffolk cheese being packed up in an iron chest and put on board a ship bound to the East Indies, the rats, allured by the scent, gnawed a hole in the chest, but could not penetrate the cheese." When complaints finally brought about a change of menu in 1758, the navy decided to lay in supplies of Gloucester and Cheshire cheese instead. These were of the "single" variety, still skimmed and inexpensive, but more successful in terms of taste and texture. The story of bang proves the longevity of a food legend, but more than that, it indicates how common standards for cheese had risen in expectation by this early date. Not even the navy would settle for a poor product.[1]

Cheese was deeply implicated in the fate of milk in the eighteenth century. Few people would drink milk in the style of Dr. Cheyne and the Countess of Huntingdon, but cheese was everyone's food: rich people ate it for digestion and poor people, for hunger, as the saying went. And in this century of war and poverty, hunger acquired an institutional face: along with armies and navies, hospitals and workhouses required large quantities of cheap foodstuffs, and cheese fit the bill. Though milk was the most fragile of consumable goods, cheese was a constant traveler. Colonial projectors sent the commodity around the globe with regiments and traders, to India, the Caribbean, and North America. "The two targets against death," bread and cheese, saved millions of common eaters from starvation. In yet another gesture of gastronomic leveling, the ubiquitous pairing propped up upper-class adventurers who depended on inns for provisions as they toured the raw wilderness of Scotland and Scandinavia, following the impulses of a new romantic movement.

The great maw of London, the generator of consumer desires of every variety, also played a part in shaping the path of milk and cheese production in the eighteenth century. Historians have credited the London market for the forward motion of the British economy at both ends of the spectrum, from luxury goods to the most basic necessities. Urban laborers ate cheese daily, relying on inexpensive varieties pouring in from Cheshire and Suffolk. In 1730, as much as 5,756 tons arrived in London from Cheshire, transported by land; this was a fraction of the unknown total consumed by the city that year, given that shipments from other areas, by sea and over land, were numerous. Agents spread across the countryside in search of consignments of cheese of all varieties, establishing warehouses in regions where they collected supplies from small and large producers. Buyers demanded finer varieties, too: dairies turned out cheeses in the shape of pineapples, fish, and trees, which were ornately colored and sometimes flavored with herbs. Despite a rise in overall production, the price of common cheese spiked by the end of the century, boosted by the demands of a growing population and war. For this common commodity, prices increased as much as a third by the 1790s.[2]

New forms of demand provoked a search for improved methods of dairying: English gentlemen and even the more humble farmer were poised to make the most of rising prices and advances in knowledge. The age of Enlightenment directed helpful beams in their direction, offering a forum for identifying scientifically tested ways of maximizing output. Not to be left behind, middlemen in charge of requisitioning cheese proved to be deft handlers of growing demand, given their contact with both producers and consumers. Their strategies are worth uncovering in this crucial period of transition into more modern structures of marketing. In earlier decades, factors had quietly colluded with one another and fixed prices offered to the dairy farms where they collected supplies. An era of agricultural improvement would inspire a more forward-looking middleman to try a different tack: he aimed his efforts at the producers, pressing them to use predictable, repeatable methods in their cheese making. These were unfamiliar expectations to bring to a workplace like the dairy. From the evidence, we can watch English farms grapple with this new world of the market and its commercial messengers.

"Sweet and neate" — this was the standard of cleanliness expected of the good English dairy. As one authority on housewifery put it, "a Prince's bedchamber must not exceed it" in such perfection. As sanctified ground, the dairy was attached, for the sake of convenience, to the farmhouse kitchen. The mistress in command subjected her assistants to the same high standard of purity. Nose and eye were alert to every detail, knowing that the slightest deviation could result in lost butter and cheese. As Thomas Tusser, the well-known versifier of household wisdom, observed:

Good dairy doth pleasure
Ill dairy spends treasure.

Good housewife in dairy, that needs not be told,
Deserveth her fee to be paid her in gold.

Ill servant neglecting what housewifery says,
Deserveth her fee to be paid her with bays [reprimands].

Nothing less than devoted diligence suited this prized arena of production.

Dairying was understood to be women's work, an expected part of the household responsibilities of women in rural settings. The household formed the basic unit of production in the eighteenth century, governed by a universally understood division of labor according to gender. Dairying was considered a by-employment, part of the tasks incorporated into a varied day, often fit between other responsibilities. Investment, once past the purchase of a cow, was relatively modest, and anyone might engage in simple production of butter and cheese for consumption or sale. The job was deemed appropriate for all females in need of employment, a liquid analogue of spinning. Local records of aid to the poor show that cows were sometimes given to single women and widows so that they would be able to generate income for themselves. In a society dependent on dairy products, these independent producers found a ready market among the villagers or townspeople nearby.

More substantial dairying promised much greater rewards. George Eliot measured its value in local Derbyshire terms in *Adam Bede:* ably demonstrated by Mrs. Poyser, "the woman who manages a dairy has a large share in making the rent." Set in Ashbourne, a fertile dairy district, the novel reflected Eliot's intimate knowledge of her native region, which she called "fat central England."[3] Her estimate of the worth of dairy products was a bit conservative for regions where cheese making was more central: a Somerset farm in the 1790s paid £90 in rent, while bringing in £175 from the sale of cheese alone.[4] The promise of financial reward no doubt inspired the actual farmers of eastern Derbyshire in the eighteenth century, for records show them converting their cropland to pasture for fair-sized dairy herds of forty or fifty cows. Wives and daughters acted as the footing for profitable household enterprises in these districts. While men of the family would tend the stock, fodder crops, and fences, women saw to the milk, butter, and cheese. Dedication to the cause was assumed, and households thrived on a healthy supply of female offspring who could provide unwaged labor. In fact, the misfortune of having only sons might force a farmer into the less lucrative line of livestock farming.[5]

Dairying on an even larger scale dominated particular areas of England, such as Gloucestershire and Cheshire in the west Midlands. Here, with all resources dedicated to the production of milk and its by-products, workforces larger than just a slightly augmented household were needed. Larger farms recruited as many as six to eight dairymaids (the average was one maid for every ten cows), several laborers to work the fields that produced fodder crops, along with a number of men and boys to clean stalls and haul milk from the pastures. The largest establishments in Wiltshire had as many as two hundred cows, but most English dairy farms averaged between twenty and forty. Advances in know-how had already brought about the cultivation of crops like turnips, the valuable item in winter feeding, and clover. These would enhance the quality and quantity of milk, and more importantly, enable farmers to defy the natural calendar of calving. Cows could be kept in milk through the winter, beyond the months of autumn when lactation would normally cease. Year-round milk production required a carefully orchestrated calendar of field use, as the Dutch had demonstrated. So it is not surprising that the word "management" appeared in descriptions of these larger establishments, a sign that dairying for the market was taken very seriously.[6]

Who were the managers? The dairy remained a woman's place even on large establishments, where "the superior dairywoman is so highly spoken of, and so highly valued," according to one account of Gloucestershire, "that one is led to imagine every thing depends upon MANAGEMENT. Instances are mentioned of the same farm, under different managers, having produced good and bad cheese: even changing a dairy *maid* has been observed to make a considerable difference in the quality of the produce."[7] Some of the largest farms relied on male managers, but these men probably handled a combination of tasks associated with fields, equipment, and haulage, similar to managers on the big estates. The nature of decision making in the dairy, as we shall see, could not have rested in the hands of such individuals. While men might adjust certain factors involved in producing milk — conditions of pasture or winter feed, for example — they would not milk the cows or examine and handle the milk or cheese. The outcome of any attempt at improving a dairy farm would be apparent only to the mistress and her assisting maids.

The years of rising food prices after 1750 worked changes on this interesting corner of the rural world. As the cow's benefits became better understood, farmers envisioned rich returns from cheese and butter production. Many turned their oat or barley fields into pasture, increased their herd sizes, or simply put more effort into producing better dairy products. We know this not just from the records of individual farms, but from signs in the social life of ideas. Among the many learned societies that sprang up across the country, agriculture societies blossomed like so many crocuses in spring. Curiosity assumed a gentlemanly demeanor, as in a letter from Sir Digby Legard, Baronet, of East Yorkshire, dated January 1770. Noting the traffic of observers and writers on the topic of agriculture, Sir Digby objected to inaccurate generalizations made about his neighborhood (probably unflattering ones, given its isolated location). "Every farmer can give the best account of his own management," Sir Digby proposed, so that the public may determine what works best from such "authentick" information about "ways of managing Pastures" and "the artificial Grasses" like sainfoin ("healthy hay" in French), which not only fattened livestock but also restored soil fertility.[8] In the affluent town of Bath, gentlemen farmers from five counties came together to form the Bath and West of England Society in 1777. Their objectives: "the diffusion of useful information" about "what is most likely to be of publick utility." One report brimmed with enthusiasm over the successes scored with sainfoin in fattening dairy cows. "The prodigious increase of milk which it makes is astonishing, being nearly double that produced by any other green food," crowed an Essex farmer. "The milk is also better, and yields more cream than any other."[9]

There is no obvious reason why a deluge of writings—called "Book-Husbandry" by contemptuous observers—should have had a particularly noticeable impact on customary dairying in England. The debate between "mere theorists" and "mere practisers" had been going on long before this particular explosion of information without rendering anyone obsolete. Yet the environment of the dairy exhibited characteristics that agents of the newest age of reason would attack with powerful ridicule. The champions of certifiably reliable methods of producing valuable food products would have little tolerance for the typical practitioners of dairying skills. The production of milk, cheese, and butter followed along lines that appeared, at least to men of science, decidedly unscientific. As professors of enlightened agriculture attempted to penetrate the secrets of the dairy, two opposing systems of thought came into view. Because the perspective of improvers was to gain considerable force in the coming century, we would be well advised to investigate their quarrel with the dairy.

The "*fair* professors" of dairying, William Marshall, a well-known agricultural writer, announced, "tho' they may claim a degree of NATURAL CLEVERNESS . . . having tried their skill, *alone,* without obtaining the requisite degree of excellency, can have no good objection, now, to let us try our joint endeavours. And I call upon every man of science, who has opportunity and leisure, to lend them his best assistance." Marshall's tour through Gloucestershire had brought him face to face with the best of that region's dairywomen; yet the itinerant investigator remained unconvinced that the leading practitioners had fully conquered their profession. His information proved, without doubt, "the imperfectness of the art" because dairywomen acknowledged that they were not able "to reach any degree of certainty, much less perfection" in what they practiced. Marshall was fairly exultant when he announced that "the art is evidently destitute of principles. So far from being scientific, it is altogether immechanical." He added some final words of damnation: "It may be said to be at present, a knack involved in mystery."[10]

Marshall hailed from yeoman stock of the North Riding of Yorkshire and claimed a long ancestry in farming. Yet he had also seen another side of life in the 1760s when he moved to London, where he tried his youthful hand first in the linen trade, and then in the insurance business, which took him to the West Indies. Returning to England in his mid-twenties, Marshall emerged from a serious illness to experience an epiphany: he decided to give up business and devote himself totally to what had been a private passion, the study of agriculture. An inheritance coincided with this moment of illumination, and so Marshall was able to throw himself into a new way forward. He gained practical experience by becoming the manager of one farm, then the agent of another, all the while energetically promoting the new science of agriculture. Proposals issued from his pen, for government-sponsored colleges of agricultural study, for an in-depth study of agricultural districts (authored by himself), and later, for an official Board of Agriculture, an idea that would bring him some recognition. But Marshall never outran his archrival, Arthur Young, who won acclaim as a more lively and competent authority on the subject of farming. Marshall's reputation for quarrelsome behavior might have trimmed his success, though one also wonders about his love affair with excessive detail.[11]

A devout believer in the empirical method, Marshall's strategy was to mount up details of practices in each branch of farming so that he might then subject the whole to a searing but corrective examination. He showed himself to be a virtuoso of the nitty-gritty, noting the composition of the cement joining the stones of the buildings, a heady mix of road scrapings and dung a century earlier. He measured out barns and dairy houses and, once inside, the length of cheese-knife handles, the diameters of vats, and the weights of cheese presses. When it was time to converse with the dairywoman about her methods, it is no wonder that the indefatigable Marshall experienced a certain sense of deflation.

"The dairy-room is consecrated to the sex," Marshall testified, "and it is generally understood to require some interest, and more address, to gain full admission to its rites." The investigator himself could not be considered particularly smitten with "the sex," as the eighteenth century euphemistically denoted females. Marshall did not marry until he was over sixty, and while in Gloucestershire, he was much the ambitious man in his late thirties, predictably impatient with what appeared to be a roadblock in the way of his survey. Like the altar of some primitive religion, the dairy represented an inner sanctum of female space. Praise be to farming, a "public employment," as he described it; cheese making, by contrast, was "a *private manufactory* — a craft — a mystery — secluded from the public eye." Not even the master of the farm would always know the methods that gave his cheeses their trademark. This last fact must have consoled the frustrated Marshall. Whatever omissions existed in his offering, he was proud of his bundle of information. He boasted that his volume on the rural economy of Gloucestershire stood as the most extensive description of that region's dairying in existence.[12]

Yet Marshall knew that his pile of pages amounted to a Pyrrhic victory. There was something highly particularized about dairy knowledge, rooted to a place and a moment in time, that defied his objectives. "Individual practice" was the rule, known to each woman and perhaps "a few confidential neighbours."[13] This was because actual work with milk, butter, and cheese appeared to defy rational explanation and systematic analysis. Cheese making presented challenges of a particularly opaque nature. In a lengthy stream of operations, each prompted by minute changes in consistency and temperature, the cheese maker made judgments that were literally incalculable without modern instruments. The correct temperature for milk at the time of adding the rennet was, appropriately enough, "milk-warm" (technically, its temperature just out of the udder), and only an experienced hand showed sufficient sensitivity to know when that moment arrived. (Thermometers were expensive, scarce, and deemed unnecessary, given the successes achieved without them.) The rennet itself constituted a uniquely mysterious substance whose properties were not fully understood even a century after Marshall. Produced from an extract of the

Eighteenth-century dairy house in Bincombe, Dorset. Dairies built of stone remained cool in the summer, which was beneficial to the process of making of butter and cheese. An adjoining thatched house, no longer standing, would have enabled the mistress of the household to move easily between kitchen and dairy in the course of a day's tasks. (Photo: Debby Rose, Weymouth Local and Family History Web site, Dorset, England)

stomach of a calf according to as many methods as there were cheeses, rennet made possible the chemical reaction enabling curds to form from the milk and cream. The pages of agricultural journals found "runnet" to be an endless source of fascination and frustration. This latter sentiment was not one that appealed much to this particular man of the enlightenment.

Region and climate played important yet ambiguous roles in the production of cheese. In the first case, idiosyncrasies were everything: native soils and their interaction with grasses, local vegetation, the nature of the terrain, along with the careful calendar of calving, barn and pasture time were just the beginning of a long list of what made the difference between one cheese and another. In the second instance, each season and year dealt a different array of conditions to the farmer and the dairywoman. How they responded would affect their fields and dairy room, along with the physical condition of their dairy herds, and hence their output of milk. If a morning's gathering contained less cream than usual, or if weather conditions varied dramatically, the maid and mistress faced seemingly small but important decisions in the next twelve or twenty-four hours. If a cow had browsed among mustard weed, the maid needed to detect this in the taste of the milk, which had to be isolated and disposed of. Each pail of milk, along with each vat of cheese, proved to be a cauldron of question marks.

More fascinating to Marshall were the "minutiae" of bringing the milk through the stages of transformation to cheese. (By contrast, his high-flying rival, Arthur Young, would dismiss them with contempt: "The minutiae of dairy concerns would fill a book, and after all would not be useful to any extent.")[14] Each maker followed different procedures, some of which were local custom, others, of a more personal provenance. In this business, Marshall the visitor was allowed to observe everything and use his own thermometer in order to record the day's proceedings. Marshall offered a log of five successive mornings, which, in abbreviated form below, present a picture of custom under the surveillance of a laboratory manager.

> Tuesday, 2 September, 1783. . . . Part of the skim milk added cold, — part warmed, in a kettle, over an open fire, to raise the whole to a due degree of heat. . . . An estimated sufficiency of rennet added. The whole stirred and mixed evenly together. The exact heat of the mixture 85° of Fahrenheit's thermometer. The morning close and warm, with some thunder. The cheese-cowl covered; — but placed near an open door. The curd, nevertheless, came in less than forty minutes: much sooner than expected: owing, probably, to the peculiar state of the air. . . . The curd deemed too tough and hard; though much the tenderest curd I have observed.

Marshall was clearly holding in check his judgment of the dairywoman, though not his differences of opinion regarding the curd. Day two, however, seemed to satisfy both parties:

> Wednesday, 3 September. The morning moderately cool. The heat of the milk when set 83½°. The cowl partially covered, and exposed to the outward air, as before. Came in an hour and a quarter. The heat of the curd and whey, mixed evenly together, 80°. But at the top, before mixing, only 77°. The curd extremely delicate, and esteemed of a good quality.

Buoyed by mounting evidence, Marshall began to see a pattern in what had earlier appeared to be uncharted activity.

The investigator was open to enthusiasm on day three. Marshall's positive state of mind peeked through the interstices of his notations when he perceived a repetition of the temperature of the curd and whey of the day before:

> Thursday, 4 September. The morning cool — a slight frost. The milk heated, this morning, to 88°. The cowl more closely covered; and the door shut part of the time. Set at half past six: began to come, at half after seven: but not sufficiently firm, to be broken up, until eight o'clock: — an hour and a half. The whey, when mixt, exactly 80°! The curd exceedingly delicate.

And day four supplied more confirmation of Marshall's beloved condition, numerical consistency:

> Friday, 5 September. This morning, though mild, the curd came exactly at 80°! What an accuracy of judgment here appears to be displayed! Let the state of the air be what it will, we find the heat of the whey, when the curd is sufficiently coagulated, exactly 80°; and this, without the assistance of a thermometer, or any other artificial help. But what will not daily practice, natural good sense, and minute attention accomplish.

Marshall was duly impressed by how nature appeared to be cooperating with the experienced dairywoman; words of praise, coming from such an irascible judge, were precious indeed. Yet it is noteworthy that he withheld the word "skill" when describing the attributes of his host.

The dairywoman's natural talent was never infallible under such pressures of performance. On the final day, Marshall was granted his own victory, emblazoned on the page in capital letters:

> Saturday, 6 September. This morning, the curd came too quick. The heat of the whey (after the curd had been broken and was settled) full 85°! The curd "much tougher and harder than it should be." Here we have proof of the inaccuracy of the senses; and of the insufficiency of the natural judgment, in the art under consideration: it may frequently *prove to be right;* but never can be *certain.* Some scientific helps are evidently necessary to UNIFORM SUCCESS.

By "scientific helps," Marshall meant instruments. What had been a triumph of intuitive sensitivity to temperature the day before now moved down a notch to fallible human sensation. But how much did successful dairies desire or require this new form of knowledge? A dairywoman was unlikely to alter her methods according to fiat from an intruder armed with a notebook and a thermometer. It remained to be seen if "helps" would have more appeal if attached to monetary incentives.[15]

Those incentives—and ambitious individuals ready to act on them—were clearly on the horizon. Marshall's decade of the 1780s was indeed a time of widespread dairy consciousness, given rising demand for dairy products in England. Others besides the dairywoman stood to benefit from the "uniform success" of her efforts. First in line were cheese factors, long understood as necessary agents spread throughout the English countryside. As early as the 1680s, cheese factors were buying up cheese in the Midlands and in other dairy regions for the sake of the voracious London market. Farmers large and small depended on these middlemen, who picked up supplies and transported them to large markets, institutions, or distant retailers. Factors were never popular, despite their essential services. Though never registered as an official monopoly (which would have been permitted under certain conditions), the factors banded together in order to agree on ways of handling their trade. Justices of the peace were needed to check price setting and manipulations of the market that put farmers at a disadvantage. Petitions to Parliament pleaded for intercession in a raft of sly practices by factors, such as holding cheeses until market prices improved without giving any gains back to the farmer.[16] Consumers were not unaware and joined in protest against the middlemen. In 1766, when violence erupted over high food prices in Derby, the crowd directed its rage at the cheese warehouses there. County magistrates sent out special forces to protect the goods and received two costly silver cups as thanks from London cheesemongers. Cheese was clearly implicated in a more general struggle over the marketing of food supplies for common people. The case of Derby illustrated perfectly how broadly it was understood: as a choreography involving producers and factors, provincial consumers and London retailers.[17]

The countryside fairly throbbed with dairy travelers during these years: while writers like William Marshall and Arthur Young roamed rural England, so did many cheese factors, including one Josiah Twamley, who worked throughout counties in the Midlands. A loquacious and confident man of business, Twamley felt he had something worthwhile to teach his contemporaries, and so this town-born man published a treatise on dairying. We may smile at his bravado and register skepticism with regard to the significance of *Dairying Exemplified, or The Business of Cheese-making* (1784). Yet the doughty little volume went on to enjoy two editions in England and then did some traveling of its own. In 1796, it was republished in Providence, Rhode Island, to serve the premier early American dairy district located in the Narragansett Basin. This suggests that its advice was seen as potentially useful; for our purposes, however, the booklet is valuable for its portrait of the dairy as a territory in need of reform.[18]

The cheese factor wished for consistency, though not exactly on behalf of science: his business required a steady supply of uniform commodities rather than the inevitable variations (and failures) that occurred in traditional cheese making of the time. Twamley would have been familiar with each farm on his route and no doubt held a firm impression of his clientele and their products from years of contact. His treatise revealed how he had tried, like Marshall, to influence the way dairywomen went about their work. "I am well acquainted, how unthankful an office it is," he complained, "to attempt to instruct or inform Dairywomen, how to improve their method, or point out rules, which are different from their own, or what had always been practiced by their Mothers, to whom they are often very partial." The factor knew what these women thought of him when he made suggestions: "What does he know of Dairying or how should a Man know any thing of Cheese making?" In his own defense, Twamley cited his years of experience in "consulting the best of Dairy-women, in many Counties," though their voices are not heard in his instructional commands. Once again, empirical evidence would trump the authority of the native artisan of the dairy, this time on behalf of the market.[19]

Of course, dairywomen had been catering to the market for many years, often with notable success. Their sensitivity to the demands of customers was on display in the yellow cast of butter, an act of artifice achieved by adding marigold blossoms to the natural product. Most dairywomen in Gloucestershire colored their cheeses, too, using annatto imported from Jamaica and sold throughout the county. Though early practices had been looked upon as a form of adulteration (we can recall the accusation against Holland cheeses reportedly colored with sheep's dung), a later age elevated the practice as an art form. By the end of the eighteenth century, the reputation of many cheeses rested on their color as much as their consistency and taste. As Marshall pointed out, "If the eaters of cheese were to take it into their heads, to prefer black, blue, or red cheese, to that of a golden hue," the dairywomen "would do their best endeavour to gratify them." His hyperbole was not misplaced when it came to true acts of artistry practiced by the most skilled dairywomen. One such woman in the Vale of Alesbury had "a pretty way of making chequer'd Sage-cheese," which a pedantic but determined treatise attempted to explain to the reading public. Factors like Twamley would have known how adaptable dairywomen could be, especially because many middlemen eagerly assumed the business of selling dyes to their cheese makers in an effort to support their creativity.[20]

But Twamley chose to present dairywomen as an unenlightened tribe, bound together by rigid custom and shared opinion. His pages interlarded instruction with tales of his interactions with dairywomen in markets, fairs, kitchens, and dairies. Twamley's authoritative presence occupied center stage in these dialogues, while dairywomen appeared as naive accessories to his campaign for progress. Most peculiar to him was the dairy community's lack of interest in improvement. "Dairy-folks will complain that there is not proper encouragement for making good Cheese, as Factors give for all Dairys in a neighbourhood nearly the same Price, though some of these Dairys are not so good as others," the trader reported. Rather than seeing this as local resentment of the factor's power to fix prices, Twamley painted it as apathy and even antipathy toward the market. He confided that a middleman could not afford to criticize a woman's cheese, for fear of alienating her and all her neighbors. Twamley unwittingly exposed the tightly knit fabric of a dairying community and its internal prohibitions against competition.[21]

Making matters worse, dairywomen adhered to neighborly values that might easily trump business interests. A vivid illustration of this conflict of perspectives emerged from a visit to "as bad a Dairy as I ever met with," where Twamley's probing questions exposed a tale more fit for local gossip than a manual of cheese making. Though he missed the significance of what he heard, except as evidence of a dairywoman's incompetence, the assiduous author reported the encounter with his customary attention to detail.

"As soon as I came into the Dairy-chamber," he recalled, "I saw, and told her it would not suit me." "Why not," she replied, "I am sure tis every drop New Milk, and nobody can take more pains with it, nor work harder at it than I do." Twamley then caught sight of a cheese on the far side of the room "very blooming in appearance, handsome in shape, well-coated, firm, fat, and much larger than the rest." He asked, "How came that Cheese there — I should be glad to know the History of it." "Why truly," she replied, "tis a strange one." Twamley informed her that "if you will make such Cheese as that, it would be worth five shillings, or even ten shillings a hundred weight more than the rest." It was at this point that the dairywoman revealed the fact that she wore another occupational hat, that of nurse and possibly midwife.

"One night," she began, "when I had rendled my Milk, a person came running to me, and said, neighbour T—— is groaning & you must come immediately; I said to a raw wench I had to help me, now be sure you don't touch this Cheese till I come back, I will be sure to come to you when I see how neighbour T—— is. But it happened she was worse than I expected, and I could not leave her till after midnight. I said, my Cheese will be spoiled, but the poor Woman shall not be lost for a Cheese. When I came home I found it not so bad as I expected, put it into the Vat in a hurry, saying, it may possibly make a Cheese that will do for ourselves, but I little thought it would ever be a saleable Cheese."[22]

"Well now," said Twamley, much like the schoolmaster who has closed his eyes during the pupil's recitation, "and is not this Cheese a proper lesson to you? Don't you thereby plainly see that you have made the rest too quick?" "Why yes," said she, "It might, if I had thought at all—but I declare, I never once thought about it." Twamley's final verdict on the dairywoman and her dairy was peremptory. "Profound stupidity! thought I to myself, and left her." The consummate businessman, he nevertheless capitalized on her mistake, relating her story as an illustration of how independent cheese making awaited liberation from a crippling irrationality.[23]

Twamley's rhetoric was a signpost of things to come. The ability to respond regularly to the factor's commands would determine the success or failure of a dairy venture. A dairywoman had no voice in determining where her cheeses would be sold, given the loss of control to the middleman who came to her door and carried her products away. How else could the isolated cheese maker in the Midlands know what the London consumer preferred, in terms of taste, as she adjusted the conditions surrounding her cheese making? Twamley knew that his most important job ultimately rested in conveying consumer preferences to the dairywomen most capable of satisfying these demands. Ideally, in his view, the fittest cheese makers would triumph, and so would he. The trader would "lead and command the opinion & interest of the best customers," just as he would lead dairywomen out of darkness. "In this happy track," Twamley enthused, "I confess I should be glad to meet many of my old friends & neighbours, in whose service I have laboured many a long day." Out of the fray of competition would emerge the factor and his followers, adjusted and restored by an encounter with the demands of the market.

The renovation of the dairy would occur slowly over the next century, prompted by shifts in attitude that brought more farmers closer to Twamley's line of thinking. After the lessons of the age of enlightenment, little nostalgia was lost on the customary practices of women. Female practitioners of the art inhabited a raw realm of nature, where they were locked in unintelligible ignorance. The combined pressures of improvers and traders ushered in a new attitude toward such work in the dairy, viewing it as a realm of a benighted female craft. In the place of custom and secrecy, renovators promoted carefully calibrated routine and transparency. Dairywomen, particularly those on larger farms, would not be opposed to such objectives in coming years. As the wives of successful farmers, their voices were occasionally heard in public, though they were in a decided minority. But the bigger business ventures in dairying forced changes in management and capital investment that modified the work of women. Even before the Victorian age advocated a less vigorous role for the farmer's wife, the association of women with the dairy underwent dramatic changes in the public eye.

Among the cries of vendors heard on the streets of London, the distinctive call of the milkmaid, sounding like the syllables "mi-ow," short for "milk below," was well known. Some contemporaries described the cry as shrill, while others thought it was "not unmusical." A familiar image by William Hogarth, *The Enraged Musician,* includes a dignified representative of the trade among the mélange of street people outside the window of a music master, suggesting that her voice was adding to the clamor that forced him to cover his ears. With her gaze openly directed at us as we survey the comical scene of chaos, she appears both quintessentially plebeian and pristine, a curious contradiction hanging over her stateliness. Hogarth went to some trouble to depict details of the pail on her head and the musculature of her arm. He must have been aware of the legendary strength and endurance of milkmaids: they were said to begin their days at three or four o'clock in the morning, carrying as much as seventy pounds of milk from the outskirts of the city to their customers. In the words of a nineteenth-century memoir, many were "robust Welsh girls or Irish

William Hogarth (1697–1764), The Enraged Musician, *1741. Hogarth's engraving combines the comic chaos of a London street scene with a statuesque dairymaid of considerable dignity. Perhaps the activity of the little boy beside her is an oblique reference to the quality of most milk sold in the streets of London.*
(© Yale Center for British Art, New Haven, USA/Transfer from the School of Music/The Bridgeman Art Library International)

women" who could be heard "laughing and singing to the music of their empty pails." In fact, Hogarth's treatment recognized two commonly held images of their character, for these denizens of the streets, veritable slaves to the rhythms of bovine production and household consumption, were also idealized in the popular imagination. Contemporaries annually celebrated them on May Day as flowery mediators between bucolic pastures and the city. Hogarth's hint is there in his milkmaid's right hand, which clutches her apron decorously as though assisting in a dance step.[24]

Milkmaids, like dairywomen across rural England, called up associations with a realm of nature governed by time-honored custom. In annual May Day rituals of an earlier time, the milkmaid occupied center stage as the primary representative of the seasonal rhythms of the calendar. Garlanded with flowers plucked from lush meadows, where cows obtained their daily nourishment, the milkmaid's body acted as the showcase of the season's promise. Latent with reproductive potential, the maiden and her flowers symbolized sexuality in its most fulsome and positive light, combined in a suggestive way with the phallic symbolism of the Maypole. Games of courtship and "unrelieved gluttony" in consuming milk products (including a potent mixture of milk and rum) would cap off a holiday ushering in a season of nature's bounty via cows, sheep, and goats — and, of course, amorous human beings.

These aspects of May Day had disappeared from London streets by the 1700s, replaced by a peculiar custom that spanned the better part of the century. Described by English historian Charles Phythian-Adams, the milkmaid's garland and pail metamorphosed shortly before the 1690s: in their place, the maid carried a headdress of silver plate — pitchers, plates, and cups — decorated with ribbons and flowers. Groups of maids, accompanied by one or two musicians, danced from door to door collecting donations from their patrons. An American observer, in obvious wonderment, recorded the details of one such scene, growing ever more elaborate over the century, in 1776:

> In Ave-Mary Lane saw the milkmen and maids again with a garland so called; being a pyramid consisting of seven or eight stories, in the four angles of which stood a silver tankard, and on the sides, between each, lessening in height as the stories rose, stood a silver salver, the top crowned with a chased silver tea-kettle, round which were placed sundry small pieces of plate; the whole adorned with wreaths and festoons of flowers, gilt papers, etc., carried on a bier and hand-barrow, it being a custom amongst them to collect of the customers a yearly contribution. The wrought silver appeared worth many hundreds of pounds, and is borrowed for the occasion.[25]

Phythian-Adams described the change as a turn to "ritually licensed begging by an occupational group," a clear departure from the rural overtones of the former dance of milkmaids. The timing of the appearance of the new headdress was significant: by the eighteenth century, the consumption of milk was no longer bounded by the calendar of nature, because cows could be kept in lactation all year long by the use of artificial grasses

Marcellus Laroon (1653–1702),
The Merry Milk Maid, *ca. 1688.
Laroon's series of seventeenth-
century London streetsellers
can be seen as part of an
iconographic tradition stretching back to the Renaissance,
in which tradespeople were
depicted in meticulous detail
as part of a catalogue of
humanity. This dairymaid
displays the arms of a working woman, but her decorous
headdress bearing silver plate,
along with her ornamented
footwear, suggest a lively world
of custom and ritual. Laroon's
milkmaid was probably the
prototype for Hogarth's version
in the previous illustration,
judging from her posture and
dress. (London Metropolitan
Archives Print Collection,
City of London)*

(such as clover and lucerne) and turnips. As the inclusion of silver tea service items in the milkmaid's towering edifice suggested, milk was now necessary for more than just cooking purposes and the occasional healthy restorative drink: it was in demand for fashionable tea drinking and coffee consumption. Milkmaids, according to Phythian-Adams, were no longer "fetching nature" for city dwellers; instead, they carried domestic emblems out of private spaces into public view. Their representative bounty — silver — rested on the goodwill of their customers and, occasionally, the local pawnbroker.[26]

London's market for milk, like its demand for cheese, represented a leading edge in terms of the consumption of goods and services. It was also a harbinger of future commercial difficulties, a euphemism for the problems associated with dispensing milk in an urban environment. Milk is a product we now understand to be a veritable petrie dish for bacterial growth. Novelist Tobias Smollett's famous description of the milk in Covent Garden, offered up in *Humphry Clinker* in 1771, turned the idea of purity on its head:

> The milk itself should not pass unanalyzed, the produce of faded cabbage-leaves and sour draff, lowered with hot water, frothed with bruised snails, carried through the streets in open pails, exposed to foul rinsings, discharged from doors and windows, spittle, snot, and tobacco-quids from foot passengers, overflowing from mud carts, spatterings from coach wheels, dirt and trash chucked into it by roguish boys for the joke's sake, the spewing of infants, who have slabbered in the tin-measure, which is thrown back in that condition among the milk, for the benefit of the next customer; and, finally, the vermin that drops from the rags of the nasty drab that vends this precious mixture, under the respectable denomination of milk-maid.[27]

For those who could pay, supplies of good, fresh milk restored the overwrought and overindulged. Pepys mentioned "drinking a great deale . . . to take away my Heartburne." He was later seized with much pain, reporting "Winde griping of my belly and making me shit often, and vomit too." Could it be that the lover of dairy had gotten some bad milk? What about the masses, who regularly depended on Smollett-like marketers?[28]

The eighteenth century can be seen as a cradle of a highly active consumer culture, one that had important implications for the future. Londoners inhabited a dense and omnivorous environment: by 1800, their city was the largest in Europe and second in the world to Beijing, China. More to the point, the city was the metropole of a global empire that connected distant regions by means of material needs and desires. Coupled with an increasingly wealthy industrial north, Britain's economic might made itself felt worldwide. The challenge of supplying inexpensive, portable food would continue to ignite interest in this industrial nation, particularly as the principles of future profitable activity seemed patently clear. Dairies must flourish and milk must travel. These two lessons could be gleaned from an enlightened eighteenth century.

CHAPTER SEVEN

An Interlude of Livestock History

"Why *cow's* milk?" is a persistent question in the path of liquid history. If goats were ubiquitous and cheaper to keep, offering a milk that was more easily digestible by humans, why did they not triumph over their bovine competitors as providers of human sustenance? Certainly many areas of the world favored goat's milk for alimentary purposes, and it may well be that in terms of household consumption, especially in rural areas, Europeans and North Americans drank and processed more of it than we can detect in recorded food history. Yet even before the nineteenth century, when scientific deliberations over the benefits of milk helped it along, the sluggish cow emerged as a frontrunner in the contest of Western food engines. By then, cow's milk was already a fixture in Western diets and, more to the point, food production. Why was this so?

Part of the answer lies at a crossroad of bovine and human history that proved critical to the future of milk: the landing of dairy cows in colonial North America. There, the evolution of animal husbandry converged with the migration of Europeans and their distinctive food cultures in the seventeenth and eighteenth centuries. The milch cow henceforth moved from success to success, partnered with beef cattle and equipped with its own trademark assets. Transported across the Atlantic Ocean, its substantial supply of basic needs, along with its symbolic baggage, hardly seem an accident. While male cattle promised to pull plows and sacrifice their hides and flesh, females offered annual calves and, naturally, their milk. Here, too, was an animal representing property incarnate, not only movable, but transferable: colonists used their cattle as ready cash, given that a serious shortage of currency impeded their commercial transactions. More valuable than a pig and less prestigious than a horse, the cow heralded an American middle way of yeoman self-sufficiency.

The importance of this fact extends far beyond the history of North America: the successful transplanting of this "portmanteau biota" (the term is Alfred W. Crosby's) began a longer process of the expansion of European attitudes toward the land and livestock. Particular property laws, arrangements of labor, and, perhaps most important of all, a high regard for profitable enterprise and trade went along with the milch cow. Such values supported the dietary regime that fueled the colonial experiment in North America. And the outcome — namely, robust Americans — stood as proof, according to some observers later on, of the validity of the project.

Not yet an especially celebrated liquid, the enthusiastic adoption of milk into American foodways can only be described as dietary opportunism. Like a theatrical understudy, milk succeeded in shining at critical moments of need during the colonial era, when the challenges of settlement drove home lessons about basic food and the natural environment. The rise of milk to a place of prominence took several centuries to play out through the rise of innovations in processing, new scientific knowledge, and entrepreneurship of the nineteenth century. But during the first two centuries of the colonial experience, essential aspects of the story are clearly visible. It was here in North America that a modern Western approach to liquid nourishment ultimately fell into place.

When western Europeans embarked for the Americas, they brought along their cattle. The second voyage of Columbus conveyed assorted domestic animals and plants, packing in additional calves, pigs, and chickens, along with oranges and lemons, on a layover in the Canary Islands. Cattle were reportedly suspended in slings because the listing of the ship would have caused them to topple over and break limbs. Little dairying came of Spanish ventures in the Caribbean and Mexico, given the dietary habits of the invaders and the hot and often dry climate they encountered. Early Spanish arrivals appear to have valued the hides and meat of their cattle, but bold conquistadores were unlikely exploiters of opportunities

Edward Hicks (1780–1849), The Cornell Farm, *1848. An iconic image of an American farm with plenty of livestock on display. This color lithograph is inscribed at the bottom: "An Indian Summer View of the Farm & Stock of James C. Cornell of Northampton, Bucks County, Pennsylvania." (Private collection/The Bridgeman Art Library International)*

for dairying. Not until English, Dutch, and Swedish settlers landed in northern parts of North America did word of milk enter into accounts of the New World.

Cattle stood as harbingers of prosperity in colonial America. In 1619, the new secretary of state of Virginia, John Pory, identified three essential things that "in fewe yeares may bring this Colony to perfection": "the English plough, Vineyards, and Cattle." Such was the vaguely biblical trinity of man-made contrivance, Edenic plant life, and essential companion of the settler that seemed to promise a destiny heavy with rewards. Pory probably had in mind the indispensable oxen needed to turn the rich soil of the land he surveyed. "We shall produce miracles out of this earthe," Pory predicted. His vision of prolific returns entailed a direct transfer from a teeming natural environment to mammalian fruitfulness. "For cattle, they do mightily increase here, both kine, hogges and goates, and are much greater in stature, then [*sic*] the race of them first brought out of England," he happily reported.[1]

The story of the early American vision of effortless abundance is a familiar one. Accounts of flora and fauna presented lavish pictures of North American sustenance, ripe for the picking, and soon enough, cattle wandered into the picture. Wild animals, in "such numbers" as to be "rather an annoyance than an advantage," pressed in upon the settlers, according to Lord Baltimore, so plentiful were the marauding swine and deer, along with legions of wild goats, birds of prey, and curiously non-invasive weasels.[2] John Hammond made a catalogue of the rich potential of Maryland and Virginia. "Diet cannot be scarce, since both rivers and woods affords it, and that such plenty of Cattle and Hoggs are every where, which yeeld beef, veal, milk, butter, cheese and other made dishes, porke, bacon, and pigs, and that as sweet and savoury meat as the world affords," he figured.[3] In the absence of hedgerows and fences, Lord Dela-ware wondered at the ability of cattle to browse and flourish, even in the snow, as "many of them [were] readie to fall with Calve; Milke, being a great nourishment and refreshing to our people, serving also (in occa-sion) as well for Physicke as for Food."[4] Here was a means of transform-ing the fecundity of the land into a liquid stream of sustenance useful to needy settlers. European invaders eagerly inserted themselves into the food chain, living off the "vast herds of cows, and wild oxen, fit for beasts of burden and good to eat."[5]

In actuality, the first generation of colonial Virginians staked their first hope on easy riches from tobacco; it was only after the boom and bust of prices in the 1620s that they turned to livestock, to which they applied hap-hazard attention and effort. Inhabitants of the Chesapeake region were, in the words of historian Edmund S. Morgan, "a crude society, peopled by crude men." Fixated on making quick fortunes, they "would not grow enough corn to feed themselves, but they grew tobacco as though their lives depended on that."[6] Their practice of relying on trade with the In-dians for cultivated crops proved disastrous. Captain John Smith's journal offers a glimpse into the deprivation and starvation experienced by settlers of Virginia; incidents of cannibalism suggest that the suffering was far more dramatic than his pages admit. The narrative of humiliation reads like a biblical test of moral fiber, as the colonists were reduced to depen-dence on charitable gestures of Native Americans and one another. Upon

Captain Nuse, Smith bestowed great praise, owing to his sharing of "some small quantity of Milke and Rice that he would distribute *gratis* as he saw occasion; I say *gratis,* for I know no place else, but it was sold for ready payment."[7]

Milk constituted humble fare indeed in the midst of the bestiary of the New World. Animal protein was always the subtext of early colonial braggadocio, particularly coming from English settlers with an attachment to beef. Yet here were the heroic pioneers of Jamestown, reduced to milk and rice. In Smith's account of the chastened community that survived to continue the colony, he celebrated the rewards of husbandry that came to Virginians who raised corn (that is, *zia maize,* or Indian corn) as well as tobacco. His description of a new native dish revealed that, despite a hierarchy of foods cooked and eaten according to social rank, milk had colonized all palates. "Their servants commonly feed upon Milke Homini," he reported in the 1620s, "which is bruized Indian corne, pounded and boiled thicke, and milke for the sauce." He added, "but [when] boiled with milke the best of all will oft feed upon it, and leave their flesh." Perhaps milk had earned extra points for helping out in hard times, for things hadn't changed one hundred years later. Robert Carter recalled from eighteenth-century Chesapeake, "Honey and milk and mush might well content [Virginians] in these summer months . . . [as they] have no meat this time of year." The American taste for corn porridge with milk had been established, for better or for worse.[8]

Virginians did in fact import cattle from an early date, either 1610 or 1611, three or four years after the founding of the colony, and by 1612, one hundred more "kine" (probably a combination of oxen, cows, and heifers) arrived on six ships. The cost of passage for an adult male settler was on average fourteen pounds; for a heifer, ten to twelve pounds. It is difficult to imagine how the animals' voracious appetite for feed and water could be met on board a seventeenth-century ship, yet it proved possible again and again. The warmth and safety of a ship's hold in fact surpassed the challenges of North American wolves, winter, and outbreaks of resistance by native inhabitants. Once let loose upon the land, cattle made good on the American promise: they gave way to simple multiplication. By 1627, Captain John Smith guessed (or boasted) that as many as two thousand cows, bulls, and oxen populated the colony. Two years later, the estimate was up to five thousand. Bovine immigrants also served as inadvertent vehicles of English grasses by means of their manure, which was shoveled onto American shores. It took several generations before colonists succeeded in sowing sizable amounts of their favorite fodder crops for winter feeding, but a beginning was launched along port town coastlines.[9]

Up north, in the imagination of New England settlers, the prospect of infinite stretches of land combined explosively with a belief in this familiar four-legged capital investment. Contractors scrambled to send good English cattle stock to the colonies, mostly females, as it turned out, because the supply of bulls and oxen was quickly satisfied by natural reproduction. Whole boatloads arrived, sometimes sold and in other instances contracted to the colony itself and then loaned to new settlers. Northerners' hunger for the beasts seemed insatiable, though not for want of shipments from home. In 1634, for example, the arrival of two ships in Massachusetts Bay brought "about 200 passengers and 100 cattle." John Winthrop's *Journal* reveals repeated disputes instigated by livestock. Settlers reported "straitened" circumstances in Watertown and Roxbury, where "cattle being so much increased" that settlers felt crowded. As time went on, fences went up. The wanderings of errant cattle repeatedly ruined fields and gardens, yet in disputes, the law came down on the side of the cattle: landowners were responsible for protecting their holdings from animals, not the other way around.[10]

Yet the English failed to appreciate fully the dairy potential of their livestock. Cattle back in England were often poorly cared for in winter and, if we can trust quotidian visual evidence, generally scrawny. The settlers were no better at husbandry: they simply let their cows roam and browse freely, even in the snow. When an English clergyman wrote home from Jamestown in the 1680s, he conveyed a certain contempt for the negligence of Virginians:

> They be coveteous of large ranges for their stock & let them run over a vast tract of land & the beast being pend in pens as our sheep are at nights in the morning they run themselves out of breath traceing 2 or 3 miles & spend themselvs before they fall to a settled feeding whereas were their pastures divided & one part preservd whilst another were eating the cattle would feed to m[u]ch more advantage.[11]

Virginia farmers approached the cow as they did the pig: they assumed that the animal was a four-legged forager with a sound instinct for self-sufficiency. Proof, so to speak, was in the pudding: their less hardy milkers simply died or dried up in the early years of settlement. Those that survived must have registered low output, unless they chanced upon some good wild rye or broomstraw. Settlers were said to throw them some hay during the winter months, which would not have been enough to produce rich milk in the spring.

True, settlers labored under an elementary grasp of biology and ecology, but their trust in the vegetation of American wilderness was extreme. Testimony to this lies in one of the few diseases actually named for milk, which appeared in North America in the years before the American Revolution: "milk sickness," or "milksick," as American settlers called it, originated in North Carolina near a mountain ridge which earned the appropriate name of Milk Sick. The entire area west of the Allegheny Mountains, from Georgia to the Great Lakes, recorded incidence of the disease. First called "trembles," it afflicted cattle and the settlers who drank their milk; it brought on listlessness, shakiness, anorexia, and sometimes coma and death. In the first half of the nineteenth century, whole settlements in Indiana and the Midwest were abandoned because of epidemics of the ailment. Its cause, known to Native Americans, was the white snakeroot, a plant growing wild in the forests of those regions. Until cattle were given adequate pasture, their roaming inevitably led to deaths caused by their browsing among poisonous plants.[12]

The New World proved to be a scientific laboratory writ large: the colonial years witnessed a Darwinian experiment in cattle breeding, as different settlers deposited their favorite cows from home on North American soil. The dairy-conscious Dutch, establishing a foothold on Manhattan Island in 1621, four years later landed their famously high-quality Holsteins. The herd was "removed upwards to a convenient place abounding with grass and pasture," most likely the pastures that became part of present-day Central Park. An additional boatload of cows arrived in 1626. Yet the Dutch, too, witnessed cattle falling from unknown causes, sometimes twenty at a time. Over the course of the next twenty years, losses drove the settlers to purchase cattle from the English, not just because the latter "have plenty of them," as contemporaries reported: the inferior English stock, not nearly as demanding of care and proper nutrition, were faring better than Dutch Holsteins, which apparently depended on more "trouble, expense, and attention." Given the arduous, snowbound winters and the shortage of winter feed, English cattle were winning the contest for survival.[13]

Despite an obdurate faith in New World abundance, settlers soon learned that nature required a helping hand. In the invasion of Europeans into North America, a rather primitive typology of gender applied with regard to food consumption. Where settlements consisted of primarily men, the practice of hunting made meat a central part of the diet; where there were women, families, and careful husbandry, milk, butter, and cheese joined other forms of protein on the table. The absence of dairying among the Spanish might well have been the result of too few female dairy experts among their adventurers. In northern regions populated by the English, French, and Dutch, women and cows made all the difference. The "comfortable subsistence" finally achieved by industrious New Englanders demanded careful attention to climate and terrain, along with considerable skills in agriculture.

Cattle — and those who manufactured their by-products, women — represented a technological and moral indicator of utmost importance to the colonial venture: they belonged to a mode of civilized agriculture, bounded by fences and furthered by labor-intensive strategies. As William Cronon showed in his study of the ecological transformation of New England, the English mounted a legal defense to their claim to the land (recast as private property) on the backs of their cattle. John Winthrop pointed out that Native Americans never "possessed anything more than a 'natural' right to property" because "they inclose noe Land and had no tame Cattle to improve the Land by." The argument could be extended to the pursuit of dairying. Native Americans did not practice any form of the art: while men hunted and trapped animals, women cultivated the land. Colonial settlers viewed this particular division of labor as primitive compared with their own. ("Their wives are their slaves, and do all the work," wrote one New England settler; this observation was based on seeing women working in the fields during the months of summer.) Though dairying was clearly viewed as laborious by the colonial women who engaged in it, the work itself was classified as household labor, part of the myriad tasks of the good housewife. Note that the invisible hand of improvement in these accounts belonged not to the divine maker, but to colonial women, who deserved a lot of credit for the survival of their households.[14]

Two developments came from these basic facts of colonial life, both impinging on milk history: one was the establishment of milk as a nearly ubiquitous element of the colonial diet. A penchant for lavish use of the products of their ever-present cows surfaces repeatedly in memoirs of the era. Some of this can be attributed to a dairy-rich English heritage, as a perusal of early English cookery books will confirm. Yet a study of every-day eating habits suggests a difference in degree in America, where readily available animal products exerted a gravitational pull in the diet of a society populated by small farms. American colonists ate like peasants of Scotland or France, but with one difference: they enjoyed the advantage of more widely distributed land and livestock. This must account for the greater abundance of dairy (as well as meat) in the diet of New World immigrants.

And repeated abundance it soon was: over time, colonists learned how to best exploit their cows for their dietary largesse. Cows grazed more systematically when New Englanders and Virginians passed laws requiring landowners to erect fences. Barns and cowhouses were erected across the landscape. Careful labor was devoted to extracting milk from cows and, eventually, to delivering it to towns and cities. From memoirs, we know that many New Englanders and some inhabitants of the Chesapeake consumed milk twice a day, at breakfast and supper, where it often took center stage. The first meal usually consisted of hot milk and bread, with tea or coffee in the later part of the century. Bowls of milk and bread often served as food given to those who came in from the fields at the end of the day during the growing season. Butter and cheese added to this supplied necessary protein in the spring and early summer months, when supplies of salt-meat were exhausted.

Recall that dairy products in the seventeenth century had served as harbingers of the new growing season, providing a necessary bridge, when most other forms of sustenance were depleted, to the more plentiful summer months. Given the ubiquity of cows and the determination of New Englanders, it wasn't long before colonists enjoyed milk even through the winter months. We know from individual accounts that milk was available to drink in December and January. When Sarah Kemble Knight found shelter from a storm near New Haven on a December night in 1704, she complained that her hosts "had nothing but milk in the house."[15] This was probably not just a matter of working out problems of storage, which a brisk trade in pond ice would have addressed later in the century. Philadelphia Quaker Elizabeth Drinker left a spare but telling entry in her diary: "1777, Janry, 31 parted with the Calf: and used our Cows milk for the first time."[16] Here was intentional manipulation of a cow and newborn calf for the sake of milk in the winter. Moreover, milk would not be relegated to second-tier status, if we are to accept a suggestion from the diary of Philadelphia socialite Ann Livingston. On a cold January night of 1784, she recorded enjoying a supper of "hominy & milk, & mince pies — the Eveng was realy [sic] delightful." Her love of "a fine Syllabub" proved that dairy products were put to more elevated uses, too. With milk and meat in winter months, there was simply more of everything.[17]

Successful adaptation to native plants meant incorporating milk into new foodways. New Englanders came to enjoy quick bread made of Indian corn, which cried out for some form of lubrication in order to go down well. "It is light of digestion, and the English make a kind of Loblolly of it to eat with Milk which they call Sampe," explained the author of *New-Englands Rarities Discovered* in 1672.[18] When Amelia Simmons's *American Cookery*, the first cookbook published in North America by a North American woman, appeared in 1796, it featured recipes that later became icons of the new cuisine: "A Nice Indian Pudding," "Johny Cake, or Hoe Cake," and "Indian Slapjack" (pancakes), all of which required copious amounts of milk. The addition of molasses and other spices would have made these dishes tasty and special in their own way. A new edition of Susannah Carter's *The Frugal Housewife* (first published in 1772), more or less a reprise of English cookery, announced an appendix "containing several new receipts adapted to the American mode of cooking." Here, again, milk played a prominent part, though Carter sympathetically added that "Water may be used instead of milk in case you have none." Her recipe for mush — cornmeal and water — showed that not all colonial diets were equal. But in ways similar to recipes in *American Cookery*, milk appeared repeatedly by the pint and quart, recommended for cold-weather drinks and invalid dishes. No household could do without it for long.[19]

American hospitality required recipes, too, as we immediately perceive in Amelia Simmons's inclusion of syllabub in her *American Cookery*. Domestic cows were a visible presence in party preparations. Her subtitle fused culinary luxury with homely equipment: "To make a fine Syllabub from the Cow." Not unlike high-class recipes of today, her instructions indicated that fresh was best. "Sweeten a quart of cider with double refined sugar," she began, "grate nutmeg into it, then milk your cow into your liquor," she instructed. The mistress was in charge of this operation, given the judgment needed to decide "when you have thus added what quantity of milk you think proper." The final touch came from slathering "the sweetest cream you can get all over it." Perhaps the most curious trait of the new American cuisine was its reprise of a medieval English manor, with providence unbounded, unembarrassed by the rather inelegant bovine guest.[20]

The multiple Protestant contributions to American culinary traditions placed milk on a pedestal; here was a food that "nourished without indulgence," in the words of one food historian. The cuisine of country Quakers was the perfect example: "morning and evening repasts were generally made of milk, having bread boiled therein, or else thickened with pop-robbins — things made up of flour and eggs into a batter and dropped in the boiling milk." (Thus, we can pinpoint the origin of the expression "Quaker food" used to refer, somewhat condescendingly, to boiled puddings and dumplings.) "We eat so moderately," wrote one Quaker of Pennsylvania, "that the whole day seems like a long morning to us."[21] Perhaps the best advertisement for milk in America has remained tucked away in a letter between wife and husband early in the revolutionary war. Abigail Adams signaled her support for the colonial boycott of tea through her testimony to the goodness of milk. "Why should we borrow foreign luxuries?" she wrote John Adams in September 1777. "Why should we wish to bring ruin upon ourselves? I feel as contented when I have breakfasted upon milk as ever I did with Hyson or Souchong."[22]

European visitors were quick to note the American affection for milk. When the Marquis de Chastellux toured America during the War of Independence, he described the custom of drinking coffee with milk alongside meals, which he found strange yet, in the long run, agreeable. He noted that in places where grain surpluses were scant, no liquor or cider could be found; the only beverages were water and milk. But even in times of plenty, the trusty dairy drink remained popular. The French Marquis du Pin, who settled in upstate New York, seemed rather amused by his neighborhood's fondness for a party punch made of boiling milk and cider, the American adaptation of English syllabub. "To this is added five or six pounds of sugar, if one is inclined to do the magnificent, or if not, the same quantity of molasses, then some spice, cinnamon, cloves, or nutmeg, &c." He reported that his "industrious guests consumed, to our delight, an enormous cauldron of this mixture, along with much toasted bread, leaving us at five in the morning, saying (this in English) 'famous good people, those from the old country'!"[23]

The second equally important development launched by cattle in the New World was the boost they gave to geographical expansion. Cast upon the American landscape, colonial pasturage demanded much more space than tillage. The desire for more land for cows lay beneath the legendary split of the community at Cambridge, where some settlers petitioned for permission to move into Connecticut. John Winthrop's *Journal* spoke of a majestic entourage that accompanied a minister on his way in the spring of 1636: "they drove 160 cattle and fed of their milk by the way."[24] A similar wish for greater pasture impelled Watertown and Roxbury residents to petition the colony. "The occasion of their desire to remove was, for that all towns in the bay began to be much straitened by their own nearness to one another, and their cattle being so much increased," Winthrop recorded. As William Cronon put it, "Regions which had once supported English settlements came to seem inadequate less because of *human* crowding than because of *animal* crowding. Competition for grazing lands — which were initially scarcer than they later became — acted as a centrifugal force that drove towns and settlements apart."[25]

If one charts the progress of the production of cheese in early America, the gradual westward movement of leading regions comes into view. The cradle of large-scale dairying appeared in the Narragansett Bay of Rhode Island, where operations housed more than a hundred cows and employed slave labor. Soon intensive cheese production arose in Vermont and expanded next in the valleys and lake regions of New York, finally arriving in the Wisconsin and Minnesota region of the Middle West. This spatial rearrangement sprang from complex forces, some of which we will examine in the next chapter. A key shift occurred in agricultural geography: the appearance of "milksheds" around major northeastern cities. As land around cities became more valuable, large-scale dairy farmers moved their operations westward, while a new form of liquid milk production evolved around urban markets. As a result, to cite just one example, farmers of New England took up new residences in the Western Reserve of Ohio. By the 1840s and 1850s, this area became known as "Cheesedom," owing to its massive production of that staple item. The landscape of dairy farming was thus reconfigured by the liquid milk market, repeatedly thrusting milk and butter operations into the hinterlands, which stretched westward as far as the edge of the Great Plains. (Only after the invention of modern refrigeration could present-day large-scale wonders of American dairying, California and Texas, take the lead.)[26]

Cattle, male and female, were most likely involved in another interesting feature of the colonial past: the outstanding height of Americans. Focusing on male height, demographer Richard H. Steckel has argued that "native-born Americans were the tallest people in the world in the eighteenth century," which he attributes to plentiful sources of protein (fish and game, in his estimation) in the average diet. Any study of colonial consumption would be pressed to add milk and dairy products to the list of staple foods with protein. Using models of height and income for the twentieth century, Steckel proposed another way of looking at the average male height of five feet eight inches: "colonial Americans were 10 centimeters [four inches] taller than estimates of their per capita income would suggest." In other words, the environment must have offered conditions capable of promoting healthy survival on less income than in Europe. When one conjures up images of the colonial landscape in the eighteenth century, replete with barns and cowhouses, it is difficult to escape the impression that the milch cow had something to do with this momentous skeletal achievement.[27]

The cow completed a resplendent vision of America as an overstocked Elysium, captured perfectly in J. Hector St. John Crèvecoeur's *Letters from an American Farmer* in the 1780s. In visiting an "outsettlement" in Pennsylvania, he exchanged greetings with a farmer of some one hundred and fifty acres, who was learning, day by day, how to make the most of his new way of life. "How do you go on with your cutting and slashing?" Crèvecoeur asked the man. Back came a characteristically optimistic response marked by an acknowledgment of generous bovine endowment:

> Very well, good Sir, we learn the use of the axe bravely, we shall make it out; we have a belly full of victuals every day, our cows run about, and come home full of milk, our hogs get fat of themselves in the woods: Oh, this is a good country! God bless the king, and William Penn; we shall do very well by and by, if we keep our healths.

Mighty engines of sustenance, milch cows seemed to be volunteering their service for the historic colonial project. Or so it seemed in the eyes of North Americans, who, with time, found ways to market such plenitude to the rest of the world.[28]

PART 3

Industry, Science, and Medicine

CHAPTER EIGHT

Milk in the Nursery, Chemistry in the Kitchen

Nature dictates that women, as good mammals, must nurse their young. But Western culture, accustomed to flouting the laws of nature, was busy challenging the practice of breast-feeding from very early on. This contradiction is one of the great dilemmas of history and justly deserves more attention than the brief treatment here. The subject certainly belongs in the larger history of food. As Marilyn Yalom aptly pointed out, the breast "offers a psychoanalytic paradigm for the Garden of Eden." We might add that in the history of milk, the sort coming from the breast is the apple itself, an object of endless curiosity and conflict for well over a millennium. It has also provided the template against which all other milks have been measured.[1]

While all medical authorities agreed that mother's milk was unquestionably the best nourishment for infants, popular practices adopted animal milk (mainly from cows and goats, modified according to local custom) as a reliable alternative. "Hand feeding" or "dry feeding," as the alternative was known, differed across cultures, but most techniques depended on animal milk as a key ingredient. (This was, after all, how the infants of wet nurses were kept alive while their own mothers' milk was redirected to others.) Recipes varied according to geography and season, but most "pap" consisted of some kind of meal made from bread crumbs or wheat flour, milk (or ale or water), and sugar. Techniques of preparation were as idiosyncratic as the choice of ingredients. For example, George Armstrong's explanation of 1771:

Crumb of bread boiled in soft water, to the consistence of what is commonly called pap, or a thin panada. The bread should not be new baked, and, in general, I think roll is preferable to loaf bread; because the former is commonly baked with yeast only, whereas the latter is said to have alum sometimes mixt with it. . . . If the infant is to be bred up by hand from the birth, it ought to have new cow's milk mixed with its victuals as often as possible. . . . The victuals should be made fresh twice a day . . . and three times in summer, and the milk must never be boiled with the pap but by itself, and added to the pap every time the child is fed; otherwise it will curdle, and grow sour on the child's stomach. It can hardly be necessary to mention, that when new milk is made use of, it must not be boiled at all.[2]

A wide range of recipes shows that milk was deemed essential insofar as it was available and fresh. Boiling milk was commonly held to be a reliable method of purification; failing that, boiled water or broth stood in as an acceptable medium for the "affusion" that would be piped or spooned down an infant's throat.

The history of artificial feeding in western Europe suggests that two separate problems were at work: the fact of a real, admittedly sometimes medically puzzling, inability to nurse on the part of some mothers; and an antipathy to breast-feeding powerfully shaped by cultural and environmental factors. The female body's ability to produce breast milk presented a challenge to scientific and medical authorities of the past. To begin with, colostrum, the yellow, watery fluid secreted in the first days after childbirth, did not win approval from most physicians until the eighteenth century, though not all of them were convinced of its benefits. Until as late as the nineteenth century, some mothers were instructed to hand over newborns to another lactating woman until true milk came in. (Today's experts point out that precisely the opposite strategy — immediate and frequent nursing — results in an abundant supply of breast milk.) Once established, a mother's milk could vanish in times of hardship: inadequate diet easily depleted the ability to produce milk, a law of nature with which the poor were intimately familiar. It is just as likely that some aristocratic women consumed too little food to sustain a sufficient supply. And tight corsetry also had a way of pinching, quite literally, the flow from breast to baby.

Crucially important was the advice women received on how to nurse during the hours and days directly following childbirth. The early weeks of nursing usually brought on an array of breast ailments, from lacerations to infections, which would need to be endured and cured while nursing continued. In elite households, the intervention of medical men probably brought more failure than success in such times. Add to this a set of surrounding factors — anxious husbands and relatives — and the likelihood of successful nursing became slim indeed. Advice books acknowledged the challenges of establishing the milk supply; one celebrated eighteenth-century French authority suggested the use of newborn puppies as a way of easing into regular nursing.[3]

Antipathy to breast-feeding sprang from many sources, including important psychological and cultural forces. As Chapter 1 pointed out, the milk of breasts served as a common bond between humans and other mammals, one that was shared with a good deal of ambivalence. The dilemma of infant feeding forced Europeans to face the profoundly unsettling fact that milk connected humans to their animal natures, which, in turn, tied them to a realm opposing civilization. The association between breast-feeding and "uncivilized" manners made the practice of seeking a wet nurse — a living substitute for the mother's breast — all the more logical.

A contradiction lay at the heart of breast milk: though it was uniquely personal and perfectly suited to infants, it was, like any other exceedingly potent substance, perfectly transferable. Even while medical authorities continued to believe that the fluid represented a form of the mother's blood, this "fact" did not prohibit redirecting the milk of one woman to the child of another. The practice of wet-nursing was commonplace across Europe and America in the eighteenth century. "Selling mother's milk," to use a phrase of historian George Sussman, represented a mutually agreeable arrangement: the social and sexual demands of upper-class womanhood required it, while the financial needs of other women, most often from poorer circumstances, benefited from it. Such relationships were almost always limited to financial terms, unlike in the Middle East, where "milk siblings" — children nursed by the same breast — would be forbidden to marry. Europeans and Americans meanwhile remained loyal to an ancient belief that the milk of the wet nurse would convey unwanted personal qualities to the recipient child until at least the 1860s.[4]

Attitudes related to gender, as well as social class, underscored a more general antipathy to breast-feeding among male authorities. Regular use of the "pan and spoon" appeared in the late seventeenth century, when royal physicians expressed preference for the method over the use of undesirable wet nurses. Spurred on by husbands, who, according to historian Valerie Fildes, "rarely contemplated or allowed their own wives to breastfeed," the aristocracy and gentry of northern Europe established a tradition of resorting to pap or panada, artificial alternatives preferable to putting infants to the breasts of strangers.[5] According to a well-known treatise on the subject by Flemish chemist Jan Baptista van Helmont (1579–1644), three reasons lay behind the use of pap. Breast milk was to be avoided, van Helmont argued, because it was vulnerable to spoiling; worse, it transmitted diseases, most notably syphilis. And because milk was another form of blood, which was thought to carry the traits of its producer, artificial feeding would prevent the transmission of unwanted qualities from women of low birth to babies of the upper classes. Fildes pointed to this "experimentation by wealthy fathers in the late 17th and early 18th centuries" as a generator of the rapid "social acceptance" that artificial feeding enjoyed at that time. Van Helmont's recommendation for successful infant feeding: a mixture of "bread, slightly boiled in smallbeer and sweetened with clarified honey or sugar."[6]

Other means of feeding infants became generally known through the writings of Pierre Brouzet, physician to Louis XV of France, from the middle of the eighteenth century. Broaden your outlook, the learned doctor appeared to be saying. Approaching the subject much as an anthropologist would, Brouzet reported that breast-feeding and wet-nursing were unknown in certain parts of the world, such as Muscovy and Iceland. "Soon after they are born [infants] are left all day, by their mothers, lying on the ground, near a vessel filled with milk or whey, in which is placed a tube, the upper extremity of which the infant knows how to find, and putting his mouth to it, sucks whenever he is oppressed with hunger or thirst. . . . [I]t is certainly true that their method of feeding children with the milk of animals is evidently not dangerous, and that it is, at least, attended with as happy effects as that of giving them women for wet nurses." This report from barbarian territory must have shocked good Parisians, who might have wished for a more decorous suggestion coming from the frontiers of nature.[7]

The romantic movement of the eighteenth century, fortified by its patron saint, Jean-Jacques Rousseau, exalted the relationship of the infant and mother at the breast. In his hugely popular book on the education of children, *Émile* (1762), Rousseau urged a restoration of natural and sincere ties between parents and children. Offended by what he viewed as avoidance of responsibility by elite women, he made the mother-infant bond the foundation of his plan for happy families in a good society. His advice became famous for its extravagant faith in the power of the maternal breast: "When mothers deign to nurse their own children, then will be a reform in morals," he contended; "natural feeling will revive in every heart; there will be no lack of citizens for the state; this first step by itself will restore mutual affection." (It should be added that Rousseau deposited his own five common-law children in orphanages.)[8]

Of course, this paean to the breast had little to do with milk or nutrition, but other equally emphatic men weighed in on the subject with those aspects in mind, too. A French lawyer named Fourcroy cited the robustness of babies he had observed on a visit to America as evidence of the effectiveness of maternal breast-feeding. Without a hint of irony, he recommended that every household embrace this natural way of infant feeding under the strict supervision of a wise patriarch. Prussians took patriarchal breast-feeding one step further by enshrining the practice in law: the code of 1794 ruled that all healthy mothers must put their children to the breast; fathers would determine the proper time for weaning. The need for strong and numerous citizens mandated vigilance on the part of statesmen, newfound champions of mothers' milk.[9]

The degree of attention poured into publications about the breast and its milk was a sure sign that much was amiss in this carnal wellspring of nature. Problems lurked in the fact that the flow of breast milk remained mysterious and irregular; these were true strikes against it in an age with diminishing tolerance of uncertainty. Debates raged over ideal size and shape of breasts, depositing evidence of difficulties caused by small and large specimens, along with various nipple formations that prevented efficient expression of milk. A wide array of devices existed for assisting breasts in their arduous task. Of particular use were custom-made *étuis*, little caps formed from wax, wood, or other materials, worn as prophylactics against voracious infant mouths. For centuries, midwives had addressed the well-known hazards of breast-feeding. By the beginning of the

James Gillray (1757–1815), The Fashionable Mamma, or The Convenience of Modern Dress, *colored engraving published in 1796. This elegantly attired woman finds the latest fashion of loose-fitting bodice design perfectly suitable to the equally fashionable practice of breast-feeding her infant. Gillray was of course poking fun at the coincidence, but the image signals changing attitudes, owing in part to the popularity of Rousseau's writings earlier in the century. (© Courtesy of the Warden and Scholars of New College, Oxford/The Bridgeman Art Library International)*

nineteenth century, a formidable chorus of medical opinion added its voices and even its apprehensions about this path to infant nourishment, ridden with obstacles.[10]

Confronted with suspect wet nurses, along with mothers unable or unwilling to nurse their own babies, what was left? The peasant practice of suckling from the udder of a beast, a known recourse in France and Italy, presented a fresh alternative. Described or pictured in classical mythology, suckling from a goat, mare, or ass came with a reputable pedigree, if not an appearance of up-to-date civility. Animal nursing rescued European societies more than once from widespread infant death. When the "new plague" of syphilis appeared in France in the 1500s, mothers bypassed infected wet nurses by employing goats. Similarly, eighteenth-century abandoned children thought to have syphilis were not put to the breast (where they would have communicated the disease to the nurse), but rather received animal milk at the udder. Brouzet described the use of four-legged wet nurses from the countryside, regarded as pure and healthy alternatives to urban milk. "I have known in the country . . . some peasants who have no other nurses but ewes, and these peasants were as strong and vigorous as others."[11]

A German book published in 1816 advertised the virtues of this approach in *The Goat as the Best and Most Agreeable Wet-nurse*. We can only wonder if the use of "agreeable" referred to the milk or the disposition of the beast, which seemed to present fewer problems than lower-class women in the estimation of wardens of foundling homes. In France, these institutions made a regular practice of keeping goats on their grounds so that nurses might carry infants, one by one, to the milk source. One account of a hospital in Aix gave the impression of goats as regimented and eager recruits. "The cribs are arranged in a large room in 2 ranks. Each goat which comes to feed enters bleating and goes to hunt the infant which has been given it, pushes back the covering with its horns and straddles the crib to give suck to the infant. Since that time they have raised very large numbers in that hospital." The story may well be apocryphal, given the notorious restiveness of goats as milkers. Yet the practice of employing goats and asses continued in Paris and London throughout the nineteenth century.[12]

Direct udder nursing was unlikely as a universal solution, given the obvious practical difficulties and, perhaps more important, the rising

Direct udder nursing offered infant hospitals a convenient way of ensuring that milk supplies were safe from contamination and disease, providing that the livestock proved disease-free. The French institution pictured here served as one of several models in a "scientific treatise" written by an Englishwoman in 1895 at a time of international alarm concerning high rates of infant mortality. (Wellcome Library, London)

temperature of debate over infant feeding. Doctors became key agents of a more modern professional sensibility in their dual role as medical practitioners and public servants. As the number and size of institutions for sick and abandoned children grew, particularly in Catholic cities in France and Italy, physicians found ample opportunity to display their expertise. In Paris of the 1840s, they busily supervised all alimentary activity at hospitals and orphanages. New medical knowledge pointed to more sophisticated approaches to infant feeding, employing exactitude in chemistry and physiology that few lay persons thought to challenge. A strict regimen emerged through the mist of custom: milk for those under six months (mother's, goat's, or cow's) and *bouillie,* or pablum, for those beyond six or seven months. Cow's milk required an additive — bicarbonate of soda — to retard spoilage. "Milk soup" appeared as another favored entrée, sometimes made with semolina or rice flour. Older infants were allowed bread with butter and jelly, but no fruit. The physiology of the immature stomach, according to medical principle, was incapable of digesting it.[13]

The middle decades of the nineteenth century present an interesting problem: even though it was well known that infants who were fed by hand were much more likely to die than those breast-fed, evidence suggests that a majority of mothers nevertheless avoided breast-feeding their infants. The extent of hand-feeding varied according to country, region, and social class, but widespread criticism of mothers who had abandoned their "natural role" suggests that hand-feeding was fairly common. One estimate reckoned that three-quarters of the infants in New York City were fed by hand in the 1840s, a method widely employed in urban settings elsewhere. Not just working mothers were searching for alternatives to breast-feeding. In the 1860s, judging from the enthusiastic response to commercial infant formulas, "insufficient milk" plagued even prosperous and well-fed women.[14]

Advice manuals and anecdotal evidence give only hints of what the causes of this complaint were, particularly in the case of bourgeois women. Medical knowledge of physiology exerted much influence in determining how the dilemma of infant feeding was addressed, not only in public institutions but in private homes. As the status and authority of physicians rose in the public eye, that of midwives, armed with knowledge gleaned from practical experience, went into eclipse. Conventions of social class also intervened. As Melanie DuPuis has pointed out, the nineteenth-century cult of true womanhood isolated middle-class women from the casual support systems of communities and custom, which had enabled women to nurse successfully. Ever fearful of not supplying enough nourishment to their offspring, mothers in doubt would have been easily persuaded to search out supplements and replacements for breast milk.[15]

It would be a mistake, though, to assume that doctors and husbands were solely responsible for a turn away from breast-feeding in the nineteenth century. Women may well have chosen to avoid the effort and discomfort that the task required; it is difficult to know how these demands might have meshed with other aspects of life for women in the past. Childbirth and nursing became more a matter of private decision-making and less a topic of broader social discussion over the course of the century. This suggests that the requisite knowledge for successful breast-feeding would have been harder to come by. Changes in material life, including the availability of new commercial alternatives to breast milk, would alter the basis for women's decisions about how to feed their babies. By the end of the nineteenth century, many women were opting to hand- or bottle-feed their infants at or before the third month of life. From the 1890s, according to Jacqueline H. Wolf, physicians worked hard to persuade mothers of the benefits of breast-feeding in order to improve the survival chances of infants across every social class. Common assumptions about infant care had clearly incorporated a belief in cow's milk as an alternative to human milk, even for infants in their early weeks of life.[16]

To fill out our picture of how milk was understood in the nineteenth century, we need to examine another new forum of discussion that shaped the future: the chemical laboratory. Though remote from the front lines of infant feeding, the privileged space of the laboratory generated new considerations having to do with milk that changed the vocabulary and convictions of everyone involved in its consumption and distribution. By midcentury, doctors and physiologists claimed greater knowledge of human nutritional needs and digestion. With the aid of chemistry, modern societies linked this scientific understanding of milk to a specific notion of infant feeding calibrated to suit the human body.

What was milk made of? Until the early decades of the nineteenth century, the laboratory provided a less satisfying answer than the dairy. But farmhouse knowledge of the properties of milk — how it behaved when heated, for example, or how it changed in form when combined with other substances — differed from the investigations of nineteenth-century science. The new field of chemistry won legitimacy by revealing the "world of the infinitely small," which its practitioners aimed to boil down (sometimes, literally) to the essential elements of nature. As both a bodily product and an aliment, milk presented unique challenges. What made milk so perfectly and powerfully nourishing? And was its perfection a sign of some larger, hidden design in nature? For the first generation of researchers, this scientific quest remained colored by a familiar, subjective understanding of the milky white liquid.[17]

"Chemistry," a then obscure English physician wrote in 1816, "in the hands of the physiologist, who knows how to avail himself of its means, will, doubtless, prove one of the most powerful instruments he can possess." William Prout (1785–1850) was busy treating patients for stomach and urinary ailments, complaints that inspired him to ponder the process of digestion in his spare time. "Animal chemistry," the study of the conversion of food into body tissue, ranked as a choice topic of investigation among elite scientists in Germany, France, and England. Prout investigated the contents of blood, urine, and that great alimentary prototype, milk, in his quest for what he later titled "the Ultimate Composition of Simple Alimentary Substances." Aiming for admission to the exclusive scientific community in London, the ambitious physician offered a series of lectures to the public on digestion and metabolism in 1814. His performances brought him professional respect, influential connections, and, ultimately, admission to the Royal Society in 1819. For the next twenty years, he offered up one theory after another, including "Prout's hypothesis," an argument about atomic weights that provoked debate for the rest of the century.[18]

Prout displayed a Victorian affinity to grand ideas, so perhaps it is not surprising that milk figured into his machinations. Here was a physician-scientist intent on limning the bare essentials of metabolism, who spent vast sums on equipment to boil down organic substances to their simplest common denominators. His method proved too expensive for general use and was quickly superseded by a cheaper and easier way devised by the renowned German chemist Justus von Liebig (1803–1873) in 1830. Yet Prout succeeded in identifying three elemental units of human sustenance, a kind of sacred trinity of survival. "The principal alimentary matters employed by man, and the more perfect animals, might be reduced to the three great classes, namely, the *saccharine,* the *oily,* and the *albuminous,*" he asserted. These classifications, not so different from the descriptive vocabulary provided by Galen over a millennium earlier, now found empirical proof in the laboratory. (A later generation of chemists would identify the categories as carbohydrates, fats, and proteins.) Here were the building blocks of flesh, bones, and human energy, and all three happened to be present in milk.[19]

As it turned out, Prout's trinity served as an emblem of more than just chemistry in the age of Victorian science and religion. In 1830, a committee of the Royal Society assigned eight men, including Prout, the task of wedding modern science to contemporary theology in what would become the prestigious and lucrative Bridgewater Treatises. Defined by the bequest of the eighth Earl of Bridgewater, Francis Henry Egerton, the works were to demonstrate the "power, wisdom, and goodness of God, as manifested in the creation." Prout's interests, as it happened, were not far from a pet interest of the late earl: evidence of divine design in the process of digestion. (Victorians always provide grist for wits: the cantankerous theologian Ronald Knox's infamous nickname for the series was "The Bilgewater [sewage runoff] Treatises.") Though no theologian, Prout the physician-chemist adopted an inspired tone in order to communicate the inherent beauty of the organization of digestive activity on earth. The title of his treatise — *Chemistry, Meteorology, and the Function of Digestion Considered with Reference to Natural Theology* — laid bare the curiously hybrid mental world of the Victorians.

In an omnibus of images not unlike a Victorian painting, Prout crammed every detail of existence onto a vast interconnected canvas. He offered chapters on the ocean, the atmosphere, plant and animal life by means of the principle of perfection in design. Nowhere did this seem more evident than his coverage of human nourishment, in which milk appeared in a starring role. "Of all the evidences of design in the whole order of nature," Prout argued, "Milk affords one of the most unequivocal."

> It is the only aliment designed and prepared by nature expressly as food; and it is the *only material* throughout the range of organization that is so prepared. In milk, therefore, we should expect to find a model of what an alimentary substance ought to be — a kind of prototype, as it were, of nutritious materials in general.[20]

Reiterating his three classifications of the aliments — the saccharine, butyraceous, and albuminous — Prout emphasized their universality, which provided more proof of the inevitable purpose of milk. The organs designated to produce milk underscored the point:

> No one can doubt that the apparatus by which milk is secreted has been formed specially for its secretion. No one will maintain that the apparatus for the secretion of milk arose from the wishes or the wants of the animal possessing it, or from any fancied plastic energy. . . . In short, it is manifest that the apparatus and its uses, were designed, and made what they are, by the great Creator of the universe; and on no other supposition, can their existence be explained.

Prout's paean to milk (and udders and breasts) may seem to us rather circular, and somewhat peculiar as the product of a distinguished scientist, as it did to some of his contemporaries. (In a later edition of the treatise, he issued a statement of his unwavering allegiance to science, probably in response to criticisms.) Yet his book offered meaningful explanations of the natural world to a general audience. And its particular argument about milk would also breathe life into milk reform efforts as far away as America.[21]

Within the annals of chemistry, Prout provided a moment of clarity. Fellow scientists immediately embraced the three categories of aliments as written law. In France and Germany, chemists strove to determine precise quantities in different kinds of milk, measuring protein and sugar content by milligrams to ascertain how near or far the milk of animals approached that of humans. This represented only a first step, however, toward understanding the chemical composition of milk in all its varieties. With the benefit of hindsight, we can say that Prout's celebration of the universality of milk threw up an obstruction that remained standing between the laboratory and scientific truth for another twenty-five years. Not until 1875 did chemists grasp the essential differences among milk proteins. Only when they identified distinct milk albuminoids in the milks of different animals could they move beyond their erroneous assumption that all milk was the same.[22]

Prout's inspirational power traveled more successfully. Across the Atlantic, *Chemistry, Meteorology, and the Function of Digestion Considered with Reference to Natural Theology* deeply impressed Robert Milham Hartley, a religiously inspired reformer in New York City, who was eager to improve the supply of milk for infants of the poor. A treatise or two cannot change the course of history, but agitators like Hartley, who had the means and influence to back his convictions, could mobilize considerable forces. The father of ten children, Hartley no doubt had personal acquaintance with the problem of infant feeding. He was also a committed temperance reformer, determined to undermine the symbiotic relationship that existed between dairies and distilleries on the outskirts of New York City. Repulsed by the state of the milk supply of the metropolis, Hartley set out to connect the wretched conditions of urban cows fed on distillery "slop" — the spent grains discarded after fermented liquid had been extracted — and the compromised state of infant health in the environs of New York. In a series of articles published in 1836 and 1837, he made the first case for pure milk in the United States. And in 1842, he published *An Historical, Scientific, and Practical Essay on Milk as an Article of Human Sustenance,* a formidable fusion of current knowledge on the subject, the first book-length treatment of milk in the English language.

The trusted cow clearly won pride of place among other milk producers. Alternatives were not an option in America, in Hartley's eyes. He
dutifully mapped a variety of milks across distant areas of the globe to
show that only exotic peoples drank the milk of exotic animals. "That of
the camel is chiefly confined to Africa and China, and of the mare to
Tartary and Siberia," he explained. "In India, the milk of the buffalo is
preferred to that of the domestic cow." Laplanders drank reindeer milk,
while Italians and Spaniards preferred goats' milk. In America, cows produced the milk of choice: it ranked as "the best and most palatable aliment for the young," "suited to nearly every variety of temperament" and
"adapted to the nourishment of the body in every age and condition."
Perhaps just as important was the fact that it was "most abundant and in
general use."[23]

Figures on the production of milk for urban areas illustrated how great
this demand had become. Cows in the vicinity of New York produced
"nearly *five millions of gallons*" of milk for residents each year, according to
Hartley. He reckoned there were roughly ten thousand cows in New York
City in 1842, housed mainly in Brooklyn and Queens. The likelihood
of spoiled, contaminated, or adulterated milk was invariably high. Were
European cities "afflicted with bad milk and its consequences?" he asked.
Not to the same extent, he offered, because the manufacturing population of England depended more on "potatoes and wheaten bread washed
down with tea or coffee." Hartley missed the fact that Londoners were
also drinking "swill milk," though with fewer complaints. The cow population of London was estimated at twelve thousand in the 1830s, when
James Fenimore Cooper remarked on seeing cows assiduously grazing in
Green Park, which was "nothing but a large field cropped down like
velvet" by the animals. Hartley mobilized the latest statistics on infant
mortality in both countries to argue that rising rates in America could be
explained by a flawed milk supply.[24]

According to Hartley, New York City's milk problem originated with
the obstruction of the spirits trade from the Caribbean during the War of
1812. As distilleries appeared at the periphery and as pasture diminished
with urban expansion, "evils of the [present] system" appeared: the use of
stalls and spent distillery grains as the primary form of cowfeed. The
"tartarian fumes" of these establishments smothered nearby residents even
before the substandard liquid reached consumers. Hartley described the

progressive decline of a child fed slop-milk when "the delicacy of the mother's health rendered it improper, and, indeed, perilous both to herself and infant, to discharge the duties of a nurse." Believing that cow's milk was "most nearly resembling the nourishment designed by nature," she obtained local milk and added water and sugar before feeding it to her child. The food resulted in immediate evidence of decline: at first "irritable, restless, and unmanageable," the child soon began to look and act sickly. His eyes became sunken, "his appearance unnaturally pale and haggard; he lost strength and vivacity; gradually fell away in flesh; so that at the age of fifteen months, his weak and emaciated body would scarcely sustain itself without bolstering." When milk "produced from natural food" replaced distillery-generated milk, the health of the child quickly returned to normal. "Who, indeed, could have imagined, that under the disguise of so bland and necessary an article as milk, was lurking disease and death?" Hartley asked his readers.[25]

Purity and danger: even while Hartley's argument rested on an extensive scaffolding of animal chemistry, he alerted his readers to the powerful dualism presented by milk. Given the poor diet of urban cows, "most inhumanly condemned to subsist on the residuum or slush of this grain . . . reeking hot from the distilleries," few if any nutrients could come from their digestive activity. Hartley reasoned that processed grain lacked the necessary saccharine matter required of healthy lactation. Kept from fresh air and exercise, consuming roughly thirty-two gallons of mush per day, the pathetic bovines faced inevitable decline. Most important, they produced bad milk.[26]

But the fault lay not in milk as nature made it. Quoting liberally from "philosophic Prout," Hartley made the case for milk as the perfect food, not just for infants of the poor (the object of his wider philanthropic efforts), but for everyone. Humans unconsciously strove to produce approximations of milk through their culinary efforts: "Even in the utmost refinements of his luxury, and in his choicest delicacies, the same great principle is attended to; and his sugar and flour, his eggs and butter, in all their various forms and combinations, are nothing more or less, than disguised imitations of the great alimentary prototype *milk*, as furnished to him by nature," the English doctor had argued. Hartley's little treatise, fortified by the author of the Bridgewater Treatise on milk cosmology, threw down his own gauntlet for posterity.[27]

Chemists acted as important interpreters of bovine benefits. Throughout the early part of the nineteenth century, chemistry shone its penetrating light into the darkness of the food supply, illuminating the fact of adulteration and stimulating a nascent sense of consumer awareness. In 1820, an appalled British public had snatched up copies of Friedrich Accum's *Treatise on Adulterations of Food and Culinary Poisons,* a racy exposé of the true ingredients of typical articles of consumption. Nearly every common item of food and drink, according to the German-born chemist, suffered from fraud. Bread acquired its texture and whiteness not from high-quality flour, but from alum (aluminum potassium phosphate). Beer tasted hardened not from age but from sulfuric acid. And much of the tea that sold as imported goods (possibly two-fifths of the ten million pounds purchased by the public) was culled from English shrubbery, treated on metallic plates to achieve the right shade of green or black. Accum's printer knew how to market such revelations: the book's title page boasted a skull and crossbones and a line from the Bible: "There is death in the pot."[28]

Although Accum achieved very little in the way of actual food-marketing reform, he did manage to introduce the literate public to the practical value of a young but important field of science. His best-selling *Culinary Chemistry* outlined the chemical dynamics of baking and brewing for those who wished to understand how these age-old processes worked.[29] "The kitchen is a chemical laboratory," he proclaimed to his readers. Intent on bringing his favorite science to the people, Accum sold chemistry sets, along with basic handbooks, out of his shop in Soho as a way of making extra income. No need to feel intimidated, he coaxed, given the familiarity of the setting and its accoutrements. "All the processes employed" in the kitchen were, in fact, chemical reactions at work. The kitchen and laboratory even depended on strikingly similar equipment: "the broilers, stew-pans, and cradle spit of the cook correspond to the digestors, the evaporating basins, and the crucibles of the chemist." And though recipes in cookery were numerous, any denizen of the kitchen knew that "the general operations (like the general process of chemistry) are but few." Here was a science intended for everybody — certainly the title invited women to purchase the book — full of wondrous and rewarding transformations.[30]

Combining meat with acid and heating the concoction on a stove amounted to a simulation of human digestion: this was not only the subject of William Prout's research, but also the focus of a great deal of energy on the part of the recognized titan of laboratory science, Justus von Liebig. Liebig's contributions ranged from the fundamental (his method of determining the content of organic substances is still in use today) to the incidental (he invented, among other things, an improved method of making mirrors). During his time, he was known for establishing an institutional model for research and instruction of laboratory science at the University of Giessen, a "small, sleepy" place when he began teaching there in 1824. Virtuosic in many branches of chemistry, Liebig was the perfect publicist for science in the industrial age. His discoveries in the field of animal chemistry led to many breakthroughs in the study of food and nutrition, along with several commercial ventures sponsored by the great man himself. Amidst all of this activity, Liebig was instrumental in promoting a scientifically informed interest in milk.[31]

Liebig's name (and even his face, thanks to the advent of modern advertising) would have been recognized in homes across more than one continent during his day. Beginning with the publication of a paper on meat juices, and then with his *Research on the Chemistry of Food,* published in English in 1847, the chemist made a foray into domestic science by analyzing exactly how much nourishment lay in the runoff of meat. Liebig's conclusions stood "centuries of traditional cooking procedures on their heads." His point was relatively easy to grasp: meat, upon being boiled, released many of its important nutrients into the water in the pot. To get the most out of your meal, Liebig's findings implied, you ought to eat both the meat and its cooking juices. Though much like common sense, particularly to poorer households, where meat juices were often consumed without the benefit of meat, Liebig's method won the day on the grounds of science, not folk knowledge. His celebrity status serves as a marker of the shifting knowledge base of the European bourgeoisie.

Another shift had to do with how this class of Europeans came to regard their diet. According to Liebig's biographer William H. Brock, we can see a change "from food being seen solely as the means for the assuagement of hunger to a more scientific concern with food as nutrition and as a key to health, efficiency, and racial strength." Mid-Victorian cookbook writers in England and America were poised to absorb his message, given the arrival of a body of knowledge understood as domestic management in the two countries. Weighty responsibilities lay in the kitchens of prosperous homes. As the eminent cookbook author Eliza Acton put it, "It is from these classes that men emanate to whom we are chiefly indebted for our advancement."[32]

More important for the history of milk is how Liebig's career exposes an enormous hunger for therapeutic advice felt by a middle-class public. They welcomed the potent and appetizing curatives evolved from Liebig's stovetop experimentation. His recipe for "beef tea," which called for soaking raw chopped meat in cold water for several hours before adding hydrochloric acid (a digestive juice), was a favorite remedy among the cognoscenti in the 1850s. Patients could obtain the precious — and expensive — liquid from doctors and pharmacists across Europe.[33] The next decade witnessed the spectacular success of his "extractum carnis," popularly known as "Liebig's Extract of Meat." The chemist himself was at the helm of assiduous and shrewd marketing activities on three continents, building a commercial network that extended all the way to cattle herds in South America. The unctuous brown liquid with its intense bouquet of cooked beef, compact and needing no refrigeration, answered the prayers of more than just the sick. Victorian armies, institutions, and households scooped up the product. Florence Nightingale pronounced it "the best thing I have seen," suggesting that Liebig's extract had been preceded by less convincing contenders. Even while medical authorities withdrew their testimonials, claims for beef extract multiplied, extending to the recommendation that the product could be used to feed infants. Portable, non-perishable, and hailing from the nutrition-packed steer, the liquid helped to shape expectations of fulfillment from sturdy bovines.[34]

Liebig's debut with artificial infant food, like formulas appearing later in the century, can be described as a rescue narrative involving next-of-kin. According to his account, Liebig mixed up a formula for his own daughter, Johanna Thiersch, in the summer of 1864, when she was unable to nurse. It is worth remembering that Liebig set out on this particular venture after his spectacular success with meat extract. Even his publisher expected to be running over the same tracks: after Liebig published his findings in an academic journal as "Eine neue Suppe für Kinder," the work appeared, expanded into book form, as *Suppe für Säuglinge* [Soup for Babies], in 1865. His publisher, probably misled by the use of *Suppe*, misunderstood the project, thinking it was intended for sick children, not newborns. But just as beef tea and beef extract could go both ways, as food

Liebig's Extract of Meat business made early use of "signature" branding in order to compete with many imitators. This American advertisement included a clever jingle, which began, "With taper fingers soft and white, she stirs the steaming cup — / Is it some love-charm, sure of might, the maid is mixing up? / Ah, no, dismiss the fancy free, before it runs to waste — / She's making Liebig's Extract broth to suit her Grandma's taste." (Holdings of the Ohio State University Libraries; used by permission)

and as medicine, so, too, might this new mixture for babies. Whether for the healthy or ailing, the infant or toddler, the product claimed to deliver first-class sustenance.[35]

We now know that another leading chemist of the age, Edward Frankland, proved the effectiveness of his recipe for "artificial human milk" more than a decade earlier, in 1854. (His mix, however, did not reach the market until the 1880s.) Frankland came to the rescue of his own baby and wife, Sophie, who was unable to breast-feed any of their five children. Described as frail and vulnerable to illness, she fits the model of Victorian delicacy mysteriously incapable of meeting the physical and psychological demands of this particular maternal role. (Sophie Frankland later died in Davos, Switzerland, of tuberculosis.) Frankland's sexual demands, later hinted at in correspondence with his son, might have played a part in this matter, too. Their second child, Frederick William, born barely a year after their first, in April 1854, contracted a serious case of diarrhea, which nothing appeared to help. Before he reached the age of one, illness and lack of breast milk reduced baby Fred to "little more than a bag of bones."[36] Given Frankland's know-how in the chemistry laboratory, there was only one thing to do.

Frankland set to work modifying cow's milk to make it more like that of breast milk. First he curdled skimmed milk with rennet, which enabled him to remove as much of the available casein (protein) as possible; to the remaining liquid, he added lactose and a dash of milk and cream in order to bring the whole mixture up to an approximation of the human percentages of sugar and fat. "The result was spectacular," and little Fred made a good recovery. This kind of success could not go untrumpeted. Frankland revealed his experiment to the Literary and Philosophical Club of Manchester later that year. His subject that evening was Liebig's extract of meat, yet this chemistry-inspired invention claimed a certain kinship with infant food. He added his account of artificial feed for infants as an addendum, one more example of the worthiness of modification of bovine products for the sake of nutrition.[37]

Liebig's formula for babies, like true chemistry, required a good deal of measuring and mixing. (This aspect of the product gave consumers something to complain about and competitors something to improve upon.) His recipe depended on a particular ratio between "plastic" or flesh-forming material and "respiratory" components that belonged to human milk: specifically, 10:38. So ten parts of cow's milk in the form of dried milk powder needed to be mixed with particular amounts of wheat flour and malt flour (the latter, because it would turn some of the starch into sugar), plus potassium bicarbonate in order to bring down the level of acid in the milk. The result was not, in fact, "like" breast milk at all; it was actually twice its strength in terms of carbohydrates, while lacking in fat, because dried milk at that time could be made only with skimmed milk. What Liebig didn't know and couldn't understand was how this whole operation failed to provide certain important amino acids and vitamins essential to infant health and growth.

Woe to the infant wholly dependent on Liebig's Food, one fears, looking back on the brushfire-like spread of enthusiasm for the product. Early advertisements for the product played on more historically accurate anxieties, which had less to do with what was in the formula than who was living in the servants' quarters. "No More Wet Nurses!" a banner headline of an American magazine advertisement announced in 1869. Were these the sentiments of mothers, chemists, or business agents? All three engaged in an intricate dance of midcentury consumer culture. The English branch of the Liebig "Suppe für Säuglinge" business strove to adapt to consumer preferences: when the original liquid formula, relatively expensive, failed to sell well, the company developed a cheaper dried mixture using pea flour. Liebig's English translator, the Baroness Elise von Lersner-Ebersburg, testified to the unalloyed benefits of the product, which she fed to infants at her institute for foundlings in Bethnel Green in London. "It would have seemed cruel to deny this boon of life to so helpless a portion of humanity," the baroness confided. She herself had been unable to breast-feed, rendering her philanthropic work meaningfully poignant. Lersner-Ebersburg was also, it should be added, the owner of the English patent for the product. She reported distributing over eight thousand pounds of Liebig's "malted food" to families in the impoverished neighborhood of Bethnel Green.[38]

Competitors soon followed with cheaper and simpler mixes: Mellin's Food for Infants and Invalids, developed in 1866 by an English pharmacist, could be stirred into warm milk. Horlick's Malted Milk, invented by an English immigrant in Chicago, first resembled Mellin's in composition, but later offered a dried-milk variety for mixing with water. As historian Rima Apple has shown, companies were well aware of the problem of contaminated milk supplies, so formulas like Horlick's delivered scientific advancement that responded to (or generated) yet another apprehension of the modern mother. Physicians and scientists endorsed the products in ways that appear familiar to present-day consumers. And like today, it was difficult to sort out the medical from the commercial plaudits in articles on the manufacturing process in *Scientific American* and *American Analyst* in the 1880s. Information and publicity had melded into one.[39]

From its origins in the 1860s, artificial food for babies depended on a modern emotion: a reverence for chemical science. The celebrated Arthur Hassall, author of the definitive Victorian tome on food adulteration (a considerable advance on Accum), seemed enthralled by the properties of Liebig's food. In the pages of *The Lancet,* Britain's foremost medical journal, he described how the malt "exercises . . . a most remarkable influence upon the starch, quickly transforming it into dextrine and sugar, so that in the course of a few minutes, the food, from being thick and sugarless, becomes comparatively thin and very sweet." Kitchen chemistry had not died after all! Having built his career on identifying artificial ingredients in common foods, Hassall's endorsement of artificial food for babies seems somewhat ironic in retrospect. But few would have thought so in the 1860s, when chemists enjoyed unparalleled prestige in matters of nutrition.[40]

The wonders of chemical transformation proved to be a powerful tool of seduction that businesses were not likely to discard. Twenty years later, *American Analyst* gave readers a bird's-eye view of the food laboratory. No longer claiming kinship to the kitchen of Accum's day, the research laboratory found a home in the same building in which Mellin's Food was manufactured. Here, one could find the "most ingenious and perfect machinery," supervised by "skilled employees." The product itself was "not in any sense a mere compound where the several ingredients are simply mixed together, but . . . the result of great chemical skill, and many years of experience, and . . . much money expended." The language was indistinguishable from that used by a certain Dr. Hanaford in "The Care and Feeding of Infants," a booklet provided free of charge by Mellin. "Men of the highest scientific attainments of modern times, both physiologists and chemists, have devoted themselves to careful investigation and experiment in devising a suitable substitute for human milk," came the reassurance. The fact that they were in the employ of infant formula companies redounded to the credit of the manufacturer; no one seems to have imagined a possible conflict of interest.[41]

The infatuation with chemically designed infant food rested on more than just effective sales pitches. If we look carefully at what was considered cutting-edge analysis of milk in 1878, we can see how current thinking about human breast milk had become deeply enmeshed with chemical science and nutrition. When the medical privy councilor of Switzerland, Henri Lebert, investigated the subject for the thriving new company led by Henri Nestlé in Vevey, Switzerland, he laid out all the necessary (and fully accepted) facts. First, he explained the "nutritive wants of the human body" according to the latest findings. Then he presented a grid of six different kinds of milk, including that of woman, according to their respective constituents of water, casein, albumen, butter, sugar, and saline substances. This prepared the ground for his discussion of how woman's milk measured up to the needs of human nutrition and, in particular, the needs of infants in the first months of life.

In the final unveiling, nutritional science trumped nature: woman's milk, perhaps predictably, came up short in nutrients and in plenitude. Lebert's strategy was to compare each constituent of human milk to another kind that had more of it; to do this, he had to shuffle rapidly among analyses of mare's milk, asses' milk, and cow's milk. Thus, he repeatedly employed a vocabulary of deficiency to the milk of mothers. "The extraordinary poverty of the milk of the woman in saline substances," Lebert pointed out, "instinctively leads the mother to give the child, at an early stage, other food in addition to her own milk, especially such as contains flour and is rich in hydrates of carbon." Here, a sleight of hand helped along the case for infant formula: never mind that "saline substances" meant salt, not "flour" and "carbon," which happened to be the components of Nestlé's Infant Food. Lebert estimated that even in the best of circumstances, the supply of mothers' milk proved sufficient for only six or eight weeks, after which, he implied, the body of an infant yearned for something more.[42]

The star of the whole production, we discover, proved to be the cow, not the company. Lebert's ode to bovine milk reflects the nutritional lessons of the previous half century and foreshadows the convictions of the decades to come. "The milk of the cow is, of all kinds of milk, the one that comes next to that of the woman," he averred. "It is, however, richer in matter, 14 per cent. being solid constituents." Solidity counted for a lot in the Victorian age. "The plastic power of this milk is therefore very considerable," Lebert continued, "and with its richness in butter, 4.3 per cent.," along with "richness in saline substances," cow's milk was "very serviceable to nourishment." "It is thus a matter of fact," he concluded, "that the milk of the cow is in every respect more nourishing than that of the woman and hence it is not strange that there are whole populations who nurse their infants with cow's milk."[43]

Business enterprises like Nestlé's were in the process of consolidating their power in large-scale concerns of international stature. But more important, shared cultural values made possible the expansion of popular interest in infant formula. A scientifically endorsed knowledge of nutrition spread alongside a new notion of convenience. As findings about milk's nourishment stoked anxieties about nursing mothers not having enough, the commercial product offered relief from worry and (just as much a selling point) measured predictability.

Women's magazines, Rima Apple has shown, voiced a virtual chorus of concerns that were peculiar to the age: an alertness to health, an expectation that one must locate the best product that money could buy, a thinly veiled recognition of other matters on a woman's mind, present in the way in which advertisements gave women permission to opt out of breast-feeding. The *Ladies' Home Journal* of 1888 advised the need to look for supplements. "If fed from your breast, be sure that the quantity and quality supply his demands. If you are weak and worn out, your milk cannot contain the nourishment a babe needs, and good cow's milk, or some food that contains the same elements as human milk, should be at least partially substituted. You will soon feel the advantage yourself, and see it in the child."[44] By the end of the nineteenth century, women of privilege in Europe and America believed that their own breast milk could not supply what newborns needed in order to thrive. They were not the first generation of women to think this way, but they were the first to feel supported in their conviction by the findings of chemical science and powerful business interests.

Chemistry, at least, still labored under certain humbling limitations: scientists at this time could not quite comprehend all that was present in milk, whether human or cow. But manufacturers of infant food were not scientists, so they did not share the skepticism of some members of the scientific community. A prophetic voice rose above the clamor over artificial food for infants in 1871, shortly after the siege of Paris. A leading chemist, J. A. B. Dumas, had survived the trials of hunger experienced by the French and offered his observations on what was, in effect, an unprecedented nutritional experiment inflicted on human subjects. Shortages of all kinds, especially "the scarcity of milk and eggs," led to "the premature decease of a great number of young children." What, if anything, Dumas asked, could "the forces of science alone" provide in such a situation?

The French chemist's particular interest was this: "Was it not possible to come to the assistance of newborn children by replacing the milk, which could no longer be got, by some saccharine emulsion? In this case, there was no question of creative chemistry, but only of culinary chemistry. Recipes were not wanting, he reported, "all reproducing an albuminous liquid, sugar, and an emulsion of a fatty body." But the compulsory experiment resulted in infantile deaths and diseases, which, from a twenty-first century perspective, were clearly nutritional deficiencies. At the time, the results proved beyond a doubt that real milk—human or animal—contained something that the artificial emulsions did not. But no one at the time knew what was missing.

Many years would pass before academic credit came to Dumas for discovering, in effect, vitamins. Though the word was not coined until 1912, his report was the first to uncover their existence by means of showing their absence in artificial infant food. It is ironic that his article included words of advice to his profession. "The power of synthesis of organic chemistry in particular, and that of chemistry in general, have therefore their limits. The siege of Paris will have proved that we have no pretension to make bread or meat from their elements, and that we must still leave to nurses the mission of producing milk." He added, "Sometimes there was such a conviction in the authors of these propositions, that one was forced to dread for the future the effects of their faith." For Dumas, the quest to reinvent milk served as a humbling reminder of the scientist's imperfect grasp of nature.[45]

The exact nature of breast milk was only one of several mysteries resistant to the efforts of the laboratory scientist. By the end of the nineteenth century, knowledge of microbes raised yet another flag of concern over the ramparts of pasture and cowshed. Milk became known as a dangerous carrier of disease by the 1880s. It was up to reformers and scientists to come up with a way of reducing risks so that this now widely embraced source of nutrition could be made available to urban populations, infants and all.

CHAPTER NINE

Beneficial Bovines and the Business of Milk

"An army marches on its stomach," according to an old saying. And as one letter from the American Civil War reveals, food was the first line of treatment for injured soldiers. A volunteer nurse, Eliza Newton How-land, described her first two days on a hospital ship in 1862. Supplies were limited when the first one hundred and fifty wounded arrived. "With two spoons, and ten pounds of Indian meal (the only food on board) made into gruel, G. and I managed, however, to feed them all and got them to bed," she wrote to her husband. Soon, however, the crew mustered famil-iar remedies. "Down in the depths of the Ocean Queen, with a pail of freshly-made milk punch alongside of me, a jug of brandy at my feet, beef tea on the right flank, and untold stores of other things scattered about," Howland recorded, "I write a hurried note on my lap, just to tell you that we keep well." For the sick and wounded, restorative liquids of this kind were as necessary as soap, water, and bandages. Milk punch delivered a much-desired relief from pain, because it was flavored with a generous shot of brandy or rum. The large quantities impressed Howland: "Beef tea is made by the ten gallons and punch by the pail." This may have been her first experience cooking for a crowd, so to speak.[1]

Viewing the history of food, one is reminded of the Sufi tale of blind men examining the elephant from all quarters. Adjusting for Western culture, the animal in question was the cow, an object of desire for pro-vision seekers on both sides of the Atlantic. The female bovine, as we learned in Chapter 7, delivered abundance in the dairy and an aura of domestic security. Its male partner also held out promise in the form of meat, and though the two products were very different, meat and milk became leading targets of food producers and marketers by the middle of the nineteenth century. Not for nothing were dairy products called "white meats": earlier consumers comprehended their substantial nutritional value and the fact that they were satisfying and filling. The challenge of making milk as portable as its dairy partner, cheese, continued to preoccupy provisioners, particularly in an era of war and colonial ventures.

Unprecedented demands made by cities and towns served as inspiration for entrepreneurs in the business of food supplies, and fortunes stood to be made in satisfying institutional needs, particularly those of armies and hospitals. A decisive shift toward large-scale output, along with industrial techniques, became apparent in the production of food during the last half of the century. These decades inaugurated an era of new food commodities designed for the market, distinctive in their uniformity, portability, and "branded" nature. It is here that we can see milk drawn into a new arena of business that would change its future in fundamental ways.

While Justus von Liebig oversaw bovine carcasses squeezed into syrup in Uruguay, a Texas entrepreneur of a very different kind invested in a portable beef substitute of his own design. Gail Borden's beef biscuit aimed to update the centuries-old recipe for hardtack, which was plain carbohydrate (a mixture of flour and water) hardened into indestructible crackers or biscuits. Later testimony of Civil War veterans described hardtack as "sheet-iron crackers," "teeth-dullers," suitable for loading into guns or as paving stones around tents (which was actually demonstrated by one troop of artillerists). Borden intended an improvement not only on hardtack's inedible character but also on its nutritional content, or lack thereof. Derived from a nitrogenous (that is, meat) source, his own modern-style biscuit capitalized on the fashion for "flesh-formers," ideal for men engaged in long marches on near-empty stomachs. With Borden's product, three pounds of the "best beef" of Texas could be had for every two pounds of biscuit. Years of fussing with drying trays in the humid environs of Galveston, Texas, paid off in 1851, when the inventor took his biscuit to London and won an award at the world-renowned Great Exhibition.[2]

Making sense of the "pre-history" of Borden's encounter with milk is not easy. Was he an eccentric with some common sense? An opportunist driven by greed? Or a missionary inspired by a purpose? His stated goals early on were frankly ambitious: "I wish to . . . be styled the Worlds Cook, — the acme of my highest aspirations for worldly fame."[3] The realities of the inventor's life dissolve into the mists of over a century of American-style hagiography. The self-made man, the roaming pioneer, the indefatigable inventor, the natural businessman, the man of God: each aspect of Borden's experience was later inflated to larger-than-life proportions. That Borden (unlike Liebig) struggled with patent offices and sources of financing nearly all his life may seem surprising, now that his name is synonymous with corporate-style milk. Yet even this fact encouraged a heroic account of the man as self-sacrificing for the sake of progress: possessed by an *idée fixe,* so the story goes, Borden never rested until he realized his dream of making a basic food available to all at a reasonable price.

More accurately and less flatteringly, Borden's first fifty years show a form of restlessness fueled by the times: the unbridled acquisitiveness of the young nation gave permission to Borden's version of American manhood, which had a callow and rather aggressive aspect to it. Borden seemed incapable of staying in one place: he migrated from his native New York State to destinations west, eventually settling in Texas in 1829. After helping to write the Texas constitution in the 1830s, he worked as a collector for the port of Galveston under Sam Houston. But his determination to score a spectacular success in business overrode his enjoyment of local eminence. The meat biscuit became his obsession, to be followed in the 1850s by his determination to obtain a patent for a process for condensing milk. And so he periodically took up an itinerant life, not out of any particular need, but because of a determined, extractive relationship to the material world.

It was Borden's particular perspective on space and time that put his entrepreneurial genius to work. As a migratory North American, Borden appreciated the demands of traveling across great geographical distances. His job in Galveston acquainted him with the essential role of sea travel, and the dependency of sea travel on long-lasting food supplies. Shipboard and military provisions were poor demonstrations of the advanced state of civilization, according to Borden; he counted himself among the boosters of American industry and ingenuity. But more than that, he comprehended the fact that people — particularly British and American people — were imperially ambitious actors in a worldwide theater. As his crafty sales pitch to the British was soon to prove, Borden was alert to international affairs and matters of state, including the threat of imminent warfare in the Crimea. Even before the onset of the American Civil War, he envisioned a global market for imperishable, portable food.

The biscuit represented a portable and fungible meat portion, the equivalent of today's dehydrated food for backpackers. Seeing its potential in exactly this light, Borden contacted British explorers Elisha Kent Kane and John Ross in order to stir up interest. He also grasped the fact that, for full realization, he would have to educate potential customers. In a publicity flyer, he laid out all the possibilities, replete with surprisingly modern advertising adjectives: "The meat biscuit makes the finest known soup; it reaches its highest perfection when boiled with vegetables; is irresistible in a pot pie; mixed with rice and sugar gives a superb pudding; is a perfect background for a mince pie, and may be used as a custard or any other desired dessert merely varying the flavoring, etc." Then, as now, apparently, few people paid much attention to the printed matter found at the bottom of the box. According to biographers, Borden had to use something like door-to-door sales pitches to drive home the versatility of the product. The manufacturer himself "could be found in the kitchens of hospitals or the galleys of ships anchored in New York harbor, showing how to prepare the food." But sales remained sluggish because "most of the cooks" were "not willing to follow his suggestions."[4]

Borden received a patent from the British government, but correspondence between the inventor and his customers reveals some dispute over the actual reception of the biscuit. The British, in a move remindful of the native approach to sausage manufacture, attempted to persuade Borden to diminish the proportion of meat in the product, from three pounds of "best beef" to two. (Borden stood firm, defending his original recipe.) A letter from the United States Navy, however, minced few words: many of the men found the "unusual flavor" of the meat biscuit downright "disgusting." The report could see no way around widespread "repugnance to this article," despite its indestructible nature when tested in "equatorial and China climates." Wherever it went, apparently, the biscuit tasted terrible. There was no accounting for taste, even in the U.S. Navy.[5]

Borden's business sense exceeded his culinary skills: the American

*Portrait of Gail Borden (1801–1874),
inventor and businessman (Harvard
University Libraries)*

inventor grasped the important fact that a manufacturer of food products
stood to gain a great deal of profit from government contracts involving
military supplies. Borden employed an agent, along with his brother John,
to promote the biscuit in London at the time of the Crimean War, though
with no more success. Their correspondence reveals a keen awareness of
the timeliness of their offering. "I can not imagine how it is you do not go
ahead here," wrote an acquaintance from London in August 1854. "These
are the very times for such a thing and I am humbly of the opinion that
you might have sold tons upon tons to the government for the troops and
sailors." The resident team did not give up. Borden's agent forwarded a
container of biscuit to the pinnacle of taste testing at Scutari, Florence
Nightingale, along with one of those instructional circulars. At least one
order came back: "600 pounds meat biscuit in cracker form to weigh 5–
1/3 ounces each, round and half inch thick. They must be just so *big* and
no bigger." But too much suspicion and discontent with the biscuit stood
in the way of complete success. The fit was somehow not right, despite the
need for more meat (confirmed subsequently by Nightingale's widely
read *Notes on Nursing*) in the field of battle.[6]

Back in the United States, Borden had already changed horses, as it
were: he was shuttling around New York State and New England, educating himself on the subject of condensing and containing milk. Once again,

the inventor had his eye fixed on the bovine species. According to legend, inspiration for the project sprang from his experience in shipboard travel from London to America following his biscuit debut at the Great Exhibition in 1851. His hagiographer put it this way: "The story goes that cows were being carried shipboard to furnish milk for the children passengers. These cows sickened and died during the voyage and the result was that children sickened and died, and that the necessity for a pure milk product caused him to turn to that enterprise. . . . Here was a product that army contractors could not corner, a product which the dwellers in the fast growing cities of the country got with great difficulty, one that could not be transported any distance, one that in its natural state was subject to the most rapid deterioration. He reasoned all this and went to work."[7] Aware that he was entering a race to arrive at a solution to preserve milk, Borden threw himself into experimentation and, in very little time, began his siege of the U.S. Patent Office.

His distinguished predecessors in milk preservation included the Mongols, whose dried product had impressed Marco Polo, and, more recently, the famed French chef Nicholas Appert, who successfully preserved (in glass) meats, vegetables, and milk for Napoleon's forces in the first decade of the nineteenth century. Appert's milk must have been discolored and distasteful stuff, given the imperfect techniques used to produce and contain a vacuum. Later methods of canning promised some improvement in quality: invented by British metal workers, actual cans of tin-plated iron provided meat for the Royal Navy in 1813, making the Napoleonic Wars a watershed in troop supplies on both sides. Friedrich Accum's *Culinary Chemistry* explained the science of the process, backed up by first-person testimonies of its success in the second decade of the century. On a voyage to China, one Captain Basil Hall expressed contentment with milk from tins. "It is really astonishing how excellent the milk is; and indeed, every thing preserved in this way is good," the helpful captain reported. Though canning technology traveled to American shores, tinned goods remained unpopular and impractical for some time. The containers were heavy and hard to open and, more important, consumers remained wary of their contents. Lack of interest must be the only explanation for why many years passed before a practical can opener was invented.[8]

Like most entrepreneurs, Borden investigated methods of food preservation at the feet of skilled practitioners and then borrowed what worked best. In the case of condensing milk, he took himself to New Lebanon,

New York, where Shakers practiced famously successful preservation techniques on fruit. The trade secret—a vacuum pan—resulted in a long-lasting "semi-fluid liquid" that needed no refrigeration. Borden was no scientist; his formal schooling amounted to a little over a year during childhood. Louis Pasteur's work on microbes lay in the future, but this was of no concern anyway to the Shakers and Borden. One needn't be able to explain why the absence of air would safeguard milk against spoilage; one only needed to embrace the fact that it did. Using a low degree of heat kept the milk from discoloring. Simple habits of hygiene, when applied to the tins used to hold the condensed milk, could further insure the product's safety and shelf life. Fewer bacteria diminished the likelihood of dangerous growth inside the can. Sugar added a chemically correct final flourish. Though Borden persisted in claiming that his technique of condensing milk required no sugar, his product ultimately depended on sucrose to further inhibit the growth of bacteria. The technical point was later lost on consumers, who adjusted their expectations to the sweetened product. As today's marketers well know, a re-created product does not have to taste like its original to be embraced by buyers.[9]

Borden eventually established factories in Connecticut with the help of loans from ambivalent investors. The financial panic of 1857 nearly ruined the overextended inventor, but at a fateful moment, just as at every previous impasse, Borden reversed his own fortunes. On a train between Connecticut and New York City, he chanced upon the New York financier and railroad magnate Jeremiah Milbank. Milbank evidently warmed to the biblical-looking figure of Borden and offered to back his business before the train reached its destination. This would be the last and most successful of Borden's many partnerships owing to yet another twist of fortune four years later.[10]

The Civil War ultimately rescued the New York Condensed Milk Company, renamed by Borden and Milbank in their new base of New York City. Despite hard effort, the company at first met with an "indifferent public." Agents distributed samples, door to door, until they built a customer base; then deliveries by pushcart provided steady but unimpressive sales. At some point (one source names 1856, even before Borden met up with Milbank), the product appeared under the patriotic label of "Eagle Brand Milk." In what may have been the earliest jackpot won by a logo, good fortune eventually arrived at the New York office: in 1861, a requisitioning officer from the United States Army walked through the door and, "after asking a few questions, announced that he wanted 500 pounds of condensed milk" for the troops engaged in the Civil War. Demand expanded so rapidly that the company could not keep up: new plants were opened in Brewster, New York, and Elgin, Illinois. By the end of the war, Borden and Milbank found themselves amply rewarded.[11]

It is worth reflecting on exactly why the sale of canned milk took off through purchases by the military forces of the American Civil War. Though the product eventually captured a piece of the market for baby food, that success had not yet materialized. Three years before the war, Borden thought it might: he shrewdly pitched his cans as an answer to "swill milk" in Frank Leslie's *Illustrated Newspaper*, the same publication responsible for exposing deception in the milk trade. But American mothers and babies preferred "fresh" milk purchased from familiar vendors on the streets of their hometowns. The secret to the Civil War market lay in the peculiar diet of American soldiers. "Coffee was the main stay," one northern private recalled. "Without it was misery indeed." According to one historian, "Army officials worked hard to insure that if only one commodity of nourishment was issued it would be coffee." As an army surgeon explained, "It has no equal as a preparation for a hard day's march, nor any rival as a restorative after one." (Coincidentally, Liebig reported these same facts from Germany, described in his popular magazine article "A Cup of Coffee," in 1866.) And with coffee went milk and sugar: Borden's condensed brand, sweetened with cane sugar, must have solved two problems at once for the soldier. The only other things necessary for refreshment were a few sticks for a fire, a day's ration of beans, and a cup.[12]

Testimonies to coffee — not milk — abound in Civil War diaries and histories, not least in John D. Billings's *Hardtack and Coffee*. From Billings, we learn that coffee made hardtack edible, through soaking, dunking, or by washing down the crackers. American coffee drinkers absolutely preferred theirs with milk. But given the scarcity of provisions all over the country and the improbability of getting supplies to troops on the march, ordinary military consumers were forced to adapt to black coffee. "It was a new experience for all soldiers to drink coffee without milk," Billings explained. "But they soon learned to make a virtue of necessity, and I doubt whether one man in ten, before the war closed, would have used the lactic fluid in his coffee from choice. Condensed milk of two brands, the

Lewis and *Borden,* was to be had when sutlers [civilian provisioners] were handy, and occasionally milk was brought in from the udders of stray cows, the men milking them into their canteens; but this was early in the war. Later, war-swept Virginia afforded very few of these brutes, for they were regarded by the armies as more valuable for beef than for milking purposes."[13] The preference for meat, which could be stretched to feed more mouths, canceled out the possibility of liquid nourishment from the cow. So coffee drinkers had no choice but to brew their bean rations with water and leave it at that.

Witnessing coffee history from his office in New York City, Borden saw military consumer habits developing in the other direction: demand for his milk was going up. Perhaps officers at government headquarters were the first to enjoy coffee with sweetened condensed milk. At least some of the stuff made its way to the front lines. Gift boxes from home, according to accounts, also helped to distribute the novel product. Word of mouth must have influenced choices and, if sales are a safe measure, consumers began to adapt. Borden's records reported that "not only would the Army take his [Borden's] milk but furloughed and discharged soldiers spread news of its advantages among the civilian population." During this time, Borden entertained the idea of marketing cans of liquid coffee combined with milk, though apparently he had no time to bring the plan to fruition. By 1863, his factories were turning out 14,000 quarts of condensed milk daily and were unable to satisfy a rising tide of orders. Borden established new factories and, with greater financial means, purchased stock in new competitors, granting licenses to them so that they could use his milk-condensing process under his patent. This would not be the last time that a contract from the government propelled a struggling company into the front ranks of American business.[14]

An honest account must note that Borden was not the first to produce condensed milk successfully. Twenty years earlier, an Englishman with the promising name of Newton obtained a patent for a concentrated, "honey-like" form of milk produced by indirect and "gentle" heating and evapora-tion. But the product was never marketed. At least two other Englishmen and a Russian came up with other condensed or evaporated milk products by the time of Borden's success. Apparently, the times called for it: one British producer boasted that he provided fifty thousand pint and half-pint tins of condensed milk on an annual basis for "Government Arctic

Expeditions." The case of Borden, then, proves the decisiveness of the personality factor: his entrepreneurial drive, a euphemism for his dogged commercial pushiness. And perhaps the willingness of American consumers to adapt to the product after the Civil War enabled Borden to move from one success to another. The history of condensed milk illustrates how many variables affect the ultimate success of a commodity.[15]

For a true picture of that history, we first need to recognize an important, rather counterintuitive fact: the production of cow's milk in the Western world increased exponentially even before milk as a drink became truly popular. One reason for this lay hidden in the identity of milk as an industrial commodity from the 1850s. As Gail Borden proved, milk in cans, condensed or powdered, gained the portability and long life that the fragile liquid had always lacked. While his process remained widely used, other patents were issued worldwide, as determined experimenters added and subtracted ingredients (mainly sugar) to come up with a better product. Powdered milk, first developed in England in 1855, also took off as a market commodity, first as infant cereal and then as the basis for the manufacture of chocolate and confectionary products. In North America and Switzerland, the milk industry became just that: an enormously lucrative business that pushed the output of the product to extraordinary levels. The popularity of canned milk raises important questions, beginning with the obvious: What were consumers using canned milk for? Were most buyers pouring it into coffee? Did mothers concoct personal recipes for infant food from their cans? What proportion of cans sold was kept on hand for cooking purposes? All of these possibilities may suggest that processed milk was paving the way for a future form of market milk, a fresh liquid deemed safe and priced within reach of ordinary people. In any case, the enormous and growing demand for milk products had a collective impact of its own: it brought into being a modern attitude toward dairy farming, visible by the time of World War I.

A serious boost came from North America, where dairy agriculture had expanded beyond midwestern states into areas of the far west, such as Oregon and Washington. The rate of expansion in the condensed milk industry provides some insight into the growth of dairying at this time. Between 1890 and 1900, the amount of condensed milk produced in the United States grew by nearly five times, from 38 million to nearly 187 million pounds; by the outbreak of the First World War, the figure had reached over 875 million pounds. Demand from Europe during the war kept production climbing, so that in 1919, the United States was producing over 2 billion pounds of condensed milk, along with 44 million pounds of milk powders. And this absorbed only half of the total milk supply. With the ratio of cows to people growing to roughly 1 to 5 in the United States, the capacity to produce huge surpluses existed as early as 1900.[16]

Many forces converged to make this explosion of production take place, not least the rise of corporate business arrangements at the sale and distribution end of the industry. Big business emerged triumphant from the reorganization of western European and American capitalist economies at the end of the nineteenth century. The milk-processing industry was among the most successful food industries of the age, benefiting from the global network of food exports linked across oceans and into hitherto untouched markets. Companies utilized new features, such as cheap long-distance transport and the technologies of refrigeration, to compete in a world market for milk. Just as important, international business interests began to reorganize producers in foreign countries. Swiss agents settled in Norway and Spain; American entrepreneurs settled in Switzerland; Swiss immigrants transformed areas of the American Midwest. The landscape of milk production for big business scrambled the map of dairying, unsettling the idea of milk as a known and native product.

By 1920, for example, the Borden's Milk Company amounted to "one of the largest industrial concerns in the United States and Canada." Its vertical integration looked very modern even at that date: Borden owned and operated thirty-one condensaries, eleven "feeders," eleven tin can factories, two confectionery factories, two malted milk plants, and two dry milk plants. The company's subsidiary enterprise, Borden's Farm Products, delivered "large proportions of the fresh milk consumption of New York [C]ity, Chicago and Montreal," supplied by its "eight certified milk farms, 156 country bottling plants, and receiving stations, 70 city pasteurizing plants and distributing branches."[17]

Clever marketing strategies found ways of packaging milk that spoke to the times, an age of urbanization and anxieties about modern life. Talk in the board rooms of advertising companies was not overly concerned about debates over milk's purity. (At Borden's in Chicago, the problem was settled simply by pasteurizing their supply, a gesture that combined public service with a business sensibility.) Familiar themes of milk history were struck when the Carnation Milk Company gave birth to its legendary phrase, "the milk of contented cows." Around 1906, the company was searching for a hook for its annual advertising campaign. The chairman of the board, E. A. Stuart, met with members of the Mahin Advertising Company in its offices in Chicago. A new copywriter, Helen Mar Shaw Thomson, joined the group and listened to Stuart soliloquize on the company's favored natural resource, placid bovines. As she recalled some twenty-five years later, the chairman delivered a performance equal to a bard of pastoral poetry. He spoke of "ever verdant pastures of Washington and Oregon, where grazed the carefully-kept Holstein herds, which yielded the rich milk used for Carnation" and described "the picturesque back-ground of these pastures — mountains often snow-capped, from which danced and dashed the pure sparkling waters which were to quench the thirst of the herds and render more juicy the tender grasses of the well-watered pastures." "The delightful shade of luxuriant trees, beneath which the herds might rest during the heat of the day" completed Stuart's rhapsody to the dairy.

Thomson came to the job having grown up in Vermont and later in Boston, where she graduated from Boston Latin Girls' School with honors. Her real ambition had been medical school, but family circumstances (including a father who didn't envision his daughter as a doctor) required that she turn her mind to more suitable work for women. The fact that she was present in a meeting with the chairman of the board, along with the owner of the advertising agency, stood as tribute to her valued intelligence. Whether through associations with her Vermont girlhood, or perhaps from some quirky principle of biology she had learned, she was able to summon up the notion of the digestibility of "milk produced under conditions of mental and physical ease." "Ah," she remembered exclaiming, after Stuart finished his reverie, "the milk of contented cows!" The executives in the room recognized a winning alliteration. "TAP came the pencil of John Lee Mahin on the glass table-top," she recalled. "THERE is our slogan!" The words became a virtual banner of celebration of twentieth-century condensed milk in America.[18]

A modern industry of international dimensions thrived on legends of bucolic origin and alliterative slogans. The picturesque landscape around

Vevey, Switzerland, studded with amiable brown cows clinging to the hillsides, inspired Henri Nestlé, a German émigré, in exactly this way in the 1860s. Riding a wave of enthusiasm for new prepared infant formulas, Nestlé translated local color into a universally seductive (and physician-endorsed) product. A secret blend of powdered Swiss milk and roasted grain, the product first sold in pharmacies across Europe. "Believe me, it's no small matter to market an invention in four countries simultaneously," Nestlé boasted to an associate in 1868. By 1873, shipments reached across two oceans to sixteen countries, including Mexico, Argentina, the Dutch East Indies, and Australia.[19]

With affordable Swiss milk as his goal, Nestlé revealed his own brand of heartfelt, vaunting ambition. "It's not the moneybags who buy the most from us," he observed. (The company history did little to improve on Nestlé's gruff maxims and mercenary calculus.) "We must try to bring the price of our baby food to within everyone's reach. It is better to sell two cans at SFr. 3.60 than one can at SFr. 2." The appealing containers pictured a mother bird feeding her nest of newborns. When an agent advised using the famous cross of the Swiss flag as his trademark, the founder said absolutely not. The bird image constituted his "coat of arms," after all—Nestlé means "little nest" in Swiss German—and so the personal trumped the national symbol. In an era of growing nationalist fervor, Nestlé's egoism probably proved better for sales. After 1871, areas around the company's factories began to blame Nestlé for pushing up the local price of milk. This was no light charge in a nation that viewed cheap and plentiful milk as something like a birthright.[20]

Serious competition was brewing in the hills nearby, indicative of the pressures evident as early as the 1860s. The Anglo-Swiss Condensed Milk Company in Cham, Switzerland (not far from Zurich), adopted its name not because its founders were "Anglo" or Swiss, but because the two countries presented the most promising market for their processed product. Charles A. Page, American by birth, had inside knowledge of Swiss agricultural production, owing to his role as United States consul in Zurich. With his brother, George, he spent some time learning the business of milk condensing from Borden in America and then returned to Switzerland to found the company in 1866. The plan worked: their best customer was England, a prize for the international salesmen, given that it came with a network of colonial dependents. Their worst customer was Switzerland: consumers there preferred the fresh article. The Page brothers showed a kind of flexibility that Nestlé's one-product company could not match until the turn of the century. Because the supply of milk waxed and waned with the seasons, periodically leaving factories with excess supplies, the brothers diversified their line of products. In some places, they used extra milk to make cheese and innovative products that sound more like the age of Starbucks than the era of Queen Victoria, such as containers of coffee and milk, cocoa and milk, and chocolate and milk. After a tumultuous and lucrative forty years, the Anglo-Swiss Company merged with Nestlé, part of a wave of corporate consolidation in 1905. It was only at that point that Nestlé moved into producing its now famous milk chocolate.[21]

With the right marketing strategies and business savvy, milk companies even tried to turn the age of "bad milk" to advantage. Swiss historian Thomas Fenner documents an intricate international dance of business interests in the example of Caesar Ritz, hotel magnate from the Upper Valais, who learned from the impact of a cholera epidemic in Cannes that disease was an enemy to be fought at all costs. Ritz utilized a network of connections in Switzerland, France, and England to produce a state-of-the-art germ-free milk according to a new German technique. His company, the Berneralpen Milchgesellschaft, founded in 1892, bore the figure of a bear, the symbol of Bernese strength and goodness. Ritz even engineered publicity through a French newspaper, reporting an Emmantal "paradise for cows" as a way of promoting the product. The project eventually failed, despite Ritz's muscular efforts to train farmers and factory workers in the art of "meticulous cleanliness, punctuality and diligence," enforced with "strict penalties." The economic downturn of the end of the nineteenth century prevented an expansion of the market for this high-end, pristine-clean milk.[22]

The age of empire at the turn of the century yielded a vast network of markets for milk in canned form. Canadian farms and factories, as part of the British colonial constellation, supplied consumers near and far. Nova Scotia, home of the Reindeer Condensed Milk Company, served fishing and lumbering communities, where manly cups of coffee cried out for the canned cow. Mining stations in Alaska and the Yukon bought up the product, too. Canadian factories shipped milk to the far ends of the British Empire: Hong Kong, South Africa, and even outposts in China and Japan found a use for condensed milk. (Australia had been conquered by Nestlé already and ranked as that company's second largest customer by the turn of the century.) Canada's biggest buyer in 1912 was Cuba, where torrid-zone temperatures made dairying a difficult undertaking. The sun never set on the need for canned milk, which poured forth in ever increasing volume from North American and European cows.[23]

But not just from North American or European cows: the condensed milk factory eventually sprang up in the Far East, utilizing the milk of transported cows. After 1868, the Meiji Restoration in Japan introduced Western dietary and cultural influences, which paved the way for a native dairying industry. In particular, the introduction of coffee drinking provided an important avenue for exposure to milk and a considerable source of demand. By 1890, the nation was devoting scientific research to the development of a native condensed milk industry. Meanwhile, a familiar pattern emerged: when the Japanese occupied Taiwan in 1897, they required all the comforts of home; hence the transport of dairy cows to Taiwanese soil. (The habit of using milk in China was slow to spread, however, and herds there dwindled after World War II.) Meanwhile, production of condensed milk in Japan exceeded that country's needs; China and the South Sea Islands became Japan's customers after World War I. Judging from fragmentary evidence, it is likely that many of these cans ended up being used as infant food.[24]

Dairy farmers themselves found ways of tapping the market for cheap, portable food through industrial methods. Cheese "factories" offered a means of producing the product in bulk and funneling it into urban markets via commercial offices. In this case, the term was not as industrial as it sounded: though utilizing equipment suited to large-scale production, factories achieved advantages not so much by mechanization but through economies of scale. By pooling the milk and cream collected from many farmers, factories could work toward selling in bulk a product that had uniform quality, which urban dealers preferred. Farmers in New York State in the 1840s were among the first to employ such methods, and they enjoyed remarkable success: soon they were undercutting English cheeses on the London market, their biggest competitor. In terms of quality, they held up well enough in the mass market to force dairy farmers across the Atlantic to turn to liquid milk for a living. Butter factories based on the same principle sprang up in England to absorb supplies. Some places combined their production with piggeries, which fattened livestock on the runoff that came from butter making, known to us as buttermilk.[25]

Milk evolved according to laws of entrepreneurial know-how and transnational migration. Vital developments in the history of consumer culture follow from this simple point: in the final decades of the nineteenth century, several highly successful milk-based products came into being. One such commodity was a mixture of malted grains and dried milk, a kind of baby food for adults. James Horlick, a pharmacist from Gloucestershire, England, moved to Chicago in the 1870s to work for the Mellin Company, which had pioneered the business of selling food for infants. Horlick soon came up with an idea for a modified product suitable, he thought, for dyspeptics. With the help of his brother, William, an accountant living in America, he established a brisk business in cans of roasted grain and dried milk, which they marketed to the general public as a health food. Today, we think of "malted" as a sweet, distinctive flavor, rather than a nutritional boost or a therapeutic compound. (In fact, malted drinks derive most of their caloric value from sugar, though they do deliver a generous serving of riboflavin, known also as vitamin B2.) In the late nineteenth century, malted grain would summoned up associations with heartiness (thanks to barley) and also beer brewing, yet Horlick was offering a product enhanced by a whiff of temperance. His strategy met with instant success.

Horlick eventually developed a dried milk-base product, giving rise to "malted milk" drinks with claims to both health and convenience. The "high digestibility, nutritive value, dietetic virtues and health-protective properties [of malted milk] render it most valuable as a wholesome food for infants and invalids, and its compactness and keeping qualities facilitate its transportation to and use in all parts of the globe," reported a late-nineteenth-century assessment of the industry. Roald Amundsen, the celebrated polar explorer, gave the product added cachet by bringing it along on his expeditions in the early twentieth century. Its portability also made it ideal for military use during the First World War. By the 1930s, the company marked out a new retail frontier: nighttime comfort food for the modern ailment of insomnia. The company opened factories around the world, taking advantage of imperial opportunities in India, Canada, and the Caribbean. By the 1960s, a factory in Punjab was making Horlick's from buffalo milk.[26]

It was a short step from Horlick's "healthful" product to the confectionery business: soda fountain drinks and chocolate-covered sweets appeared at the turn of the century, expanding the venues in which milk might make a disguised appearance. One of the earliest products, the "Malteser," was "made by sawing the dry chunks [of malted milk,] as they come from the vacuum pan, into bars, dipping the bars in chocolate syrup and wrapping them in tin foil."[27] With rising expendable incomes among prosperous urban populations, the demand for milk forged a close and important relationship with sugar during this phase of nineteenth-century food history. Sugar consumption soared in Europe and America during the mid-nineteenth century: in Britain alone, annual consumption rose from roughly 17 to 60 pounds per capita between 1844 and 1876. Add to that the arrival of powdered and molded chocolate, both products of important processing innovations, and you have the formula for a massive rise in demand for milk as a star ingredient of universally desired commodities.[28]

The history of chocolate, a fascinating subject in its own right, developed with little input from milk until the late nineteenth century. Van Houten's Dutch-processed cocoa, invented in 1828, gave consumers access to a more palatable product, a powder that could be mixed with water. Within the next decade, French and German manufacturers were able to devise a way to make "eating chocolate"—that is, solid forms of chocolate that needed no preparation by the consumer and delivered a novel and satisfying sensation when eaten. A long artisanal tradition of making chocolate lay in many of the western towns of Switzerland, where confectioners trained in France and Italy had set up shop in the first decades of the nineteenth century. Chocolate, it should be added, was always imbued with exoticism and associated with luxurious elite French culture. So when the English Cadbury Brothers first sold their block chocolate in 1842, they labeled it as "French," and five years later, Fry's (another English Quaker company) called theirs "Chocolat Délicieux à Manger." Such products were expensive and limited to a small clientele of buyers, though that fact only served to enhance their desirability among the have-nots.[29]

Milk promised to make chocolate the treat of the people in the late nineteenth century. A fortuitous convergence of locale and technology made this possible: in Vevey, Switzerland, the home of Nestlé's famously successful milk company, a former candlemaker named Daniel Peter de-

This Cadbury's Chocolate advertisement from 1929 revealed the commercial savvy of the company by graphically demonstrating that a glass and a half of "English full cream milk" went into every half-pound chocolate bar. The fact was sure to impress milk-conscious consumers in an age of nutrition. (Mary Evans Picture Library, Picture No. 10035180)

vised a successful way of combining chocolate with dried milk in 1876. Without Nestlé's processing innovations on the side of milk, the winning recipe could not have worked in quite the same way, given the mysterious alchemy involved in the preparation of chocolate. The combination turned out to be a brilliant stroke: milk made chocolate less concentrated and more digestible, meaning that one could eat more of it, an added benefit to both the manufacturer and, presumably, the consumer. And just as important to some purchasers, particularly the English, was the promise of nourishment that came from the milk. Cadbury's adoption of the title "Dairy Chocolate" advertised the fact, pure and simple, and packaging enhanced the point by picturing references to the crucial ingredient. The inclusion of chocolate in troops' rations in the First World War must have given the product another advertising boost, though it's unlikely that much help was needed in the first place. More important, prices fell as manufacturing costs fell, once technological advances and international competition between large companies came into play. By the early twentieth century, chocolate had arrived as a mass-produced, relatively inexpensive commodity, and much of what was consumed in America (as high as 86 percent in the late 1930s) and Britain (85 percent) was the milk chocolate variety.[30]

One other commodity placed tremendous pressure on the production of milk: ice cream, regarded as "one of the most spectacularly successful of all the foods based on dairy products." Its recent history reveals an interesting paradox that directs attention to the process of commoditization that this chapter has described. Despite the multicultural dimension of ice cream's long and venerable past, the product acquired an identification with American culinary practice, which in turn made ice cream "American." Part of this had to do with the already well established Anglo-American attachment to dairy products, transplanted in vigorous form to the North American continent. Yet the American passion for ice cream went beyond this, locking ice cream in a resolute embrace. Love for the product was universal in a way that mattered. As an American magazine declared in 1850, "A party without ice cream would be like breakfast without bread or a dinner without a roast." And with universal love came marketing strategies galore. By 1900, even the English ceded precedence to the Americans, though they, too, had always preferred cream-based frozen desserts over the flavored water ices of the continent. "Ices [made from cream] derive their present great popularity," wrote an Englishman at that time, "from America, where they are consumed during the summer months as well as the winter months in enormous quantities." Steamers carried American ice cream to India, China, and Japan. By 1919, American annual output was reported as 100 million gallons, valued at roughly $140 million. That figure would translate to $1.74 billion today using the Consumer Price Index, or as much as ten times the extent of Ben & Jerry's profits in the year 1997.[31]

What made American ice cream so successful? Perhaps simply the knack of producing sheer volume, which began as early as the 1840s, when Nancy Johnson devised a hand-cranked churn equipped with a dasher and surrounded by a bucket of ice and salt, suitable for easy home use. The technique of using a dasher, or paddle, created a product that was distinctively "light and soft" owing to the volume of air incorporated into the cream as it froze. According to one source, "this increase in volume, known as 'swell' or 'overrun,' may be as great as 80 per cent" beyond the original amount of cream in the American variety. Somehow, ice cream had managed to replicate the national trademark of abundance in its very processing.[32]

But not all credit should go to the quality and quantity of the product itself; clever marketing accounted for a good share of the success of American ice cream. The various stratagems devised by American vendors, such as the "I scream, Ice Cream" street sellers of New York, circulating as early as the 1820s, deserve note. A distinctly new mass market appeal seems to have appeared in the early twentieth century. All accounts pay homage to the legendary Syrian immigrant Ernest A. Hamwi, one of several claimants to the title of inventor of the ice cream cone, who made ice cream uniquely portable by wrapping it in a waffle at the St. Louis World's Fair of 1904. Other innovations soon joined the North American competition: the I-Scream Bar, coated with chocolate, appeared in 1919, the Good Humor Bar reached the market in 1920, and ice cream sold in Dixie Cups ("the sound of it was patriotic, musical, snappy, clean, and modern," Margaret Visser points out) were a success by 1923. Visser's entertaining account of the history of ice cream in America depicts a culture of consumption that was busily tying milk to a spirit of expansive capitalist enterprise.[33]

Beneath this seemingly unstoppable multiplication of dairy products lay fundamental changes in Western agriculture. By the 1870s, staple food crops such as grain from North America and beef from South America were entering an expanding global economy. Carried by steam transport, the products of such vast monocultures (areas devoted to only one crop or type of livestock) both enriched and threatened all parts of the world by their cheapness and plenitude. So while more and better foods became available to ordinary and even poor consumers, the arrival of affordable foreign foodstuffs posed a perilous challenge to the livelihoods of many European farmers. Those who worked small plots of land or had existed on the margins of rural economies could not survive the changes happening after 1870. A worldwide economic downturn, known as the Great Depression because of its severity and persistence, began in 1873 with what Eric Hobsbawm has called "the Victorian equivalent of the Wall Street Crash of 1929." The malaise lasted until 1896, and by this time, the demographic map of Europe had begun to look more as it does today: a steady migration from rural to urban areas, as well as across oceans, left the countryside in some areas dramatically depopulated. By the beginning of the twentieth century, the number of people living on the land had fallen to a fraction of what it had been in the nineteenth century.[34]

Some farmers who remained found salvation in producing liquid milk, butter, and cheese for near and distant markets. But the playing field was not level: economies of scale enabled large-scale farmers to prosper, while small, independent farmers struggled to survive. The most successful, particularly in Britain and Denmark, found that dairying cooperatives provided a solution. By pooling milk supplies and capital, small and large farmers joined together in what one Dane described as "a piece of living democracy." That nation emerged triumphant by specializing in dairy products, particularly butter: by 1924, Denmark produced 308 million pounds of butter, 90 percent of which was exported. And yet enough liquid milk remained to supply an annual 70 gallons per capita consumption, a remarkable quantity at that relatively early date.[35]

The history of the business of milk is a complex subject, shaped by regional and local factors that have not been touched upon here. Although customary uses of milk and dairy products persisted, perspectives on milk were changing with the times. Medical knowledge about the liquid played an important role in creating a distinctly nineteenth-century profile of milk. The following chapter provides a window into yet another world of milk, its therapeutic role in treating a disease that was very much a part of the modern era of business.

CHAPTER TEN

Milk in an Age of Indigestion

Perhaps the most famous upset stomach of all time resided in the body of Thomas Carlyle, Victorian writer, "the sage of Chelsea" in the middle of the nineteenth century. Given his prolixity on the subject, it is safe to say that Carlyle put dyspepsia on the map of Victorian maladies. In distant New York City, a newspaper story about the peculiarly widespread ailment felt obliged to cite Carlyle as the quintessential sufferer. (The article then pinned the blame for his dyspepsia on his habit of eating oatmeal.) Dyspepsia spoke volumes about the age and its beliefs about the body and food.

In private war against the ailment, Carlyle's main weapon was milk. From Scotland, he reported details of his attempts to quell the riot in his stomach to his wife back in London. "Dear Goody," he confided, "my *inner man* is far out of order; I have dined today on boiled rice and milk, — declining chicken-broth till the morrow."

Bland food and cow's milk — the couple agreed that no other milk would do — became Carlyle's mantra against the enemy. Suppers of hot milk and porridge, rice and milk, or just milk, warmed on the hob, formed the centerpiece of his life when his stomach was low. His condition improved only sporadically. Some years later, again from Scotland: "I got here in the hollow midnight between Thursday and Friday. . . . I have done little ever since but sleep; and drink milk in various forms. I will write you farther tomorrow."

In Chelsea, Carlyle demanded his supper of bread soaked in sweetened hot milk, which his wife, Jane, called "breadberry" and their servant Ann dared to call "Master's PAP!" The stuff comforted him, offering relief from the tribulations of work and the noise of the train across the river. As he confessed to his brother, "Residence in the country, with meal and milk *ad libitum,* fresh air, and *quiet,* quiet: — this is my constant daydream at present." The press of modern urban life only enforced the coupling of milk and rustic salvation.[1]

Whether in country or city, the "demon of dyspepsia" found and pinioned Carlyle. "Dyspepsia sits heavy on me at present," he wrote to his friend Monckton Milnes. "On Friday I cannot come to breakfast." To another friend, "I get so dyspeptical, melancholic, half-mad in this London summer." And in a more reflective mood, he devoted a philosophical air to his account: "Dyspepsia . . . it is a hard case, in fact, this distemper. A malady of soul one can embellish and dignify a little by enduring; but this carries with it the indellible [*sic*] stamp of nastiness and lowness; do what we may, it seems to pollute the very sanctuary of our being; it renders our suffering at once complete and contemptible."[2]

Carlyle *was* contemptible at times: his irritability and finickyness remain legendary, and a whole field of sympathetic scholarship has grown up around the spousal miseries of Jane Welsh Carlyle. Ever charitable friends (and biographers) excused his worst behavior as the unavoidable expression of his stomach disorder. Yet he was far from alone in his pain: a list of distinguished contemporary victims included George Eliot, Charles Darwin, and Charlotte Brontë. Whatever the dimensions of this inner plague of the nineteenth century, we can be certain that dyspepsia was everywhere in need of a cure.[3]

Was milk a legitimate treatment for this bodily and psychic scourge of the nineteenth century, or simply comfort food? Today we know that milk may have an initial soothing effect on some forms of upset stomach, but its slightly acidic nature only adds to the distress of excess digestive juices. Medical expertise in Carlyle's time expanded daily, but despite this fact, groaning patients still banked their hopes on the milk remedy used by Platina in the fifteenth century. In view of one doctor's prescription of biscuits soaked in milk, we are led to suspect that Victorian medical professionals, powerless in the face of this mysterious ailment, relied on custom rather than hard science. For the patient, milk at least represented an

Thomas and Jane Welsh Carlyle in their home in Cheyne Walk, London
(Mary Evans Picture Library, Picture No. 10059522)

alternative, even in our modern sense of the word, to the more conventional cures. Regular draughts from the cow proved more affordable than a histrionic retreat to a fashionable spa. There, one would down saline mineral water, sit in salted bath water, and wait for the stomach and bowels to respond to the many assaults inflicted upon them. A more time-saving refuge lay in auspicious-looking bottles of bitters. A random perusal of newspapers of the time offers up testimonies of "remarkable" cures of dyspepsia from mysterious formulas and pills. By the end of the nineteenth century, Beecham's reported selling over six million boxes of their pills, which claimed to relieve dyspepsia, among other ailments. Given their ingredients (aloe, ginger, and soap), their main effect was laxative. Medical and popular opinion appeared equally wedded to the opinion that this was a valuable first step to conquering almost any stomach upset.[4]

Some patent medicines proved certifiably deadly. Charles Darwin dosed himself regularly with Fowler's solution, a mix containing potassium arsenite, gaining at least temporary relief from his dyspepsia. But in the long run, a medical biographer has hypothesized, Darwin's continued suffering probably sprang from chronic arsenical intoxication, which caused symptoms similar to the vague signs of dyspepsia: nervousness, lowness, sleeplessness, along with intense stomach misery. In fact, given how frequently the medical establishment in London prescribed Fowler's potion, a whole swath of prosperous male clientele could be said to have suffered from "Fowler's disease" rather than dyspepsia. Darwin died at a relatively premature seventy-three, while Carlyle groused about his stomach until the riper age of eighty-five. Of the two treatments, clearly milk won the contest.[5]

Dyspepsia in all its vague and varietal symptoms served as a mirror of an age of industriousness. Stomachs rebelled, it was thought, because they were overworked, or because they were housed in bodies made anxious by too much to do. And if stomach turmoil sprang from such busy behavior, then milk could offer a solution with modern-day legitimacy. From the 1830s, milk became associated with a scientific battery of nutritional advice, winning a place among the latest dietetic discoveries. No longer under the fog of humoral explanations for how ailments came into being, nineteenth-century authorities now couched matters in contemporary metaphors: all creatures were "locomotive furnaces" (Liebig's phrase) in constant need of fuel to work properly. Not only "heat-making" material was needed for this biochemical power box: thanks to a German theorist named Mulder, contemporaries came to understand the concept of "protein" as a particularly necessary component of food. To "repair the waste which tissues are constantly undergoing," one doctor explained, food with protein was a must; this particular component enabled people to undertake hard physical labor. Decomposition and degeneration must be fought at all times, particularly with patients entering their later decades. Milk could "perform the duty" of replacing whatever bodily matter was spent, whether by work or worry.[6]

But dyspepsia fit poorly into the new, fashionable models of nutritional theory. Degeneration was not very evident among dyspeptics, who were often overweight, nor did protein seem in short supply; most victims probably erred on the side of providing their systems with too much rather than too little food. In truth, most people — including physicians — didn't understand the disease in scientific terms at all, but instead chose to fasten on its social and moral dimensions. Dyspepsia, they believed, was punishment for intemperate eating. Its true cause was "overloading the stomach," according to a leading physician popular on both sides of the

Atlantic. The French called it "the remorse of a guilty stomach." The American reformer Mary Peabody Mann read dyspepsia as a sign of "ignoble appetites" gone awry, which made the ailment "disgraceful, like *delirium tremens.*" Once "the gospel of the body is fully understood," she predicted, dyspepsia will vanish and the process of "redeeming the race from its present degradation" can begin.[7]

Mary Peabody Mann would have endorsed Thomas Carlyle's milky suppertime ritual. Though she championed scientific diets, her recommendations had the air of priestly abstinence. She condemned "melted butter, lard, suet," and anything in a fatty, congealed state. Wedding cake and plum puddings, "masses of indigestible material," ranked as evils to be avoided at all costs. On the good side of the dietetic ledger lay "innocent" cream and potatoes. The cow, she deemed "the most valuable possession for a family of children," for obvious dietary reasons. In a confiding tone, she recommended that "it would be better to have a less expensive carpet, or chimney ornaments, or even bonnet and cloak, and have a friendly cow in the shed or barn." The American homestead as a bulwark against stomach disorder: here was an idealized haven that might withstand the advancing tide of modern eating habits.[8]

Medical authorities of the day were not so very different from Mann in their judgments about dyspepsia, though they relied on different methods of treatment. A division of good food and bad food was beside the fact, according to physicians. Such talk about dietary choices opened the way for delinquent eaters to deflect attention from their behavior. If our stomachs and bowels "feel uneasy after a heavy meal, it is not *we* who are to blame for having eaten it. Oh no!" mocked Andrew Combe, noted authority on stomach diseases, "it is the *fish* which lies heavy on the stomach, or the stomach which is unfortunately at war with the soup, or potatoes, or some other well-relished article. *We* have nothing to do with the mischief, except as meek and resigned sufferers." In his widely acknowledged *Physiology of Digestion,* the doctor argued that anyone in a decent state of health should be able to eat all foods without difficulty. But most complainers were also violators of all-important "laws of digestion." The luxuriant slice of wedding cake, then, served as a foil for a far more complicated web of problems.[9]

In a world unacquainted with *Helicobacter pylori* bacteria and food allergies, an understanding of stomach ailments rested on a relatively primitive paradigm. Observation was everything, focused as much on the world at large as the internal workings of the stomach. In one signal exception, a soldier named St. Martin who suffered a gunshot wound to the stomach agreed to allow a doctor to peer through the unhealed cavity in order to watch what went on inside when he ate. But even this diorama reenacted a predictable drama: physicians reported a virtual factory bustling with muscle action and acid baths. The stomach, like everybody else, had work to do, and its job was best accomplished if given peace and quiet before and after meals. Eating (especially the act of chewing) ought to be leisurely and the mind should not be distracted while doing so. Meals should be taken in a state of calm, deliberate enjoyment.

Combe condemned the English habit of reading the newspaper while eating: here was a case of allowing worldly cares to penetrate the very act of filling the stomach. But he reserved his greatest disapproval for Americans, famous for bolting their food "as if running a race against time."

> Amidst hurry and bustle, and while the mind is still labouring under all the anxiety and excitement of business, the food is swallowed in larger quantity than is required, and very imperfectly masticated. The natural consequences are, that indigestion prevails to an extent unknown elsewhere; and that the thin and hungry leanness of the people has become almost as proverbial as the portly paunch, and ruddy, well-fed cheeks, of their great ancestor John Bull, from whom, in this respect, they have so much degenerated.

No wonder that North Americans outnumbered all other people in suffering from dyspepsia, according to Combe. Sounding like Mary Peabody Mann, he decreed the illness as "the punishment which the Americans bring down upon themselves."[10]

The truth was, dyspepsia served as a catch-all category for a great many disorders with distinctly different characteristics. As the century progressed, physicians carefully parsed the cases they observed, arriving at as many as ten different varieties of the ailment. Dyspepsia could arise from weakness, congestion, inflammation, kidney or liver dysfunction, disorders of the stomach muscles or stomach chemicals, or from that other catch-all condition, a nervous disorder. Yet even in cases far from the concerns of nutrition, milk retained its odd, persistent association with remedies for the stomach.

In "sympathetic dyspepsia," which sometimes accompanied the notoriously feminized disease of hysteria, physicians believed that the stomach became slave to a deranged mind and body. Obeying the commands of some unseen master (often thought to be the uterus), the stomach would assume roles completely alien to its normal functions. "Vicarious menstruation" sometimes appeared in cases of dyspepsia. In these reports, the stomach would vomit up blood in place of menstrual shedding of the uterine lining. Victims exhibited unacceptable attitudes toward nourishment, either refusing food or vomiting up what was eaten. In one severe instance, a nineteen-year-old woman was admitted to hospital, motionless, moaning, and refusing food. Her doctor suspected that an attempted abortion lay at the root of her condition, but attention focused instead on the more immediate problem of her refusal of food. The first solution was, in a word, milk. The imagination recoils from the brevity of the report: "Milk was poured into the mouth, and she was *made* to swallow it; in this way a considerable quantity of food was taken." Following this, a "blue pill with colocynth and henbane" cleared the woman's bowels, leeches were applied to her groin, and "sparks of electricity were taken from the spine." Doctors considered the results successful when her stomach retained food and "the patient became able to walk." How different this was from the treatment of a male manual laborer, suffering a hemorrhaging stomach from alcoholism: his regimen, delicate by comparison, consisted of "infusion of roses with acid, and milk diet."[11]

With such a legion of body organs casting a shadow on the stomach, victims could breathe a sigh of relief. Perhaps habitual gluttony wasn't the cause after all. Wise doctors hypothesized that some unsuspected agent might be at work, too. No matter what the source of the disorder, the treatment that stood out as most logical was the "regulative" approach. And this was where milk played a very important part.

Every generation of medicine since Galen had its promoter of "the milk cure," and for the nineteenth century, the mantle appears to have fallen on Philip Karell, the physician to the Russian emperor in the 1860s. His pronouncements on milk filtered to the West by way of a British doctor stationed at the embassy in St. Petersburg, who heard Karell deliver a paper to the local medical society. In his preamble, Karell noted a long legacy of milk cures throughout history, entertaining his audience with the fact that a Roman emperor had awarded Galen a medal in the shape of a cow in gratitude for the remarkable results of a regimen of milk. Further successes had piled up since the third century. Particularly in France, doctors were achieving outstanding results in cases of dropsy (edema, or the collection of fluid in parts of the body) by administering milk throughout the day. (One doctor, adding a personal touch, administered raw onions after each dose.) German doctors claimed victory over Bright's disease and typhoid fever by using the milk cure. An epidemic of typhoid fever in Poland and Lithuania was snuffed out with milk and Hungarian wine. On a visit to military encampments in the Russian steppes with the emperor, Karell toured a fever ward and happily glimpsed "a bottle of milk at the bedside of each patient." The gospel of milk was spreading.[12]

If this report made milk seem like some kind of hokum or, worse still to the students of science, an all-purpose panacea, Karell begged to differ. His claim to originality lay in a carefully laid out method applied to ailments of a systemic nature. Unlike a Moscow colleague who cured "nearly 1000 chronic cases" of miscellaneous illnesses by simply feeding his patients milk, Karell distinguished his approach from such permissiveness. "If we allow milk to be taken *ad libitum,* the patient will likely soon suffer from indigestion," he warned. The best approach was a gradual one, applied "with great caution." He ordered his patients to begin with two to six ounces of skimmed milk "three or four times daily, and at *regularly-observed intervals.*" All other food was forbidden. "They should not gulp it all at once, but take it slowly and in small quantities, so that the saliva may get well mixed with it." Milk from "country-fed cows" was best, served tepid in the winter and at room temperature in the summer. If the patient's feces afterward were solid, then the doctor increased the dosage. In week two, that would mean a pint of milk every four hours, amounting to two quarts each day. Regular intervals, preferably of four hours, must be enforced. Here, Karell understood the psychological power of routinization: "No confidence can be inspired, and no cure expected, if the physician says to his patient, 'Drink milk in whatever quantities, and whenever you wish.'" The doctor admitted that the first week was the hardest, "unless the patient has a strong will and a firm faith in the cure."[13]

Gaining the patient's faith was half the battle. Karell's caseload groaned under the weight of chronic sufferers, patients who had exhausted every clinic and spa. Their complaints were often of a gastric nature: diarrhea, frailty, obesity, liver disorder, stomach and abdominal distress of every variety. Resistance to milk ran high: patients insisted that milk was "repulsive to them, or that they are unable to digest it." Some confessed fearing they would starve to death on the treatment. How could it work in a situation in which all medicine had failed? Karell mustered the scientific argument: "in milk are united all the elements necessary for the nutrition of our body, and besides," he added, "this substance is easily assimilated" when given in careful doses. Patients were reminded that milk "is the first food of man" and "to die of hunger, even when taking nothing but milk, is impossible."

And so the doctor imposed order on disorder. In the case of a Mrs. B., whose complaint was chronic diarrhea and vomiting, he began with three daily doses of milk, four tablespoons each. He ordered her to discontinue every other liquid, including the beef tea she was apparently fond of. (Karell was no fan of beef tea.) "From that time the vomiting ceased, and after the third day the diarrhea disappeared," Karell reported in biblical tones. "The feces acquired their normal appearance, which had not been the case for years before." Once she was consuming two bottles of milk per day, Mrs. B. made a "complete recovery."[14]

Karell scored successes on many fronts: patients watched their swellings subside, their waistlines diminish, and their headaches lift. Trips to Karlsbad Spa fell off. The market for beef tea diminished, at least by a fraction, as the demand for milk, at least locally, spiraled. Several of Karell's cases were men in their sixties who were suffering the effects of rich diets and aging organs. Some were young people who had been sickly since infancy. With the knowledge of hindsight, we may recognize particular cases as tuberculosis, poorly understood by doctors at the time. A sixteen-year-old girl displayed many signs: scrofula, diarrhea, a dry cough, intensified suffering in the evening, a general thinness, and a sallow complexion. Karell had treated her for fourteen years, to no avail, when he prescribed the milk cure. Taken away by duties, he returned to her neighborhood a year later and encountered the patient "entirely metamorphosed" by the treatment. Eventually, she had been able to add porridge, vegetables, and fish to her diet, without suffering from diarrhea. Here was a case in which the "cure" had worked, first, by staunching the loss of all nutrients, and then by supplying the basic nutrients themselves. It appears that with the help of a steady, healthy diet, the patient had cured herself.[15]

But was milk the solution to the modern epidemic of dyspepsia? If we are searching for a scientific answer to the question as the doctors might have explained at the time, it may as well lie in what Karell and his medical contemporaries believed: milk worked well as food for the sick because it was, at least theoretically and at least for some patients, easy to digest. Karell argued that it was probably the fact that milk was close to "chyle," digestive juice itself, that made it go down so well. Certain doctors recommended adding lime water to neutralize the acid in milk, thereby "*rest-[ing] the stomach,* by not requiring its solvent services."[16] We have no way of knowing how lactose-intolerant patients survived the milk treatment, given the silence of the record on that score. Doctors seemed to have viewed purgings as beneficial to patients who were overfed, so it is possible that nothing seemed amiss if some digestive tracts rebelled in that way. What can be seen with certainty was a growing body of knowledge about the digestive tract, which promised to keep the book open on the subject of milk's nutritive value. The phrase "perfect aliment" surfaced repeatedly in works on the subject of digestion, and the fact that it was a liquid put it in a distinguished family of bodily fluids. Its near-relative, blood, retained its regal position as king, or, in suitably Victorian financial terms, as "the floating capital lying between absorption and nutrition." Yet milk nevertheless won the spot as a reliable supplier of the "treasury" that in turn fed the "capital." No wonder that the stuff showed up on the bedside tables of patients as often as it did in the laboratory of physiologists and chemists.[17]

The mantra of "method" also helped boost the popularity and possibly even the success rate of the milk cure. Invoking this term amounted to an endorsement of the scientific community, breathing faith into the bourgeois sufferer of dyspepsia. As Dr. Karell well knew, faith had much to do with curative power. The authority of the doctor, the comforting refuge of his dictates, the fact that illness had the power to bend circumstances around the patient in a way that effected change: these were real benefits to sufferers of stomach ailments. Milk may have gained from a parasitic relationship to all of these factors.

Yet the benefits of milk on a schedule were not all derived from an imaginary world of time, work, and capital. Physical disorders of the alimentary canal were real and painful, and much could be gained (or shed, as it were) by paring down the forces working on the dysfunctional digestive system to a few basic elements. As one wise doctor from Tübingen put it in a congratulatory letter to Karell, there were many diseases forcing the body into "a perverse [state of] nutrition of which we are *unable to define either the extent or nature.*" The idea was to change the body's way of processing nutrients while continuing to nourish it. "How much more dangerous are other *alternative* cures," he added, such as "hydropathy, sea-bathing, saline mineral waters, etc." Consider the regimens endured by hydropathy patients at Karlsbad and Marienbad, where "a cold-rain shower" or the more vigorous "Scottish shower" (pictured as a tremendous force of water shot through a hose at the backside of the patient) might accompany compulsory draughts of mineral water and "lavements" of the intestine. In comparison, Karell believed (and we must believe with him) milk provided an "innocent" and perhaps even "efficacious" solution.[18]

The nineteenth-century milk cure for dyspepsia was distinctive in two final aspects, best described as negatives: these therapeutic doses of milk did not come under the category of self-help, nor were they simple and inexpensive. We would do well to recall the draughts of milk consumed by Platina, which were obtained, no doubt, from a nearby cow on the day of production at relatively little expense. In our own world of self-doctoring, we may know of people who have decided to drink milk as a trial run in solving a stomach complaint. But in the last half of the nineteenth century, milk suitable for sick stomachs had to be pure and fresh milk, and its benefits were enhanced by the presence of medical supervision. Looming large in the next phase of milk history is the problem of quality control. As a result of a rapidly changing marketing environment, milk was becoming hazardous as it became more widespread as a commodity. Used by most people, often without much (or enough) awareness of its provenance, the supply of milk in the age of Carlyle and Darwin was destined for a day of reckoning. Given the long distance it traveled and the number of times it changed hands, the product risked contamination, adulteration, and worse as it made its way to distant city consumers. By the end of the nineteenth century, urban milk was more commonly associated with disease, not healthfulness. This darker image, oddly enough, became responsible for ushering the commodity into modern consciousness.

CHAPTER ELEVEN

Milk Gone Bad

Among the most celebrated wonders at the Paris Exposition of 1900 were two surprisingly humble entries: milk and cream imported from Illinois, New York, and New Jersey, "raw and in the natural state, without preservatives of any kind," "as pure and sweet as milk freshly drawn from the cow." On the road for over two weeks, the products even survived a time-consuming gauntlet of customs inspections between Paris and the dairy exhibition near Vincennes. Americans kept the samples coming at a daily rate — the only exposition participants to do so — to underscore the reliability of their bravura performance. Their achievement had more to do with simple cleanliness than with modern technology. The newsworthy point of the story was written between the lines: pure milk at any distance from the cow evoked surprise in the year 1900.[1]

In fact, milk was more often associated in the public mind with disease at the turn of the century. The seemingly innocent white liquid was known to soak up and incubate germs as well as any laboratory petrie dish. Continuous publicity linked it with tuberculosis and other epidemic illnesses: milk regularly played host to diphtheria, scarlet fever, septic sore throat, and typhoid fever. (The list of diseases that can be transmitted through milk is much longer than this and includes viral as well as bacterial illnesses.) Only tuberculosis originated with cows; all the rest joined up with milk from contaminated surfaces and human hands. Milk spoilage remained a worry, and consumers could now count on a whole new generation of hazards ushered in by an age of industrial chemistry. Farmers and marketers depended on additives that to us sound more like automotive liquids — "Iceline" and "Preservaline" and "Freezine" — to prolong the life of unrefrigerated milk. Their key ingredients, formaldehyde and boracic acid, killed what were vaguely understood to be "germs" present in the many impurities: bits of manure, dead flies, and generic dirt that had fallen into the product in its journey between the cow and the consumer. No wonder, then, that pure milk had the power to amaze in 1900.[2]

Sour milk was a very familiar taste to nineteenth-century people; this is quite apart from the fact that, depending on the region, soured milk products would have been a regular part of the diet for many. In the case of milk gone sour, excess lactic acid did not necessarily sicken the consumer, but often such spoilage indicated that extraneous bacteria lurking in the supply had been able to multiply by the thousand over the course of one or two days. Taste and smell provided unreliable measures of true disease pathogens, however; reformers dwelled on lurid descriptions of infected udders oozing pus when the impact of numerical data failed to impress public audiences. Tolerant or captive consumers were more likely to suspect milk supplies after the fact: repeated outbreaks of milk-borne diseases generated telltale neighborhood geographies, which eventually revealed that a single supplier or even a single deliveryman had acted as transmitter responsible for an outbreak of diphtheria or tuberculosis. No one doubted that chemical and medical researchers had a role to play in solving the problems related to milk. The nagging question was exactly how that role should be regularized by guidelines and law.

The modern consumer may be surprised by the minor role played by the larger-than-life figure of Louis Pasteur in the career of milk. He certainly did not revolutionize the marketing of milk from his laboratory in the 1860s. It is true that he identified small living bodies called microbes as the agents of spoilage in consumable liquids. Applying heat could often kill them, or at least curb their harmful effects. But consider the vagueness of these revelations in a world untutored in modern epidemiology (not to mention steeped in free market ideology), and their limited impact seems perfectly understandable. Pasteurization as we know it took sixty years to become common practice. At the outset, roughly twenty years would pass before microbes became the target of milk reformers. And because Pasteur had concentrated in the first instance on wine and beer, more work needed

"I drink to the general death of the whole table." The skeleton offers a mock toast, playing on a line from Hamlet. The cartoon, employed in publications by advocates for pure milk in the first decades of the twentieth century, had won first prize in a contest sponsored by the American Medical Association. Probably from the first decade of the twentieth century. (Harvard University Library)

to be done to figure out how to apply what was known to milk. A German agricultural chemist, Franz Ritter von Soxhlet, deserves credit for devising equipment capable of doing an acceptable job in 1886. He remains obscure in the annals of history, probably because his surname could not possibly be turned into a technical-sounding noun.

In the end, "pasteurization" was actually fairly simple. The procedure needed to retain the characteristic texture and taste of milk, so overheating was to be avoided. And because warm milk was exceedingly hospitable to the further growth of microbes, the process of killing most of the bacteria with heat had to be followed up with rapid cooling. Roughly 140 degrees Fahrenheit worked well (though some producers favored higher temperatures, later found to be harmful to certain properties of the milk) for either twenty or thirty minutes. (Today's practices use higher temperatures, often between 162 and 165 degrees, for fifteen to twenty seconds.) Then the liquid had to be chilled immediately and poured into sterilized containers with clean and effective stoppers and kept cool. Another method, usually referred to as "sterilization," involved holding milk at the boil for specified periods of time. Trumpeted alongside white-coated bottling regimens to ensure germ-free products, sterilized milk offered the appealing bonus of an aura of the laboratory. The process enjoyed popularity in England, Germany, France, and to some extent the United States. To the general public, however, these extraordinary measures seemed too expensive and fussy. Pasteurization could be done at home with commercially designed equipment and careful attention to procedures, but this, too, was a costly affair. And yet, at the beginning of the twentieth century, it is worth remembering, every aspect of pasteurization presented obstacles of expense and know-how, not to mention the prevailing attitude of skepticism.[3]

It wasn't for lack of advice of chemists, apparatus, or even resources that the advance of pasteurization marched to a very slow beat. Passions ran high one hundred years ago, just as they do today, over reasons for and against the process. Some of the most ardent advocates of clean milk opposed pasteurization because of its power to forgive dirt and condone slovenliness. Others questioned its effect on milk, arguing that "ferments" (natural enzymes) were destroyed by the process of adding heat to the product. (They were correct in this regard, but given that vitamins had not yet been identified, could not add that some of these had been destroyed, too.) Still others objected to the taste of pasteurized and sterilized milks, which lacked the fragrance and fullness that true milk lovers readily identified in the natural product. Many farmers and suppliers continued to deny the need for treating milk, claiming that their supplies were acceptable in the raw state. And chemists themselves disputed what constituted the right procedure for pasteurizing milk. Perhaps most important, local battles in several countries revealed underlying financial obstacles: who should bear the cost of providing a disease-free milk supply? How should such production be regulated and supervised? Until the 1930s, the general populations of Europe and America were not guaranteed safe milk at the market because of lack of consensus on all these matters.[4]

These were critical years in the emergence of modern milk. The threat of disease drew the commodity and its producers into the new and important arena of debates about public health and the responsibilities of the state. Most notably, the connection between cow's milk and the widespread feeding of infant formula placed the issue of pure milk squarely in front of all other food safety issues. Science and pediatric medicine would play an important role in the determination of the future of the precious white liquid. And an unforeseen consequence of these years became evident in the marketplace: reformed measures in milk production would lead to the expansion of milk as a universally available commodity. As a result of solving the problems of purity, refrigeration, packaging, and transport, the production of market milk was made complete. The campaign for purity carries us into the twentieth century, when milk becomes a very modern commodity.

A peculiar paranoia colored the last forty or so years of the nineteenth century, summed up in a single, meaning-laden word: hygiene. It may be only slight exaggeration to say that if Pasteur hadn't existed, hygienists would have invented him. According to historian Bruno Latour, Pasteur's discovery of the threats in microbes crystallized a pervasive uneasiness with physical surroundings long before anyone learned of the inner life of a drop of milk. Europeans of an earlier generation had responded to waves of infectious diseases with sanitary reform: this amounted to an attack on "filth" present mainly in poorer districts of urban areas. Pasteur's refutation of the Aristotelian theory of "spontaneous generation" of disease organisms — in effect, the ushering in of the germ theory of disease — was laid on top of previous anxieties about the environment. Newfound bacteria of the 1860s and 1870s, ubiquitous in nature, prompted an attack aimed at all environments, including the ordinary home. An alliance of physicians, scientists, and governments mobilized against microbial invasion as though warring against a thousand hidden enemies. "The vague words 'contagion,' 'miasma,' and even 'dirt' were enough to put Europe in a state of siege," Latour tells us. "Food, urbanism, sexuality, education, the army. Nothing . . . human [was] alien to them." The entire world would have to "make room" for the war against the microbe.[5]

Modern industrial society had good reason to take up arms against disease: following advances in methods of statistics gathering, some nations learned that their populations were no longer growing. The French were acutely aware of dwindling human power, given their losses in the Franco-Prussian War of 1870–71 and the decades that followed: the census results of 1891 informed them that for the previous five years, the nation had experienced an excess of deaths over births. (Not all European nations were on the losing end: by comparison, "the German population had increased more than four times that of France's from 1880 to 1891.") A similar anxiety preoccupied the British, who became aware of population trends of the 1880s and 1890s: by 1897, it was known that the crude birth rate had dropped by roughly 14 percent over the preceding twenty years.[6] The matter "require[d] the consideration of all who have the well-being of the country and of the empire at heart," pronounced an editorial of the *British Medical Journal*. The reversal of attitudes in the United States led one sardonic observer to note, "From the Malthusian cry of 'over-population' to the Rooseveltian cry of 'race suicide' is an astounding transition." Modern nations were waking up to the larger consequences of family limitation among the middle classes.[7]

A falling birth rate was not the only problem contributing to "race suicide" and other population anxieties. It was clear to statisticians that populations were stymied by the shocking fact that infant mortality was actually rising in the latter decades of the nineteenth century.[8] High mortality of children under five had been a sad feature of urban life for many years: in New York, for example, the city inspector reported that the rate of mortality for that age group was 50 percent in 1840.[9] Since that time, despite environmental improvements, mortality rates among the very young were still high and, in parts of England, France, and America, still rising. Reformer John Spargo extravagantly estimated that roughly 95,000 babies were dying annually owing to bad milk. A Boston official invoked the key phrase of alarm in 1880: why, he wondered, did the "slaughter of the innocents" continue in societies of such prosperity and comfort?[10]

All data pointed to a single target. Physicians knew that the greatest number of infant deaths occurred in the warmest months of the year from a broadly defined complaint described as "summer diarrhea." It was a short step from that awareness to the medical focus on milk contamination. Bottle-fed babies were known to contract summer diarrhea in disproportionate numbers compared with breast-fed infants. Those who investigated the problem knew well the cause, as muckraker John Spargo pointed out in his *Common Sense of the Milk Question*. A health officer in Birmingham, England, estimated that the mortality of infants "artificially fed" was "at least *thirty times* as great as among those who are nursed at the breast."[11] Though exaggerated (a more exact measure from Liverpool was by a multiple of fifteen), the figure reveals the frustration felt by advocates of breast-feeding, an unexpectedly ardent coalition of doctors and urban health workers who were witnessing what appeared like a peculiar modern malady among mothers. "There is reason to believe," argued an English doctor with experience working with both rich and poor in late-nineteenth-century London, "that the function of maternity is undergoing atrophy in the women of modern civilization." Citing a study of more than two thousand women in central European cities, he added, "the inability to suckle is a symptom of degeneration . . . handed down from mother to daughter."[12] Physicians across Europe and America thus became increasingly alert to the dangers of "artificial feeding," a deadly combination of unclean utensils and questionable milk substitutes, usually from the cow, seldom fresh, and almost always negligently handled.

A French physician named Pierre Budin publicized a month-by-month graph to illustrate the point: he labeled the peak of mortality in late summer and early autumn the "Eiffel Tower" of death of mainly bottle-fed infants. The image soon became as iconic as its eponym, appearing across the Channel on posters and in journals elsewhere. "Don't wean in hot weather," advised the "Save the Babies" broadsheet issued by the Babies of the Empire Society in England. Attempting to solve the problem in a practical way, a French factory owner in Elbeuf deposited one hundred francs into a bank account for any female worker who successfully breast-fed her child until "the age of weaning."[13]

The "Milk Question," as it became known, provides a window into one of the earliest health and food reforms of the modern age. Doctors, scientists, farmers, marketers, and muckraking writers shared a public arena, harnessing the medium of print to the fullest extent. "There is death in the bottle," cried reformers, resuscitating a tagline from the earlier temperance movement. Bring an end to the "slaughter of the innocents," demanded advocates for safe infant feeding. An ad hoc international network of experts bore down on the subject, filling medical journals with articles fulsomely titled "The Great Bacterial Contamination of the Milk of Cities" and "Infant Mortality and Its Principal Cause—Dirty Milk." That a fairly arcane series of articles in the *British Medical Journal* later sold as a book stands as a measure of just how much information inundated the public in the first years of the century.[14]

Models of success were already visible in France, where the state joined hands with doctors and philanthropists, providing the funds for new forms of health care. Dr. Budin established the first *consultation de nourrissons,* a health clinic specially arranged to provide maternal education in breast-feeding and sanitary practices for the home. Since 1874, with the passage of a law sponsored by the celebrated philanthropist Théophile Roussel, mothers everywhere could obtain free advice and material support in child-birth and subsequent nursing of their newborns. Constant supervision of both mother and baby followed their discharge from the hospital and a common policy governed what happened next. Breast-feeding was of para-mount importance, and if and when a mother's supply failed, she was required to obtain sterilized milk from the hospital on a daily basis. A contemporary report described the laboratory-like procedures:

> The milk department is on the ground floor, and is very simply fitted up. In one room is a copper sterilizer. The milk, on its arrival from the country, in glass bottles with indiarubber stoppers, is sterilized, and is exposed to the temperature of boiling water for three-quarters of an hour. Each mother receives daily the number of bottles adapted to the child's age . . . the day's supply being handed through a window, which is approached by a special entrance at the side of the hospital.

Hospital attendants kept careful record of infant weights and survival rates, and the evidence over the course of a sample four years seemed clear to all: breast-fed babies exhibited a mortality rate of 3.12 percent, while babies fed either partly or entirely "artificial" food showed between 10 and 12.5 percent mortality. None among the 448 babies in the study con-tracted scurvy or rickets.[15]

Similar arrangements, called *gouttes de lait,* provided sterilized milk for households with infants among the French poor. Distribution centers run by charitable organizations would subject fresh milk from the countryside to an hour of sterilization and a "hermetic" sealing process in specially designed bottles. Some of the milk was given away; the rest was sold at varying prices according to the ability to pay. Unsterilized milk remained available through some milk depots, sold to those who wished to use it for purposes other than infant feeding. For bottle-fed infants requiring what was called "humanized" milk, the station used sterilized or pasteurized milk to compound a mixture of water, cream, sugar, and salt. Mothers could obtain a specially made basket holding several bottles at once (six for the day, three for nighttime), which they would return for replacement the next morning.[16]

In the United States, milk reform of the 1890s began as a private venture that gained a lot of publicity. By that time, Nathan Straus, a German Jewish immigrant in the china business, had become a highly successful department store owner in New York City and Brooklyn. In 1892, he turned a growing sense of philanthropic duty to the cause of milk. At least some of his zeal came from personal experience: he and his wife had lost two of their five children under the age of five, and one of the deaths occurred as a result of contaminated milk. Earlier reformers had addressed the problem of "swill milk" in the 1840s and 1850s. Frank Leslie's *Illustrated Weekly Newspaper*, for example, ran a sensational series of articles on swill-cow dairies in the spring of 1858. Straus comprehended that both disease and spoilage were plaguing poor consumers, particularly mothers buying milk for infants, who could afford no alternative to the cheapest milk available. In 1893, the Nathan Straus Pasteurized Milk Station at the East Third Street Pier in New York City became the first public infant milk depot in the United States. Straus also established a sterilizing laboratory supplied by regularly inspected farms and disease-free cows. Milk arrived on ice and moved through equipment tuned to heat the liquid to precisely 167 degrees for twenty minutes, long enough to kill tubercle bacilli and other "noxious germs in the milk," while preserving "the nutritive qualities of this most perfect of nature's foods." Some of the treated milk was then used to produce "modified milk for infant feeding," sold for one cent per six-ounce bottle, according to formulas provided by leading pediatricians. Meanwhile, Straus scrutinized the returns from the Bureau of Statistics of New York City for children under five years of age. By July, despite higher temperatures for the summer of 1894, deaths were down by over 7 percent. By August, deaths of children under two years were down by 34 percent. His project appeared to be working and his publicity campaign made sure that the public heard about its success.[17]

Straus's genius for establishing a network of influential backers was no small part of the success of his campaign for safe milk. His earliest ally was none other than the New York physician Abraham Jacobi, a longtime advocate of treated milk for babies. Recognized as the "dean of American pediatrics," a novel line of medical specialization at the time, Jacobi intro-

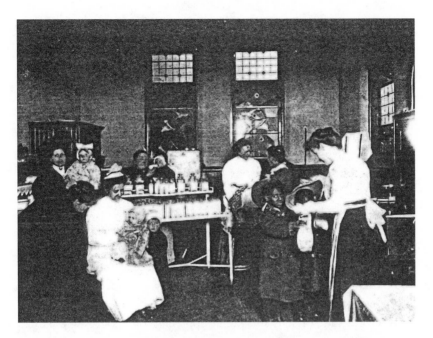

Nathan Straus sponsored milk stations across New York City, including one in the neighborhood of Morningside Heights. Photographs displayed the success of his efforts in reaching the immigrant and black populations of the metropolis. (From Disease in Milk: The Remedy, Pasteurization; the Life and Work of Nathan Straus, *ed. Lina Gutherz Straus, 2nd ed. [New York, 1917], Straus Historical Society)*

duced the process of pasteurization developed by Soxhlet to American medical discussion in 1889.[18] With his professional blessing, Straus then enlisted the help of Health Board doctors and hospitals, who promoted sterilized milk for newborns across the city. Doctors treating the poor were given booklets of order forms; the poor could use the coupons at any of the six depots situated around the city. Straus made a point of establishing a "large pavilion provided with comfortable seats" on the pier at the bottom of East Third Street, which was open "all day up to midnight," enabling busy mothers to stop in whenever convenient. While waiting to see the doctor, they could "enjoy the fresh sea air." Straus also organized free lectures by physicians "on the proper care and feeding of infants" — a means of increasing the demand for sterilized and modified milk. Finally, in an effort to change the minds of skeptics who believed that sterilized milk was somehow "medicated" and not for general consumption, Straus set up booths in New York City parks to sell the beverage at a penny a glass during the hot summer months. It did not take long for demand to outstrip supply, so Straus sold "pure" raw milk, too, proving that fresh, safe supplies could be had under the proper circumstances.[19]

English milk reformers were more interested in emulating French examples, opening "milk depots" in London and provincial cities at the turn of the century. But like Straus's arrangements, these stations sold milk rather than giving it away. Nevertheless, popular demand could support an expanding network of stations. In Battersea, a densely populated area of South London, for example, in the space of one year, a single depot expanded its service to roughly fifty gallons of treated milk per day for three hundred mothers in 1903. Stations boasted state-of-the-art heating and bottling technology, which reporters strained to describe in terms their readers would comprehend. The stoppers used at Battersea were like those used on German beer bottles. Familiarity, reformers hoped, might breed respect.[20]

After a decade of activism, public apprehension about raw milk had risen and expectations had changed: a fair number of mothers and caretakers of infants, young children, and the sick now depended on the regular availability of specially treated milk. Yet the biggest question of all had not yet been answered: who was responsible for ensuring that milk supplies were safe for consumption? Local governments had already played an important role, and so had philanthropists. The next decade of activism would raise the bar, demanding that nothing less than the power of regional and national governments work to bring safe milk to the market. A closer look at this later phase of activity reveals the surprising alliances forged between a remarkably diverse set of reformers in the United States.

"The housekeeper is abroad in her city," announced a Boston women's magazine in 1912. Well-to-do women awoke to a vast cleaning project that stretched beyond their own posh neighborhoods on Beacon Hill and Commonwealth Avenue. "Science had recently confirmed our democracy," read the article, warning readers that modern disease respected no class boundaries. "When Boston had 1,200 deaths from tuberculosis and 1,600 from typhoid, the Back Bay would have to have some of them," and that should rouse women of the upper classes to action. How could they check the spread of illness? "The housewifely mind paused to ponder," as one of their advocates recounted in a contemporary women's magazine. "Mary, the cook, who used to bake bread in the kitchen, had to be carefully watched to see that she wore a clean apron and washed her hands. Did the Superfine Baking Company wash its hands?"

Boston women organized campaigns to track down violators at every level of society. Before the city government assumed such responsibilities, the Women's Municipal League of Boston paid inspectors with bona fide degrees from Wellesley and Radcliffe Colleges to "patrol the streets and the alleys, especially the alleys," for garbage and flies. Their vigorous "Rat Campaign" posted literature to Brazil, India, and Russia and organized a citywide "Rat Day," when residents were urged to deposit captured bodies in designated collection bins around Boston. They sent confident women into the homes of rich and poor, especially the poor. One Miss Clarke visited tenements "day by day," where she "patiently interprets American civilization to the ignorant housekeepers who do not understand . . . that dirt is dangerous." To be fair, inspectors bullied not only immigrant women but Back Bay matrons, too, along with intransigent landlords. Perhaps most admirably, they bore down on businesses that handled food. On Saturday nights, sporting the latest technology, they projected images of sanitary violators in "stereopticon pictures . . . thrown against a building" in districts full of leisurely strollers. One of their favorite graphics described the route of germs via the typhoid fly, "from the stable to the baby's milk bottle." Here was something legislators needed to hear about, given that infant survival rates were currently newsworthy. So they marched to City Hall and eventually to the State House to demand laws enforcing the sanitary rules they believed in.[21]

Women of the Progressive Era, along with "maternalist" reformers across all of Europe, proved to be bold pioneers of a new era of welfare provision. Their efforts on behalf of mothers and children paved the way for innovative government policies in the early decades of the twentieth century. Through settlement houses, housekeeping cooperatives, mothers' clubs, and a general campaign for educated motherhood, a generation of women changed the expectations of literally millions of families. Despite the fact that the female protagonists in this story lacked full citizenship rights before 1920 — and despite the irony that some were active antisuffragists — we can see how they expanded contemporary understanding of what the duties of the state were: to ensure the health and safety of all citizens, including the underprivileged. Protecting babies was one of their foremost objectives.[22]

Nowhere was this more apparent than in the history of milk reform in Massachusetts. The "milk question" occupied reformers from 1909, when an indomitable Elizabeth Lowell Putnam (1862–1935) became chair of the Committee on Milk of the Women's Municipal League of Boston, which her sister, Katharine Lowell Bowlker, had founded. Like Lina and Nathan Straus, Putnam had experienced personal loss before beginning her campaign for purity: her two-year-old daughter Harriet, one of five Putnam children, died from drinking contaminated milk in 1900. Putnam's social circle of supporters included an intimidating array of influential Bostonians — pediatrician Milton Rosenau at Harvard Medical School, banker and philanthropist Henry Lee Higginson, and Richard Maclaurin, president of the Massachusetts Institute of Technology. No doubt her own brother, Abbott Lawrence Lowell, president of Harvard from 1909 to 1933, lent his name to the cause. (The social clout of the Women's Municipal League itself must not be forgotten: the *New York Times* used the adjective "aristocratic" to describe the organization when it decided to "ask poor women to join" in January 1909.) Descended from the Lawrences and Lowells, scions of the New England cotton industry, Elizabeth and her two impressive sisters (besides Katharine, there was Amy Lowell, a celebrated poet) could command power and prestige when the occasion permitted.[23]

The League exerted a surprising degree of force by way of the foot soldiers of reform: the phalanx of middle-class women who invaded shops and intimidated lax merchants so that substandard supplies could not be sold. Neighborhood groups called "Market Tens" sprouted up across Boston, pledging "to exercise watchful supervision over the sanitary condition of the stores in which they made their daily purchases. Is the food protected from dust and flies? Would the floor soil your dress?" Such were the questions printed on "report cards" that each investigator returned to headquarters. Violators would receive visits from the formidable chief inspector. If conditions weren't corrected after that, the League submitted a formal complaint to the Board of Health. Hence, the "menace to the health of the community" found itself in more serious trouble.[24]

*Portrait of Elizabeth
Lowell Putnam hold-
ing her son and
daughter, possibly
George and
Katherine, ca. 1891–
93. (The Schlesinger
Library, Radcliffe
Institute, Harvard
University)*

Yet even the cleverest stratagems of the League were powerless to affect
the problems plaguing commercially sold milk. Contamination often occurred long before the product reached the shopkeeper: diseased udders,
unclean containers, warm railcars, and infected deliverymen contributed
their fair share of problems. Many of these practices eventually became
public knowledge through muckraking journalism that characterized the
first decade of the twentieth century. But the Milk Committee of the
Women's Municipal League of Boston was incapable of reaching out to
the complicated network of contractors and farmers who supplied the
considerable urban market. When farmers shipped their milk long distances, their contributions to the market were inevitably mixed with the
milk of other suppliers, and with no coordination of regulations or treatments, one heavily contaminated contributor could spoil the lot. Of the
360,000 quarts consumed each day by the city of Boston, 324,000 arrived
by rail. Contractors purchased rail deliveries and took responsibility for

distributing the milk to retailers. As Putnam quickly learned, the problem was not always with the milk itself, but instead, with coordinating supplies at the level of the market.

Putnam obviously possessed the financial means to opt out of the ordinary market for milk. If she had wished, she might have purchased milk from certified dairies, which supplied cities with high-grade milk at high-grade cost: in the words of a contemporary, the milk was "produced by carefully selected cows, under conditions bordering on surgical asepsis, and modified in sterile vessels with boiled water, under the direction of a qualified physician."[25] But Putnam went beyond market reasoning to demand a higher grade of milk for the general public, a highly controversial stance at the time. Even pediatricians expected less from the marketplace. C. E. Rotch, one of Boston's foremost infant doctors, viewed Straus's New York project as an unrealistic model, too labor intensive and costly for wider distribution. But other leading health experts shared Putnam's expanded vision of milk safety. As J. H. Mason Knox of Baltimore argued, not just the "tiny heirs of comfort and luxury" were deserving of safe sustenance.[26]

Given the basic premise of infant welfare, Putnam and other milk reformers adopted a new attitude of public responsibility on behalf of the greater society: in Putnam's words, "the rights of the innocent consumer to protection from disease and death."[27] Conscious of being part of a more general wave of female activism across the country, Putnam made no claims to originality. Outstanding reformers like Jane Addams and Florence Kelley had already won public acclaim by calling attention to the needs of immigrants, the poor, and mothers and children in Chicago. Compared with activists engaged in the settlement movement in the United States and England, Putnam's political views were quite conservative: she disavowed any interest in women's suffrage or socialism, and she had nothing to do with the burgeoning interest in organizing women, labor, and African Americans in pressing for equal rights. Much of her energy was aimed at middle-class women, whom she believed were neglected in the effort to improve standards of living among the working classes. Yet her upper-class upbringing led her down an unexpected path of educating the public on points of health and hygiene. Before she knew it, Putnam was deep into the territory of improving the lot of all. Her career as a milk reformer stands out as a remarkable performance of patrician bravado.[28]

"Our sewage in Boston contains 2,800,000 bacteria to the cubic centimeter," Putnam reported to the Committee on Milk in 1909. "In Washington they recently found their milk supply averaging over 22,000,000, and some milk ran up to over 200,000,000. Among these germs were many of tuberculosis, one charitable institution for children buying milk which contained tuberculosis germs virulent enough to give the disease to guinea pigs inoculated with the milk." She employed a competent knowledge of sanitary conditions in other cities to the best effect. "Berlin is said to consume in its milk 300 pounds of cow dung daily and New York 10 tons of filth and refuse." She cited the professionals. "As Professor Jordan says: 'Cow dung as an addition to champagne or beer would be an efficient aid to temperance.' "[29] Her alliances ultimately were with the rising tide of pediatricians, who were gaining ground as important health authorities in the United States and Europe.

Putnam's pamphlets reveal a working knowledge of the latest research on bacteria, along with a flair for shock-value revelations. Her speeches and writings also display the inevitable town versus country snobbery that plagued the campaign for pure milk. Even John Spargo, a spirited Vermont socialist, exuded an air of exasperation in trying to capture rural attention with news of bacteriology. Putnam no doubt had read Spargo's *Common Sense of the Milk Question,* which soliloquized at length on the many pitfalls of the milking process carried out by the typical farmhand, as experienced and valued as he might be. ("Why, it would take a college graduate to be a milkman!" admitted Spargo — without irony — after reviewing the need for sanitary procedures all along the way of milking.) "Along comes Bill, the farmer's man, pail in either hand, to milk the cows," Spargo narrates with gusto. Measuring the proximity of manure piles and the inadequate way in which Bill washes his pails before milking, Spargo adds, "How could you expect poor Bill, who doesn't know a bacterium from an elephant, to know that his pails are not really clean[?]" We hear of Bill wiping his hands on his trouser legs, wiping his nose with the back of his hand, and then sitting on a milking stool "which is filthy, [and] saturated with dirt." No washing of teats takes place, no antiseptic wash cleans Bill's hands. Flies "are busy carrying bacteria from the manure heap into the milk" while the cow contributes a lump of dung to the pail now and again. A new "nickel-plated strainer which Farmer Jackson bought at the Dairy Show" removes a "round ball of dung and hair" from the milk, which sits for a time in an uncovered pail before traveling to a railroad station, where it is to sit for a while longer. Spargo explained each step of the industry to the presumably sickened reader. Comely Holsteins, their hind legs covered in manure, peered out from photographs punctuating the text. The chapter was a tour de force of literal muckraking; it also documented just how far the countryside was from the sanitary heights of urban reform.[30]

For Putnam, two decisive events occurred that placed her at the head of an aggressive campaign for state involvement in coordinating a clean milk supply. Legislative reform had been stalled for two years, and the leader of a milk reform bill in Massachusetts, Dr. Charles Harrington, appealed to Putnam for help in gaining its passage. When he died suddenly in 1908, Putnam felt convinced that she must assume leadership of his campaign. Her determination rose to the boil when a milk strike in Boston brought the issue to the attention of state legislators in the spring of 1910. A committee called witnesses to the State House, but Putnam found the question of milk purity edged out of discussions. Given the prevailing sexism of the day, which would have relegated female philanthropy to the margins, this is not surprising. Yet Putnam was perfectly aware of her own degree of expertise and practical knowledge. Bridling at such "cavalier" treatment, she now set her sights on direct lobbying of both state government and farmers. Using her home on Beacon Street as headquarters, she organized the Massachusetts Milk Consumers' Association. Within a year, the group boasted a membership of around fifteen hundred, mostly "prominent men and about ten women," more proof of the privileged circles that Putnam was able to mobilize. "The object of this body," announced the group's first report, "is to obtain legislation to assure the community a proper milk supply, putting the control of this supply in the hands of the State Board of Health."[31]

Only the Board of Health could coordinate the necessary labors of inspection and, most important, enforcement of standards for bacteria counts and evidence of disease. Conjuring up the metaphor of battle, Putnam pointed out that legislating for state supervision "puts a general at the head of the Army for the protection of disease." Her position was seconded by leading physicians in the reform effort. "Pasteurization is too important a public health measure to leave to the caprice of the individual," Milton Rosenau argued in his widely acknowledged book on milk.

"The process should be under strict surveillance, and guarded by health laws and regulations." These were modern words indeed.[32]

Putnam and the continuously expanding troops of milk reformers achieved a Pyrrhic victory in the first years of the movement: milk sales declined as the public became aware of the threats to health implanted in the public supply. By 1912, it was evident that press coverage across the state was beginning to gain traction. Repeatedly, state legislators debated and voted on bills concerning milk, including the Ellis Bill, one of the most stringent sets of regulations, drawn up by a coalition of doctors and legislators. But throughout the years of World War I, the governor of Massachusetts vetoed one milk bill after another, claiming that existing regulations were quite enough. New demands, he argued, would leave Massachusetts dairy farmers at a grave disadvantage compared with their competitors across state borders.

What was holding up passage of such common-sense measures bringing about safe milk for the general public? It is easy to forget how our present conception of milk is shaped by twentieth-century laws belonging to an industrialized food supply. Boston consumers harbored herd instincts of their own in 1912. Buyers were accustomed to existing arrangements, or resigned to them and incapable of imagining an alternative. The metropolitan region constituted the main problem, given its dependence on milk hauled long distances by large contractors. But outside urban districts like Boston, Cambridge, and Somerville, many communities relied on milk from trusted, albeit untested, dairy herds from nearby rural areas. Small dealers still dominated such towns, meaning that farmers themselves were in charge of collecting and marketing their own milk. And small contractors hauling milk short distances didn't feel compelled to resort to pasteurization, partly because the cost of equipment was prohibitive. Even dairy-herd testing remained relatively rare, possibly applied to as little as 10 percent of all cows in Massachusetts. As the Special Milk Board Report pointed out as late as 1915, if only cows tested for tuberculosis were allowed to supply the state, a "milk famine" would certainly ensue.[33]

During the first decades of the century, belief in pasteurization constituted just that for the lay person: a suspension of rational understanding and an abiding faith in a process vaguely understood at best. Physicians published learned and enthusiastic accounts of the process, but a survey of dealers, contractors, and farmers revealed a broad range of opinion and ignorance. Some believed that pasteurized milk was equivalent to "a sort of bacteria soup," full of "dead organisms" and "strictly parallel to the sale of a . . . soup made from oysters of an uncertain vintage." (Pediatricians themselves can be credited for spreading the association with oysters, allegedly the only other food consumed "raw" like milk.)[34] Sterilized milk did, in fact, putrefy and produce toxins quite distinct from the bacteria in raw milk. This, too, contributed to popular confusion. Some milk dealers denied that pasteurization could kill "germs of tuberculosis." Opinions varied on whether one could get sick even if milk *had* been pasteurized. And given the variety of methods of pasteurizing milk, the results might easily have varied. According to the experts, milk properly pasteurized should be indistinguishable from raw milk. But reality was probably very far from the ideal.[35]

Critics of reform placidly viewed milk drinking as something like an everyday test of mettle, best endured by healthy adults. Putnam fought hard to silence those who adopted a eugenic view of the matter. "An official of the milk producers has advanced the theory that it is a good thing for the vitality of the race that infants who cannot withstand the poisonous effects of dirty milk should die," she informed a state legislator. "Do you think the test of poisonous milk should be deliberately applied to determine whether your baby should be allowed to grow up or not?" Putnam proved that she was perfectly familiar with this line of reasoning. "Poisonous milk is just as likely to kill embryonic Philips Brooks' [*sic*] or Robert Louis Stevensons who are not strong physically," she continued, "as it is to kill prospective inmates of feeble minded institutions who happen to have ample powers of resistance to disease germs." Many progressive-minded people of the era subscribed to eugenicist views, replete with racist and class-bound prejudice. And so did many farmers, who viewed Putnam and like reformers as effete and ignorant intruders from the ranks of the urban elite.[36]

Practical problems existed and would continue well into the 1930s. Experts remained skeptical that pasteurization could eliminate deaths caused by milk that entered the market chain. As the Special Board pointed out, the "enormous number of people" handling the supply put milk at risk. Typhoid, diphtheria, and other diseases were still chronic across the general population, including among milk deliverers (not the saintly creatures that legend would suggest).[37] In fact, many supplies did become infected after treatment, when cooled improperly or poured into unsterilized containers. Some supplies received treatment twice before reaching the consumer, and yet, given inadequate refrigeration and ignorant dealers and purchasers, the milk still became contaminated, spoiled, or both. Roughly half of the poor households in the greater New York City area purchased "loose milk" ladled from large metal milk containers as late as 1931. The cheapest milk available, it constituted only 25 percent of the milk market, but that figure did not fully represent its importance to a considerable part of the population.[38]

The biggest challenge of all would be to impose regulations on the substantial trade in what was known as "market milk." By the early twentieth century, many farmers delivered their milk to rural depots far removed from the actual point of sale, where contractors collected cans and hauled large shipments to distant cities. The Boston milk strike of 1910 uncovered just how far this system had evolved. With contractors in the middle, both producers and consumers came up vulnerable. Farmers complained of rising costs of cows, feed, and labor, while they continued to receive the same payment for containers of milk over the course of two years. The strike originated in their refusal to accept "summer prices"— lower than winter payments owing to the flush of supplies in the summer — at the beginning of May 1910. If consumers thought they were paying too much for their milk, the farmers asserted, they should blame the contractors and not the producers, whose income had actually declined in real terms.[39]

Even before the strike broke out, the state of Massachusetts backed down from imposing regulations that would have required the use of glass bottles. The modern milk bottle served as a lightning rod for debate: it represented high costs and high maintenance, which neither farmer nor consumer wanted to assume. The wide-mouthed American milk bottle, developed in 1886 by Hervey D. Thatcher in Potsdam, New York, aimed to solve the problem of complete cleansing that thin-necked European counterparts had presented. But the specially designed container required an initial investment by the producer, along with the equipment needed to wash, sterilize, and fill it. Many poor customers purchased only a few cents' worth of milk at a time, since they had no means of storing larger quantities of milk without ice chests of their own. The new law required cold storage in shops, too, which small shopkeepers could not afford. Cleaning the bottle continued to prove problematic: consumers were expected to "scald" bottles before returning them, a demanding process seldom performed to perfection. (As Dr. Rosenau intoned, "It is nothing short of crime to place any milk, raw or cooked, in bottles that have not been properly disinfected.") Milk dealers complained that milk was too often the scapegoat for every outbreak of disease, and not surprisingly, they blamed the troublesome glass bottle when an epidemic of scarlet fever broke out at the time of the strike.[40]

In the midst of the strike, which lasted for nearly forty days, Elizabeth Lowell Putnam addressed the state commission charged to investigate the conflict. She testified that "consumers will pay a fair price for milk if they are assured that it is clean and fresh."[41] Given so much diversity among milk consumers, including those who were buying two cents' worth each day, her assessment could not have represented the whole truth. The point was moot, however, given the state's inquiry into the big business of milk. Putnam's voice was eclipsed in the Boston papers, where the big contractors occupied center stage and provided good entertainment with quips about bookkeepers and profits. (Surpluses, one contractor claimed, were like bacteria in milk, "sometimes more and sometimes less." The committee and the assembled audience were said to erupt into "roars of laughter for several minutes" while listening to the witty commentary of W. A. Graustein.)[42] The largest dealer, H. P. Hood, held out the longest against the farmers and provided much grist for later commentary on the strike. Once an agreement was signed, giving the farmers all but one month of "winter" rates for milk, the clerk of the farmers' association went on record with his own prediction. "The customer must look this matter square in the face," he warned, "and be willing to pay a price that will insure a modest profit to the producers and afford the distributors a fair compensation[,] or face a famine in the near future."[43]

The state of Massachusetts ratified additional milk laws in 1914, which rendered illegal any attempt to "sell or deliver any milk without Board of Health authorization." But without serious enforcement, the law failed to satisfy milk reformers like Putnam, who pressed on with their cause throughout the decade.[44] New York State's pasteurization law went into effect in 1916, and other cities and states joined in the effort to protect consumers from tainted milk by means of pasteurization. (Chicago's municipal law of 1908, though not mandating full pasteurization, was the first of any American city; New York followed in 1910.) European nations enacted pasteurization laws only after decades of stalemates between reformers and farmers. Voluntary standards were relatively ineffective in England, where different grades of milk substituted for a uniformly trustworthy product. Less than 1 percent of all milk was both graded and certified in Great Britain in 1926; during that decade, around two thousand childhood deaths per year were attributable to bad milk. The issue remained contentious, associated with an antipathy to market capitalism that increased during the Depression years. When consumers were warned to boil milk before giving it to babies, one member of Parliament suggested boiling distributors, too. Not until the "Save the Children" Bill of 1949 was the greater public in England ensured of safe milk. Edith Summerskill, then parliamentary secretary at the Ministry of Food, called passage of the bill a "triumph over ignorance, prejudice, and selfishness."[45]

Historians differ in their interpretations of the torturous path to pasteurization. Most accounts resist seeing its legal enforcement as the inexorable march of progress, meant to enable the expansion of a mass market for good milk. As we have seen in the case of Boston, the adoption of pasteurization lagged behind both technological know-how and the widespread demand for an inexpensive yet relatively safe product. The emergence of large-scale production of pure milk wasn't inevitable or even especially desirable in the first twenty years of the century. Yet the decisions of governments, which came out of the struggle with relatively low-cost responsibility, did have an impact on subsequent history. Once the pasteurization process had been made efficient by technological means, the road to mass consumption had been cleared. In the long run, this meant that small farmers would suffer and large ones would prosper.[46]

Perhaps most noteworthy of all was the less visible shift that had taken place as a result of the fight for pure milk: the debate over breast-feeding had quietly subsided, relegated to the margins of discussion, as pediatricians, milk reformers, and, ultimately, infant formula suppliers pressed farmers for clean milk. The rise of the "expert" — invariably someone other than a childbearing female — meant that modern societies like the United States and Britain implicitly placed their faith in what medical science had to say about the latest trends in infant feeding.[47] Most women were no longer relying on breast milk to feed their offspring for longer than a few weeks. Only the most outspoken critics, like John Spargo, dared to state the obvious. "The humble and docile cow is the foster-mother, or wet-nurse, of the modern infant," he wrote in 1908. A young health officer in London, Dr. G. F. McCleary, put the point more bluntly: "The human infant tends more and more to become a parasite of the milch cow."[48]

PART 4

Milk as Modern

CHAPTER TWELVE

The ABC's of Milk

It is worth remembering that milk reformers were demanding pure milk for a minority population — namely, infants — when they laid siege to the Massachusetts legislature in 1910. The leading pediatric authority of the day, Milton Rosenau, described the liquid as "too perfect food for the adult," meaning it lacked the necessary bulk that a healthy digestive tract required. Reformers had to bring in a second physician to advocate for a broader band of consumers. "Milk is the chief food of sick folks," the doctor testified. "The importance of milk as a food for sick people — adult sick people — cannot be overstated or overestimated." Though it was widely used as a kitchen ingredient, the profile of milk as a beverage remained strikingly low, particularly in cities. Hence, reformers rested their case for pure milk mostly on compassion for innocent children, rather than on a sense of duty to feed the masses. The argument for milk as a food for everybody emerged from a different context following the First World War, when developments in nutritional science came upon new evidence of milk's goodness.[1]

Although a global economic depression stretching from 1873 to 1896 created hardship for farmers, paradoxically, it provided a boon for consumers. As prices for agricultural commodities dropped by some 40 percent over the period, the price of food fell, too. As a result, more people enjoyed the luxuries of meat and dairy products than ever before.[2]

Agricultural output continued to soar worldwide, despite the fact that fewer people were responsible for making it happen. Milk production, in particular, was on the brink of becoming truly modernized, tuned to a high pitch of productivity and consumption.

But what exactly did the "modern" milk production look like? It wasn't necessarily a massive, mechanized affair: even in the United States, only 3.6 percent of farms owned tractors by 1920. The United States Department of Agriculture "estimated that a farmer needed at least 130 acres to make the purchase of a tractor affordable." That ruled out most farms in dairy-abundant places like New York, Connecticut, and Wisconsin. So rising productivity sprang from three factors: cheap labor (often unpaid family members); a few timely inventions, many of them adapted to draft animals; and an infusion of advances brought by scientific research in agriculture.[3]

Depending on where you looked, either science or the machine was at work, changing the labor force and the process of production. Such changes had their own internal logic in the wide expanses of United States farmland. "The shortage of labor in a country where everyone might own land made the new machines almost a necessity," wrote Wilbur Glover, longtime editor of that famous bible of midwestern dairying, *Hoard's Dairyman*. By "new machines," Glover meant McCormick's reaper, along with newly designed harrows, rakes, and threshers, all of which were in use by the 1850s, pulled by horses, oxen, and mules. These inventions "were made to produce a surplus and so to pay for themselves," Glover explained, "—and for continually increasing purchases of machinery to lighten the labor and enlarge the income of the farm." Those who started down the road of technological improvement found themselves committed to production for the market and dependent on the railroad for access to it. Despite some resistance, they came to recognize that their survival depended on keeping up with the times through collective associations and regular access to the latest research in science.[4]

This made Wisconsin a particularly hospitable environment for an agricultural experiment station funded by the federal government. The College of Agriculture at the University of Wisconsin, under fire from farmers around the state because of its pretensions, provided the logical location in 1883; with a new agenda and an annual allowance of four thousand dollars, it could hope for a better future. Its mission: "to extend by publication and by lecture the knowledge thus acquired" through scientific research in agriculture. Many farmers remained skeptical ("An educated fool is one of the most disgusting things in life that I can think of," one protested), but those who were heavily invested in dairying admitted it was worth a try. Unless they kept up with modern innovations, they knew they might meet the same fate as New Englanders, who were surpassed in productivity and output by farmers farther west. "Necessity," argued one farmer, "is urging dairymen as a class, to appreciating [*sic*] the importance of a more scientific education, and a much wider range of business knowledge than we have hitherto possessed."[5]

With an "Experiment Station," newly named, and with innovative venues like "the short course"—a program that ran for two winters, when farmers and their families would be unemployed at home—the reformed college made a deep and growing impression on the agricultural population. "I have increased the production of my herd about seventy-five pounds of butter per head as a result of the Short Course," wrote one enthusiastic farmer afterward. A laborer even reported, "It doubled my earning capacity." Increased output of corn, thanks to the university's research, amounted to roughly $20 million per year. American journalist Lincoln Steffens visited in 1909 to report on the college's growing achievements and attendance: from 19 students in 1885 to 393 in 1907, a number that included wives attending the "Housekeepers' Conference." "Madison," he announced, "is indeed a place where anybody who can go there, may learn anything"—the "old ideal of a university."[6]

Ideals were one thing; reality, another. Many farmers at first were resistant to academic assertions. One unpopular fact was the idea that disease was impossible to detect with the naked eye. The drama of introducing farmers to the threat of tuberculosis, for example, made newspaper headlines in 1892, when the Experiment Station demonstrated the latest method, Koch's lymph injection, to test for the disease on its own choice herd. "Hundreds of farmers flocked to Madison to see the results," the report stated. The lymph test revealed a shocking number of positive cases, requiring the station to act according to its own scientific convictions. "As the sleek well-fed dairy animals were led before the assembled crowd, murmurs of disapproval were loudly voiced to see such fine specimens of the different dairy breeds sacrificed. But the autopsies told another story." Twenty-eight of thirty fine-looking cows showed internal evidence of tuberculosis. The news amounted to a public crisis, especially given that the Experiment Station's milk had been marketed in

Madison. But the test achieved the desired results with farmers, who soon "swamped" the facilities of the station to obtain testing for their cows. Like dairy farmers across Europe at the time, many joined cow-testing associations to monitor stock on a voluntary basis.[7]

No discussion of increasing supplies at the beginning of the twentieth century can omit the creature most directly responsible for producing milk, namely, the dairy cow. Experimentation in breeding was not new to the twentieth century, but modern genetics would encourage more rigorous record keeping and herd culling. Farmers now understood that the all-purpose cow, used for milk one day and for meat the next, did not necessarily deliver the best product in either case. Considerable interest swirled around the Cinderella of the barnyard, the Holstein. Mottled black-and-white, not particularly pleasing in shape, producing a whitish milk that contained less milkfat (3.5 percent) than the nineteenth-century milk taster would have preferred, the Holstein nevertheless was a record-breaking producer. One of the most important projects of the early 1880s had been a long-running evaluation of fat content from Holsteins and Jerseys, valued for their creamy milk. (Cattle-breeding societies offered annual proof of the reliability of Holstein stock: in 1888, for example, a remarkable cow named Pietertje produced an astonishing 30,318.5 pounds of milk.) Breeders yearned for a genetic mix of the two, giving the Experiment Station one more goal.[8]

While work on the production side continued at the University of Wisconsin's Experiment Station, research on milk consumption was definitive in changing the future of milk, and more. The work of one researcher in particular, Elmer V. McCollum, crystallized the hypotheses of nutrition scientists in a way that enabled them to make universal claims about what everyone should eat and drink. The puzzle of diet and organic growth had always concerned farmers, who were in the business of rearing livestock for their flesh and by-products. But the same questions also extended to human well-being. McCollum's message about milk, though not entirely new, captured international attention and created a rather new job description for the humble cow. It was time for the beneficial bovine to feed the world.

Elmer V. McCollum prefaced *The Newer Knowledge of Nutrition* with bad news. "The author recently enjoyed with a friend," he confided rather promisingly, "a dinner which consisted of steak, bread made without milk, butter, potatoes, peas, gravy, a flavored gelatin dessert and coffee." The self-esteem of the biochemist's native land lay embodied in that menu, yet McCollum used it as a prime example of a nutrient-deficient meal. Such a "diet of seeds, tubers and meat" would prove incapable of "promot[ing] health in an experimental animal over a very long period," he reported. Something was missing, hiding in plain sight.[9]

Without milk, no diet was complete, and more emphatically, McCollum argued that "the consumption of milk and its products forms the greatest factor for the protection of mankind."[10] At the time — 1918 — these insights represented the pinnacle of international research on biochemistry, disease, and nutrition, reached by competing teams of scientists from many disciplines. McCollum can be credited with the most significant discovery of all: a substance present in the fat of whole milk, which acted as the key agent in enabling growth in animals and humans. He called it "vitamin A," classifying it with "amines," or amino acids, discovered in recent years. The "A," reminiscent of milk's formerly touted asset, "albumen," seemed oddly appropriate. It was as though milk were receiving a good grade for its excellent fat content.

McCollum's discovery spoke to a generation haunted by the threat of deficiency. The First World War had drained the food supplies of western Europeans and, to a lesser extent, Americans. Shortages of essential commodities persisted when McCollum wrote *Newer Knowledge,* making animal experiments at the University of Wisconsin look like odd parodies of actual wartime menus. What was life like, living on the same meal, day after day? No sugar in the diet? So what were the effects? General readers were perfectly willing to think about analogies between their eating habits and the diets of laboratory animals. As the French chemist Dumas had commented in the wake of the Franco-Prussian War in 1871, warfare renders whole populations subject to a nutrition experiment. Although McCollum's discovery of vitamin A in milk had actually preceded the First World War, a full appreciation of his research had to wait until the appropriate conceptual apparatus — what historians of science might call a paradigm — arrived in its aftermath.

"In meeting McCollum one involuntarily thinks of Lincoln," one of his associates observed years later, after McCollum had retired from a prestigious position at the Johns Hopkins School of Hygiene and Public

Health. Despite his achievements and subsequent fame, McCollum always retained a "homeliness of manner" and "rustic dignity." After sharing a meal with him, a famous Dutch pediatrician visiting Baltimore was said to exclaim, "At last I have met an American."[11] McCollum's life story certainly wears the look of archetypal. His parents were uneducated settlers who lived by their wits, moving across Tennessee, Arkansas, and Kansas in order to farm. Elmer was the fourth of five children, surprisingly unfavored by his parents, despite his status as older son. He enjoyed little formal schooling, few books, and no educated conversation until he was eighteen; most of his childhood was spent at labor, which he apparently bore without resentment. McCollum recalled memorizing poetry at the plow (he later became an ardent fan of Emily Dickinson) and devising a clever trap for rats (a system he reproduced when the dean refused to fund the purchase of laboratory rats at the University of Wisconsin). While his sisters received the privilege of being sent away to a remote divinity school for high school education, McCollum and his brother had to wait until they were older, after his mother moved the household and her disabled husband to Lawrence, Kansas, for that purpose. McCollum contributed to his own support by becoming a lamplighter (except on nights with moonlight) in the town of twelve thousand. Throughout the first half of his life, he was signally frail, suffering from dental abscesses and inflamed tonsils. Six feet in height, he weighed no more than 122 pounds. Like Lincoln, he appeared tall and not particularly well fed.

McCollum knew about deficiencies first-hand, not because of the scarcities of war, but from the circumstances of isolated rural America. Miles from resources and forced to depend on an idiosyncratic supply of wisdom, farm families commonly experienced serious dietary deficiencies without knowing it. Much rested on a mother's practical intelligence. As an infant, McCollum suffered from scurvy, though no one knew why the infant had brown spots on his skin, swollen joints, and swollen and bleeding gums. The disease could be traced to the way in which he had been weaned, four or five months before his first birthday. McCollum's mother believed that raw cow's milk would "surely" kill a baby in the weaning process, a position McCollum later gratefully applauded. (The real problem with milk on the farm, he argued in his autobiography, derived from barnyard flies, which happily circulated from dung to straining cloths hanging in the dairy.) As an alternative, she cooked up a mash of potatoes and milk, heating it to the boiling point, as her own brand of American farm pap. The heat, he realized in retrospect, had deprived his diet of the necessary nutrient. Proof of the pudding came in his sudden recovery not long after his first birthday: peeling winter-stored apples, his mother gave Elmer scrapings to chew on, and when he seemed to like them, she gave him more. Within two or three days, his condition had improved. Strawberries in May completed the treatment. Without realizing it, Mrs. McCollum had discovered, so to speak, vitamin C.[12]

Milk was inscribed in McCollum's life story as indelibly as his farmstead practicality. His parents' homestead near Fort Scott, Kansas, built in 1879, the year of McCollum's birth, was the "best house" in a six-mile radius of rural hardship. The family owed its material advantage to twenty-five dairy cows. "Milking cows, churning cream to butter, washing and molding it into one-pound cakes involved great labor," McCollum explained in his autobiography. "None of our neighbors were willing to be tied to the care of a dairy herd, so that we had a virtual monopoly in this field." McCollum's father was not the passive sort of farmer, indifferent to his herd. Recognizing that cows were money makers, he painstakingly measured output and culled his herd, selling the "poor producers." Everyone, boys included, worked at churning butter for the market, which delivered bounty every two weeks. McCollum fed calves skim milk, a lesson in nutrition he would not forget; he also remembered teaching them how to drink from a pail, the foundational act of generating a milk supply for people. Later the scientist credited the many processes he witnessed on the farm, including the making of butter, as training in observation for the laboratory. And there is no doubt that McCollum owed much to butter later in his research career.[13]

McCollum proved to be a determined student at the University of Kansas, lunching at his laboratory desk and reading beyond assigned texts so that he could excel. Admitted to graduate study in organic chemistry at Yale, he strained to master his new environment, working long hours, reading faculty papers assiduously, and competing for prizes to prove his worth. It is clear that these years were not easy for him. To earn money, he taught chemistry courses at the local YMCA, while relying on the no-meat meal plan at the Yale dining hall. Subtle signals suggested second-class status. Upon arrival, he found himself working in the undergraduate laboratory, rather than in the more private spaces allocated to other doctoral candidates. One professor suggested that he had better start working on foreign languages, assuming that he was deficient. (McCollum had already taught himself enough German and French to pass the exams.) And in his final year, on completion of his research, another professor announced that three years was too short a time for McCollum to receive his degree. (The dean weighed in, and McCollum got his degree that spring.) While McCollum's cohort pursued positions in medical research and teaching, the former farm boy found himself back in the laboratory, learning biochemistry as a postdoctoral student. One year later, his mentor, Lafayette B. Mendel, a German émigré a mere five years older than his charge, found a position for McCollum at the College of Agriculture in Wisconsin. Perhaps this is where Mendel, a highly ambitious wunderkind in his own right, thought the painfully striving student from Kansas belonged.[14]

The two places could not have been more different in spirit. Wisconsin researchers could hardly suppress their contempt for the assertions of pure scientists from their citadels on the East Coast. McCollum's former teacher at Yale, R. H. Chittenden, received a verbal beating in the pages of *Hoard's Dairyman* after the esteemed professor pronounced borax "harmless" as a milk preservative. Wisconsin researchers knew the chemical to be injurious when ingested by animals. Stephen Babcock, sporting invulnerability as a world-famous inventor, stood up to attack the Yale professor's laboratory data.[15] But the pecking order in the world of science remained an inescapable reality. Female students, for example, were often steered out of the organic chemistry lab and into the field of nutrition, where they would commandeer classrooms of women studying the field of home economics. Animal nutrition in Wisconsin may have been more marginal than that.

Events would prove Wisconsin to be a frontier in more ways than one. There, McCollum could combine his past with a future in biochemical research. Certainly his trademark method, a "biological analysis of foodstuff," amounted to a carefully applied system of feeding farm stock. McCollum must have known the farmer's proverb, "A cow milks through the mouth"—in other words, a cow's output of milk depends on what she eats. Once he arrived in Wisconsin in 1907, he encountered the textbook version in W. A. Henry's *Feeds and Feeding* (1898). Emblazoned on its title page was another proverb: "The eye of the master fattens his cattle," which Henry translated as "stock feeding is an art and not a science." In midwestern fashion, he argued that practical experience and the wisdom that came from it—the farmer's "art"—should take precedence over information handed down from laboratories. The tension between pure and applied science was as old as the American nation, but McCollum was about to initiate an unusual moment of dialogue, when the two branches would find themselves working on the same problem.[16]

The orientation at Wisconsin was more profound than its folksy antipathy to the eastern academy suggested. It embodied a kind of skepticism that McCollum shared, a sense that the cherished truths of laboratory scientists were not the whole story. Mystery—or perhaps just nature's unpredictable potential—lurked at the edges of the hallowed truths of contemporary science. Henry, a noted advocate of feeding experiments, argued that something was missing from the pure chemical analysis of nutrients, which farm animals got from foodstuffs in their original form. In feeding animals oats, for example, field researchers found that the actual plant delivered nutritional "value beyond that shown by chemical analysis." This same skepticism was cause for a local joke known across the Madison campus, originating with dairy researcher Stephen Babcock. Offering advice to W. O. Atwater, the leading human nutritionist at the turn of the century, Babcock pointed out that, according to chemical analysis of its nitrogen content, bituminous coal would make an excellent food for pigs. One can imagine the gag itself acting as a litmus test of loyalties. Laboratory or barnyard: depending on which was your true home, you might chuckle or not.[17]

Enter the cow. In his autobiography, McCollum attributed his move to Wisconsin to his pursuit of the beneficial bovine. "In my opinion the cow experiment at Wisconsin clearly indicated that something fundamental remained to be discovered," he wrote years later.[18] Here was a land that valued its dairy stock. Debate over whether or not cows should stand in stanchions, with their heads in "neck-vises," had reached the level of the state assembly in the 1880s and 1890s. The dictum "Speak to a cow as you would to a lady" could be seen on banners at meetings of farmers' institutes across the state.[19] A lot rested on milk research, which dairy interests had made a priority since the establishment of the Wisconsin Experiment Station in 1883. Wisconsin farmers added up to a forward-looking crowd, willing to join hands with a select group of laboratory scientists.

Which feeds worked best? Which milk contained the most nutrients? Some experimenters had even investigated which teats on the udder produced the highest quality of milk. The field was open for discovery.[20]

Discoveries had sprouted profusely by the time McCollum joined the faculty at the College of Agriculture. Babcock's cream separator, developed in 1890, stood above all others, though the contraption never was a priority of its perpetually busy developer. Even while other cream test methods appeared in England and elsewhere, Babcock's was, by most standards, the best of its kind. It soon became widely used in Europe as well as the United States. As a chemist, Babcock had succeeded by figuring out the correct amount of sulfuric acid and the appropriate speed of centrifugal force necessary to separate out the cream from whole milk. Here was the answer to the needs of every farmer and milk vendor: how much fat did the milk at hand contain? With a reliable cream separator, farmers could grade cows according to the quality of their milk; they could then cull their herds of "boarders," as the poor producers were known in the Midwest.[21] Milk itself was sold at prices adjusted according to fat content, so farmers and vendors would now have a reliable means of measuring their product. And finally, with general use of the invention, buyers would be less vulnerable to fraud. The problem of guessing fat content had been troublesome enough to French and Belgian consumers to compel them to buy lactometers — inexpensive and relatively crude fat measurers — from street vendors when shopping for milk in towns and cities.[22]

The gravitational pull of milk proved inescapable to everyone at the College of Agriculture at Wisconsin. Babcock seems never to have strayed far from milk and its products. One of his earliest experiments spoke to the question raised by the ancients: was milk analogous to blood? Babcock accidentally proved that indeed it was, owing to his investigation of a component called fibrin that was present in both, playing a role in whether the liquid flowed or coagulated. In another momentous discovery, Babcock and his colleague H. L. Russell identified the enzyme involved in curing cheese that was "inherent in the milk itself" and named it galactase. Revelation of its properties guided the cheese industry in determining the best temperatures at which to ripen cheese. Along the way, Babcock also ascertained that lactation occurred at the time of milking, a discovery that had practical as well as philosophical value in a state deeply devoted to bovine well-being. Such achievements even the gods would have applauded.[23]

Soon the pull would draw in McCollum. But while Babcock had initiated a method of feeding cows and heifers a single-plant diet, a "radical departure" in the field of biochemistry, McCollum's first historic contribution was the idea of employing rats as his laboratory subjects.[24] German biochemists were currently experimenting with mice, owing to their rapid growth rate and relatively short life span; but rats, according to McCollum, presented the assets of "convenient size" (they could be grabbed easily for daily weighing), "omnivorous feeding habits" (they would cooperate in the face of steady changes of menu), and the fact that they had "no positive economic value." McCollum first confided his wish to Babcock, who encouraged him in his plan, despite the veto of a disgusted dean. The rat, after all, ranked as public enemy number one in the farming community. As McCollum recalled years later, the dean worried that if "it ever got noised about that we were using federal and state funds to feed rats we should be in disgrace and could never live it down." With Babcock's secret blessing, McCollum set about catching rats in the horse barn of the Experiment Station anyway. When these proved too "savage," he spent six dollars (roughly $140 at today's value) for twelve albino rats from a "pet-stock dealer" in Chicago. His experiment began in January 1908, the first of its kind in the United States.[25]

All experimenters at Wisconsin sought, in effect, pathology induced by deficiency: if animals ate a dramatically restricted diet, could they grow and (the definitive proof of health) procreate? Babcock's heifer experiment had shown that wheat or oats alone rendered cows small, mangy, sometimes blind, and unable to produce healthy or even living offspring. But after four years, McCollum had "lost confidence" in the drawn-out project. Instead of employing Babcock's "single-plant source" method of testing nutrients, McCollum wanted to try combinations of purified substances. With rats as subjects, the effects would manifest themselves in quick time.

By now, McCollum had acquired an assistant, Marguerite Davis, a Berkeley graduate who "worked full-time for five years without any pay" ("except for Saturday afternoons and Sundays"), caring for the rat colony and running experiments while McCollum continued to do his duty with the heifer experiments. Davis's volunteer work appears unmistakably as exploitation by today's standards: she was "physically handicapped" from a childhood accident and she was female. The dean "felt she was not sufficiently trained to be placed on the staff" when McCollum pressed for a salary for her. (McCollum showed that he, too, was a man of his time when he noted that "she did not need to support herself.") In her sixth and final year, he reported, "I got six hundred dollars for her." That would be roughly $13,500 in today's currency. In his autobiography, he referred to her appearance one day in his laboratory as "nothing less than a blessing from Heaven," and he was right.[26]

Like most animals at the Experiment Station, the rat colony often dined on some kind of milk product. McCollum made mistakes at first, believing that taste mattered to the rats (he and Davis painstakingly flavored their chemical rations with bacon and cheese) and not realizing that the rodents were also dining on their own excreta. After three years of dogged effort, something came of adding butterfat and egg yolk to various foods: in these instances, the rats always showed growth. Comparing butterfat and egg yolk with olive oil and lard, McCollum and Davis found markedly different results: rats on a diet of lard or oil failed in health. Something special lurked in the two yellow-colored fats, which, even in tiny amounts, promoted well-being. By 1912, the experiments had yielded the exciting presence of what they would later call fat-soluble vitamin A.[27]

Appreciation of the weightiness of this discovery requires some effort for the twenty-first-century supplement-saturated reader. At this point, not even McCollum and Davis comprehended the full impact of their findings. They were struck, first of all, by the simple news that not all fats were alike. According to received wisdom since the time of Justus von Liebig, chemists believed that carbohydrates (sugars and starches) and fats delivered the same nutritional value, namely, "energy." The prince of nutritional substances, proteins, the "flesh-formers," reigned supreme in the tripartite biochemical universe. A half century of attention had unlocked the multidimensional character of protein, revealed through the discovery of amino acids. In the meantime, the stepsisters of the chemical family continued to strike poses as boring monoliths until the early twentieth century. Now McCollum and Davis were finding that fat, of all things, was a many-faceted substance.

As for the concept of "vitamin," the word first appeared in scientific

Portrait of Elmer V. McCollum (1879– 1967), studying his own textbook on the history of nutrition. (Photo: Journal of Nutrition *100 [*January 1970*]: 1–10, American Society for Nutrition; used with permission)*

literature in 1912, the same year in which the Wisconsin researchers had realized their discovery. Polish biochemist Casimir Funk would coin the term, though with an incomplete understanding of the mechanisms at work. Funk related necessary elements of the diet to a particular group of amines involved in preventing beriberi; hence, he combined "vita," or "life," with "amine." At the time, only twelve of the twenty amino acids of the body had been discovered. Not until 1916 did McCollum and another researcher, Cornelia Kennedy, propose using alphabetical letters to designate nutrients. By this time, he and Davis had also come up with a water-soluble "B" vitamin, which they had extracted from wheat germ and rice polishings.[28] But mysteries still remained. Researchers had not yet established exactly what function vitamins performed. It took several more years before they comprehended their role in enabling the body to absorb nutrients from food.[29]

As early as 1893, readers of *Hoard's Dairyman* learned that "there is something about milk which is nearly impossible to replace, that stimulates assimilation and digestion and promotes growth." The premise may have been imperfectly formed, but the outcome counted more than anything else. According to Theodore Louis, legendary midwestern pig-raiser, "hogs responded better to even the best grain feeding when they had a skim-milk supplement."[30] Had farmers long understood something that biochemists didn't yet "know"? Some twenty years later, McCollum's mentor at Yale, Lafayette B. Mendel, reckoned the same thing, but from the chemistry laboratory. Collaborating with Thomas Osborne, who had directed McCollum during a brief stint at the Connecticut Experiment Station in 1906, their experiments with "protein-free milk" led them to conclude that "milk food contains something that is essential for *both* growth and maintenance." Even Yale scientists were now willing to turn their attention from protein to fat.[31] (The team challenged McCollum and Davis in their claim as the discoverers of fat-soluble vitamin A, but a protracted debate in the pages of professional journals decreed McCollum and Davis the winners.)

What farmers understood by practice and some chemists knew by experiment, many others continued to resist, *tout court*. In 1917, when McCollum moved to the new School of Hygiene and Public Health at Johns Hopkins University, he found himself on the front lines of conflict. After hearing McCollum deliver an impromptu discourse on supplementary nutrition at a dinner party, a statistics professor responded flatly, "If you have long-lived ancestors *it doesn't make any difference what you eat*." McCollum recalled how the man had pronounced the last words of the statement "slowly, singly, and with emphasis."[32] At the time, controversy was also swirling around the idea of "deficiency disease," prompted by a new theory about the cause of beriberi. A Dutch doctor, drawn to Southeast Asia by colonialism, had experimented with chickens and a restricted diet of polished rice, which produced the neurological symptoms of beriberi. Mainstream medical men resisted, arguing that the human disease was either infectious or caused by an undiscovered poison in the environment. But researchers in physiology could turn the symptoms on and off by adding and subtracting the chemical contained in rice polishings from the diet of the chickens. While pathologists shunned animal diet experiments, pediatric specialists were more open minded. At Johns Hopkins, researchers working on childhood rickets later collaborated with McCollum and nailed down its nature as a deficiency disease involving a hitherto unidentified substance: they called it vitamin D.[33]

By 1912, McCollum had arrived at the idea of supplementary relations between foods: the nutritional value of one food could supply factors lacking in another when eaten together. Foods could interlock like pieces in a puzzle.[34] McCollum rested part of his case on the pioneering work of British biochemist F. G. Hopkins, who suggested the idea of "accessory" articles of diet. The primary example Hopkins used was milk, which, when added to inadequate feedings of purified substances, worked its magic by producing growth in laboratory animals.[35] As Prout had shown long ago, milk offered protein, carbohydrates, and fat (recently promoted). In *Newer Knowledge,* McCollum hoisted the tripartite banner again, this time with a different emphasis: "the composition of milk is such that when used in combination with other food-stuffs of either animal or vegetable origin, it corrects their dietary deficiencies." His specific recommendation: everyone should drink a quart a day.[36]

The gospel of milk was going places by 1918, uniting disparate fields of research while extending to a public beyond the scientific community. The growing field of pediatric medicine, now convinced that milk was essential in preventing childhood deficiency diseases, embraced McCollum's message with enthusiasm. At the invitation of Milton Rosenau, author of *The Milk Question,* McCollum delivered the prestigious Cutter Lectures at the Harvard School of Public Health in 1918. Rosenau then arranged to publish the lectures; hence, the appearance of *The Newer Knowledge of Nutrition.* Not least of McCollum's followers was Herbert Hoover, then head of the United States Food Administration, who shared the dais with him at a theater in Philadelphia that same year. Food was a concern in the final months of the war; Hoover called for international cooperation and concern for "starving people in Europe." McCollum followed with twenty-five minutes on "good and bad combinations of foods and the results of adherence to faulty diets." Included was his infamous condemnation of steak and potatoes, but there was also much positive advice in the talk. Using knowledge of complementary foods, McCollum recommended pairing foods that "make good each other's deficiencies." Above all, he called "attention to the value of milk and green, leafy vegetables as supplements to the staple foods"—the "protective foods," as they became widely known. Hoover responded with as much enthusiasm as the audience. Through the Food Administration, Hoover arranged for McCollum to take his lecture to eighteen American cities.[37]

Milk and good nutrition—the two subjects spread like wildfire across the United States in the years following World War I. *Newer Knowledge* sold fourteen thousand copies in its first three years and went into five editions by 1939. In 1922, McCollum began his long association with

Advertising by American farm machinery companies capitalized on the role that the United States played in exporting badly needed food to Europe and other parts of the globe after World War I. The Holstein cows pictured in this flyer from 1922 contributed their mite to the surge in dairy exports. (Wisconsin Historical Society, WHS Image ID 4646)

McCall's magazine, eventually becoming its nutrition editor. Interviews in the *Saturday Evening Post* and the *New York Times* popularized his idea of "protective foods," along with particular recommendations for menus including dairy products. State and local authorities sponsored milk promotions and even "Milk Weeks," when its consumption became a virtual social duty. Something about McCollum's message resonated with a public hungry for information on how to avoid sickness and achieve something known as health through diet.[38]

It is no exaggeration to say that McCollum began to see the world through the lens of milk. The first edition of *Newer Nutrition* offered a hefty dose of American jingoism, not unusual for the period, yet distinctly framed by a dairy farmer's consciousness. "Mankind may be roughly classified into two groups," the author intoned in his discussion of the urgent need for attention to "public health." On one side stood the Chinese, Japanese, and "peoples of the Tropics," who consumed plant leaves as "almost their sole protective food," with a few eggs to round out their nutrients. On the other side were Europeans and North Americans, eaters of leaves, but more dependent on milk and milk products for "a very considerable part of their food supply." (He estimated that part at between 15 and 20 percent.) The consequences of diet were momentous:

> Those peoples who have employed the leaf of the plant as their sole protective food are characterized by small stature, relatively short span of life, high infant mortality, and by contended [*sic*] adherence to the employment of the simple mechanical inventions of their forefathers. The peoples who have made liberal use of milk as a food, have, in contrast, attained greater size, greater longevity, and have been much more successful in the rearing of their young. They have been more aggressive than the non-milk using peoples, and have achieved much greater advancement in literature, science and art. They have developed in a higher degree educational and political systems which offer the greatest opportunity for the individual to develop his powers. Such development has a physiological basis, and there seems every reason to believe that it is fundamentally related to nutrition.[39]

Supplied with milk, Americans and Europeans were taking their gospel to the rest of the globe. "There is no substitute for milk," McCollum wrote, "and its use should be distinctly increased instead of diminished, regardless of cost."[40]

The timing could not have been better for dairy farmers. The war would make tremendous demands on American farms, and once European agriculture rebounded, a new abundance would be in search of markets. Production of milk expanded in America, despite the loss of European buyers. Years later, McCollum pointed out that "between 1919 and 1926 the national production of milk products increased by one-third and that of ice-cream by 45 per cent." The future of the powerful white liquid appeared bright. A tide of milk was flowing across the land.[41]

CHAPTER THIRTEEN

Good for Everybody in the Twentieth Century

Milk won pride of place as a public necessity throughout Europe and America by the 1930s, but not without concerted effort on the part of reformers, many of them women. The new nutritional intelligence about milk became part of an international campaign to improve the health of the masses. McCollum's teachings about milk as a protective food, equipped with all-important vitamins, resonated in a world preoccupied by worries about hunger and food. Like Pasteur's identification of microbes, it wasn't so much what was discovered as when the discovery became widely understood. The First World War had altered public awareness, limiting the availability of necessities and carving out a role for governments as managers of supplies. State agencies quickly absorbed the lessons of McCollum's *Newer Knowledge* and inscribed them in new policies. A modern assumption had emerged almost imperceptibly: certain commodities were now understood as essential for entire populations, and governments were ready to assist in providing them.

Milk was more than ready to assume its modern identity: blessed by pasteurization and verified by biochemists, the liquid warmed to its new profile as good for everybody. Processed and canned by manufacturers, purchased by governments for armies and navies, and transformed into butter, cheese, chocolate, and ice cream, one could say that milk had spread just about everywhere. Consumers had learned to expect easy access to what was now seen as an entitlement, made possible by the handmaid of the twentieth-century state, modern science. The campaign for safe and affordable milk joined forces across oceans. In 1943, Winston Churchill uttered his famous dictum, broadcast over radio to approving listeners: "There is no finer investment for a community than putting milk into babies."[1]

The new century was in fact witnessing a tectonic shift in Western food symbolism. The first tremblings were felt in the food protests of postwar Britain, where the pint of milk was displacing the loaf of bread as the emblem of the people's entitlement to basic sustenance. A scarcity of essentials soon schooled the British population in market operations affecting critical commodities like milk. Suspicion of market practices after the war motivated American consumers to insist on lower-priced staple foods, too. They were joined by reformers eager to address the alien eating habits of urban immigrant populations. Mixed with maternalist arguments about the rights of mothers and children, the case for affordable milk acquired an aura of invincibility. With the hand of the market so visible in bringing milk to the table, why, postwar citizens asked, was this familiar gift of nature held hostage by big business? Like no other foodstuff, milk commanded special treatment.[2]

We tend to think of the English countryside as a timeless wellspring of dairy products, but the consumer of 1900 would have known better than that. Free trade in the milk industry had brought per capita consumption of fresh milk in Britain to the lowest in Europe: for 1900–1902, 14.5 gallons per year, and for 1910–12, a modestly increased 16.3 gallons. In Saxony, by contrast, per capita consumption was as high as 46 gallons, and in Sweden and Denmark, 40 gallons. Derek Oddy has shown that the British working classes before the war consumed "less than half of what New Yorkers and Parisians consumed at the time." The reason is not hard to fathom: prices for the native treasure were prohibitively high. In Bradford, a northern industrial town, 22 percent of the population purchased no milk at all. So it is worth contemplating how a nation like Britain managed to lay claim to its own natural resource in the dairy.[3]

The First World War dealt a heavy blow to dairying everywhere; in the competition with military needs, dairy's demand for backbreaking labor, large quantities of cattle feed, and transport facilities lost out. Herds across Europe were destroyed, either for beef or by belligerence. During three months of German occupation, the number of Belgian dairy cattle fell from 1.8 million to 700,000.[4] Consumers, meanwhile, paid steadily higher prices for fresh milk, or they fell back on the canned variety. Suspicions of profiteering grew as liquid milk distributors continued to behave as businesses like any other. From 1917 to 1920, however, women purchasers wouldn't stand for it.

The time was right for guardians of the family to assert their rights as consumers. Here was a case, they argued, of profits reaped at the expense of common and indefensible human need. Women across Britain organized boycotts of fresh milk, holding out for an entire week in one town, and in another designing a "muddling strike," dramatically changing their purchasing patterns from one day to the next to confuse distributors and suppliers. In Bristol, women marched under the slogans "We Want Cheaper Milk" and "God Save the Babies." In London, hundreds of thousands gathered in that revered place of rallies for social justice, Hyde Park, to hear about and protest the big food trusts, including the alliance of milk distributors. A new sense of citizenship was being born, one that linked democracy to the right to affordable food and the assurance by government that "transparency in pricing mechanisms and profits" would apply to food, particularly milk.[5]

Women proved to be masterful international networkers during times of crisis. British women learned that mothers in New York City "had been taught that milk was a necessary and bought high-grade milk in consequence." A pamphlet circulating in Manchester put the case in plain language: "British mothers too have a right to clean milk for their children, and the government must see to it that arrangements are made, not only for the proper housing of the people, but also for the proper production and handling of the children's food—clean milk."[6] Wartime regulation of foodstuffs had revealed the logic of the state as the citizen's protector: "If men could be commandeered for the Army, surely milk ought to be commandeered for the benefit of the health of the community," a woman cooperative member insisted.[7]

Governments soon learned that they, not the free market, would be the elected guardians of a cheap and plentiful milk supply. The British government took control over the supply of milk in the last year of the war, but that did not solve the obvious problems of high prices and distribution. Vociferous members of the Women's Cooperative Guild, a British organization of working-class women invested in local cooperative stores, made headlines in their unrelenting protests. Their testimonials showed a remarkably deep distrust of wartime economic measures, which were meant to discourage too much civilian consumption of essentials. The Guild's newspaper outlined how the price of milk came to be so high: when the wartime government fixed prices at high levels, farmers incorrectly understood such prices to reflect limited demand, so they curbed production. But Guild women insisted that milk belonged to a special category of commodities and should be managed for the good of the people. An editorial remarked in December 1919: "And in the newspapers the other day we read of thousands of gallons thrown away. It makes one's head dizzy puzzling out what or who is to blame. It is heart-breaking to go amongst the poor babies and realise their need of this God-given food that is being wasted and quarrelled about. We do not want to blame the farmers unduly, but for Heaven's sake let us 'get a move on' and see to it that the price is dropped. If the Government has made a mistake can they not say so, and try some other way?" Advocates for affordable food for all pulled out all the stops, describing hungry toddlers who looked "wistful" at the mention of milk and asking if readers recalled the "creamy rice or sag puddings" of childhood. Every sentimental chord attached to milk was sounded.[8]

Tensions over food were simmering in the United States, too. Europeans viewed Americans as too tolerant of big businesses (five large midwestern meatpackers, nicknamed "the American Meat Trust" in Britain, enjoyed particular notoriety abroad), but even American consumers refused to tolerate the sharp increases in food prices at the end of the war: in Boston, New York, and Philadelphia, women and men took to the streets. Especially in February 1917, crowds called for local control over pricing in retail stores. Desperate demand for potatoes resulted in guards stationed around a shipment sitting in Boston harbor. Like British women, American mothers claimed a voice as guardians of children, organizing against the intrusion of social workers and demanding breakfasts for schoolchildren. Many buyers simply ceased to purchase milk and "accused both the producer and the distributor of profiteering." Concerned that the rise in prices was hurting the large number of poor inhabitants in New York City, the mayor appointed a committee to investigate milk consumption in different neighborhoods, comparing results to minimum standards recommended by "pediatrists."[9]

The Rushden Cooperative Society's No. 2 Milk Float, with Miss V. Reid in Crabbe Street, 1920. (Photo from the collection of Eric Fowell with kind permission, and the Rushden and District History Society Research Group)

But the milk crisis was bigger than the challenges presented by any single locality. Governments everywhere faced a lengthy string of petitioners: besides dairy farmers asking for help and milk distributors asking for regulation, there were manufacturers of condensed milk, ice-cream businesses, butter and cheese companies, along with regional and municipal authorities and, of course, consumers' leagues, all demanding protection from high prices for milk. Given "the importance of the problem and the increasing danger of the situation arising from the uncontrolled controversies," the United States government felt compelled to step in. In November 1917, the Food Administration appointed commissions across the nation to act as tribunals in determining fair prices for dairy products. This solution established an important precedent for future negotiations with dairy farmers.[10]

It was clear that a fear of "milk famines" haunted wartime governments. Yet that is precisely what occurred as war requisitions placed heavy demands on available supplies. The United States Food Administration watched high-priced milk dam up on farms and feared, quite rightly, that unpurchased milk would signal farmers to cut back precious herds. Yet how could mothers and babies do without the liquid? In Britain, the Ministry of Food increased available milk by directing supplies away from cheese and processing, toward the liquid milk market instead, and localities were empowered "to distribute milk at reduced prices to mothers and children." The product still wasn't cheap — at ten pence a quart, it cost more than twice the price of the prewar commodity — but a semblance of control coaxed buyers back into the market. Americans, meanwhile, adjusted to paying around fourteen cents a quart in cities like Chicago, which would be $6.84 in today's market, using the unskilled wage as the best indicator to measure relative worth.[11]

Cheap liquid milk for popular consumption could not be the first priority of dairy farmers, even under ordinary circumstances. The cost of producing milk was high even before they factored in labor, which they often squeezed from family members. Beleaguered farmers were forever assessing different options for their supply, each with different rates of return. Liquid milk for direct consumption competed against milk for processing into cheese, butter, or condensed milk. Prices for milk destined for cheese, for example, would vary according to what kind of production methods were nearby and, more remotely, the degree of foreign competition challenging the final product. At the same time, payments for milk for direct consumption were hostage to fluctuating costs of transport and steadily evolving hygiene regulations. To make matters more complicated, milk continued to be a seasonal product, despite the fact that cows could be kept in milk through the winter. Production reached its peak in the spring and tailed off in December, so farmers needed to find profitable outlets for surplus while surviving the inescapable dearth of winter months. And finally, there was the cost of feed for cows. This, too, fluctuated according to all of the same variables, particularly now that commercial concerns were involved in manufacturing high-nutrient "cake" for cows. It is no wonder that the revolution in transport at the end of the nineteenth century wreaked havoc on the dairying industry everywhere, simply because a winning competitor could turn up on the doorstep even in hard times, burying local farmers in debt in a hurry.

A specially inspired invisible hand nevertheless seemed to be preparing the way for a new age of milk. Even while the war was sucking up supplies, progressive-minded school reformers in New York City sized up schoolchildren, quite literally, and pushed for an expansion of a recently endowed school lunch program. Hunger was on the minds of the entire nation during the economic downturn of 1914–15 and in the critical year of 1917, when an anthropometric "craze" seized health reformers intent on weighing and measuring children from New York's poorest neighborhoods. Many immigrant children, Harvey Levenstein points out in a dietary history of America, were designated "underfed" or, in the latest semantic shift, "malnourished," a term hinting of a distinction between good and bad food. Inspired by the latest findings in nutrition, a new generation of social workers saw these offspring through the lens of deprivation — ironic, Levenstein points out, given our own stereotypes of Italian and Jewish mothers. At times, standards were laughably inappropriate: when "key indicators" devised by the Carnegie Institution in Dumferline, Scotland, were applied to Italian and Jewish immigrant children, many subjects were found deficient in height, eyesight, and muscularity, not to mention "rosiness of complexion." The findings forced health officials to conclude that over 10 percent of the school population was "undernourished." Clinics and classes in nutrition materialized to deal with the outbreak, whether apparent or real, of dietary deficiency. By 1917, fully a quarter of New York City's 1 million schoolchildren fell into the category of "severe malnutrition."[12]

The United States Children's Bureau was prepared to help out. By this time, progressive women had taken charge of the Children's Bureau, a newly created government agency, and adeptly engaged modern media to get their message to the people. Under Julia Lathrop, the bureau turned out a series of pamphlets for mothers, priced at a reasonable (at least for middle-class parents) five cents a copy, announcing the latest findings of nutritional scientists, mostly the work of Elmer McCollum. Dorothy Mendenhall's *Milk: The Indispensable Food for Children*, appeared in 1918, before the war in Europe had come to an end. The voice in the booklet was confident, passionate, and maternal. Its message, fulsomely patriotic, was simple. Children, Mendenhall argued, belonged to a special category of society, demanding attention and protection for the sake of the nation's future. Unlike adults, the young could bear no curtailment of their nutritional needs. "Milk," she intoned, "has no substitute in the diet of the child."[13]

Warning that "our own great cities" were showing deleterious effects of "underfeeding," Mendenhall lamented the low per capita consumption of milk in the United States. According to the Department of Agriculture, the nation produced 1.15 quarts per capita per day. Mendenhall argued that only 40 percent was consumed as milk, and this inadequate amount was likely to decline, owing to the inevitable contraction of the cow population following the food crisis of recent years. The rise in price by two cents per quart in 1917, she argued, had led to an alarming decline in consumption in the United States. In a rather bold move, she suggested that Americans could "profitably study the way that Germans have controlled the milk situation" by protecting "the nutrition of . . . children under 6 years of age by fixing the price of milk early in the war, and insuring the use of milk for nursing mothers, weaned infants, young children, and the sick." Mendenhall demanded fixed or controlled prices and careful monitoring of the national supply so that enough liquid milk was available to the young at all times.[14]

Like an enormous river drawing from tributaries at every turn, the cause of milk gained an ever growing number of advocates. The precise timing of this development is proof of the strange and compelling power of milk itself. Even before the discovery of vitamins, that is, before the science of nutrition had caught up with popular interest in the commodity, milk was deeply desired across whole populations. Women reformers in Europe and the United States deserve the credit for laying the groundwork through philanthropic organizations and neighborhood visiting. They had also been discussing "domestic science" for at least sixty years, long before the formal discipline gained recognition in institutions of learning. Milk occupied a privileged place in the food chain, associated as it was with mothers, nurturance, and wholesome "natural" food. But decades of practical experience had taught middle-class women that dictating domestic dietary practices to poor women stood little chance of changing limited circumstances. Determined reformers decided that providing schoolchildren with milk was an efficient, direct way to improve the health of poor families. And poor mothers often proved to be active participants in the programs.[15]

Milk given to needy children at school presented its own kind of problem, known to researchers today as "the substitution effect." If poor parents understood that their children were receiving food at school, they might withhold food at home in order to economize. At Toynbee Hall, a London settlement house project run by (male) Cambridge University graduates, however, administrators were reassured by a trial run of free school milk. The basic drink hardly registered as "food" in the minds of disadvantaged parents, according to the dispensers, so institutions could give out remedial milk without worry. Meanwhile, earnest London teachers, determined to ply their students with the product, circumvented public debate by organizing "milk clubs" at their schools. By collecting halfpennies from children, they established voluntary programs of milk at morning break.[16]

Institutions full of hungry children presented a research paradise for nutrition scientists. Once installed at Johns Hopkins, Elmer McCollum launched a dietary study at a Baltimore orphanage of black children to see if a quart of milk a day would result in appreciable gains in growth. Some children showed signs of rickets and tuberculosis; others, probable signs of neglect, even from the institution. Dividing a group of eighty-four children in two, McCollum gave half a milk supplement each day, reconstituted from powdered milk. The others lived on the regular orphanage diet of cereals and soup. The children consuming milk supplements benefited measurably from the newly anointed liquid, but the experiment seemed compromised on many methodological grounds. Looking back on the project, one researcher suspects that results were muted, probably owing to the "Hawthorne effect," the improvement of institutional feeding provoked by the presence of researchers. Nevertheless, McCollum reported the positive results, apparently with pride, to a meeting at the World Dairy Congress in 1923.[17]

Meanwhile, nutrition research flourished in Scotland, where a strong medical tradition happily converged with a tradition of diligent dairying. Scotland was home to the British Isles' own answer to Elmer McCollum, Sir Frederick Gowland Hopkins, a biochemist and trained physician who had been pointing to an unidentified nutrient in milk in 1912, while McCollum was carrying out his rat experiments. (British historical literature today continues to give McCollum short shrift in favor of the native hero, who was knighted in 1925 and awarded the Nobel Prize in physiology or medicine, along with the Dutch physician Christiaan Eijkman, in 1929 for "discovery of the growth-stimulating vitamins.") Hopkins's student Harold Corry Mann worked with a captive group that ranked as literal poster children in the world of British philanthropy: namely, the "colony of boys," ranging in age between six and twelve years, maintained by the famous Dr. Bernardo in Woodford, Essex, ten miles from London. For three years, the boys ate supplemental servings of milk and butter, along with allotments of sugar, margarine, and casein. Measurements of height and weight proved that milk and butter fostered the greatest gains in weight (averaging eleven ounces every six months) and height (0.38 inches for milk and 0.17 inches for butter). More studies showed that biscuits and milk made children grow, and even "separated" (skim) milk, usually fed to hogs, made children grow. And so the data continued to flow and accumulate.[18]

Judging from the public expression of sentiment for milk during the postwar years, researchers needn't have worried about convincing the public with data on the benefits of milk. The symbolic power of cows was hovering over Europe, ready to inspire activism among the many true believers already on the ground. When the Women's Cooperative Guild learned that the reparations treaty would force Germans and Austrians to surrender 140,000 milch cows, they quickly sent delegates to Paris to plead with leaders of state to change their minds. A "Miss Marshall," a working-class woman, reported in the Guild's newspaper on her experience there: "I was very nervous indeed, but I felt it my duty to go, and I am sure my fellow guildswomen will agree with what I have done. To those who know the conditions of child-life in Germany and Austria today, to take away the 140,000 milch cows is a most cruel act, and will mean the deaths of 600,000 babies! I told Sir John Bradbury that women realised what this condition meant as no man could. Women knew what it was to go through the pains of child-birth and to feed their children at their breasts, and I pictured what our feelings would be if we had to put our children to bed crying of hunger. I told him we thought the punishment should not fall on helpless babies." Her account, titled "Massacre of the Innocents," had the ring of recent debates for pure milk, but it also displayed the trademark of a distinct, new outlook: the conviction that commonalities among women made them suitable spokespersons for basic human needs. They spoke out beneath a banner of milk.[19]

The time was also right for cows to become veritable vessels of inter-national good will and nutrition. Edith Pye, a British midwife decorated by the French government during World War I, devised a plan to provide for milk-deprived children of Vienna throughout most of 1921 and 1922. With the help of her friend, physician Hilda Clark, Pye raised funds from British and American sources for a carefully worked out scheme. First, from Switzerland, they obtained donations of Swiss brown cows;

what better act of charity could that nation perform? Next, with their donated funds, they purchased cow feed from Croatia. And finally, they distributed the resulting milk free to Viennese children under the age of four.[20]

Amid so much sentiment and science, something equivalent to the perfect storm was happening in food industries. Wartime had fostered unprecedented consolidations everywhere. Consumers in the United States witnessed the emergence of Edward F. Hutton's General Foods and J. P. Morgan's Standard Brands, huge enterprises that controlled several major companies producing different food products. The milk industry was no exception. According to Levenstein: "Giant holding companies such as National Dairy Products (Sealtest) and Borden's used exchanges of stock, leverage, pyramiding, and other new devices of the corporate finance markets to come to dominate milk distribution in most major cities. In one week alone, Borden's purchased fifty-two concerns."[21] In London, three wholesalers amalgamated to create United Dairies in 1915; by 1920, their operations controlled 470 shops. In the wake of price mediation by governments at the end of the war, the big companies did not win the right to price-gouge the consumer. They did, however, enter into a nutritional social contract: given the evidence for a need for wider milk consumption, corporations joined hands with nutritionists and medical researchers to launch a massive campaign to promote their products. Here was victory in new form: an unprecedented economy of scale.[22]

Consumers everywhere witnessed a snowfall of propaganda documenting the miracles worked by milk. Organizations of dairy interests, staffed in part by home economists, medical experts, and nutritionists, generated campaign posters, pamphlets, and recipe booklets to get the message out. Buoyed by their own funded research, the British National Milk Publicity Council, established in 1920, united producers, distributors, and representatives of the medical profession. The NMPC generated "educational publicity," which was sent home to parents: your children will grow by drinking milk, the leaflets argued, and here is the scientific data to prove it. The pattern was repeated across Europe and North America. Not surprisingly, advertising firms were enlisted, too.[23]

American dairy ephemera used every route possible into the imagination of consumers. Pamphlets touted science and medicine as the handmaidens of dairy agriculture. Readers found reassurance of purity in learning that "graduate veterinarians" at Borden's "make frequent physical examinations of the cows in all herds" supplying the company. Users of the "Treasured Recipes" booklet of Sanitary Farm Dairies found photos of the company officers inside the back cover, with an invitation to visit the "beautiful and modern plant" in Houston. ("You are always welcome here.") An accompanying photo showed Holstein cows wading across a brook in a landscape that looked decidedly un-Texan. Yet, as we know, the idea of milk thrived on contradiction. Twentieth-century milk strove to fuse the bucolic and the modern. The rhetoric of a New York State Bureau of Milk Publicity recipe book reminded readers that their daily delivery came from "one of the most elaborate and efficient distributing systems in the world," from "millions of farms to the very doorsteps of millions of consumers." Milk traveled "many miles at maximum speed," delivered "every morning, regardless of storms, floods, or cataclysmic events." (An image of a milkman in a hurricane comes to mind.) "The story of the milk industry," it pointed out, was "an epic of modern times."[24]

All of these efforts targeted mainly women, so why not trot out the age-old allusion to Cleopatra's secret? "Milk Dishes for Modern Cooks" promised that milk contained "*elements that preserved the characteristics of youth.*" (The same publisher offered "The Milky Way," described as "a diet and beauty book," free of charge.) Milk was both valuable protein and end-lessly protean. While cribbing shamelessly from McCollum's dicta about protective foods, the booklet's recipes incorporated milk and (less often) cream into vegetable and meat dishes of every variety. The culinary econo-mies of the Depression era are evident in numerous meatless dishes, in spite of the relentless optimism of advertising lingo.[25]

A more subtle message had to do with a social contract holding together the alliance of milk producers and distributors. Readers heard about the distinction between "fluid" milk meant for liquid consumption and the rest of the farmer's supply, which fetched a lower price and ended up in manufactured products, such as butter and cheese. This lesson presumably meant to encourage consumption of the liquid, which, the reader was told, was the farmer's primary source of income. Producing milk for society was a labor-intensive, year-round commitment, one with moral overtones. Purchasers needed to know this. "*The marvel is that milk can cost so little,*" the booklet emphasized. Even if consumers, on their side of the issue, believed that milk ought to be low-priced, they should marvel at the fact that it was.[26]

The cardinal duty for mothers in this new child-friendly world was to make milk palatable to children. Whatever it took to coax the liquid down throats, sugar notwithstanding, was fair play. Home economists advised mothers to mix chocolate into milk, and additives like cocoa and malt were vaunted for their own health properties. Horlick's found a perfect niche in this environment. Clever advertising campaigns, using the jingle below, capitalized on a British pedigree:

Little Miss Muffet,
Who sat on a tuffet,
With nothing but curds and whey
Each day for her dinner,
Grew thinner and thinner,
And soon would have pined away.
While her mother sat crying,
Afraid she was dying,
A lady in satin and silk
Said, for nutriment ample,
Just send for a sample
Of HORLICK'S MALTED MILK.
There was no time to waste,
She wrote off in haste
To 34, Faringdon Road,
And a bottle post paid,
For the poor little maid,
Came swift to Miss Muffet's abode.
She liked it so well,
Both the taste and the smell,
That her joy was delightful to see;
She took it each day,
And the parents both say: —
She's now just as strong as can be.
When the news spread around,
It was very soon found
To be good for sick folk of all ages;
The relief which it wrought
And the comfort it brought
Are a source of surprise to the sages.[27]

Colored flyers offering free samples made their way under doors and into shopping baskets. Who could resist such entreaties?

The milk business and popular culture entered a phase that resembled a delicate courtship ritual. No better evidence of this appeared in Britain in the form of the milk bar. Replete with contemporary chrome fittings and a black-and-white art nouveau look, the bar offered all the service aspects of a pub, including a barmaid and cocktail-like drinks. The first establishment caused a sensation when it opened on Fleet Street, the legendary neighborhood of alcohol-girded newspapermen, in July 1935. Around fifty nonalcoholic drinks were on offer, "taking the place of 'sun-downers' and 'elevenses.'" Bars served "malted milk, yeast milk, milk cocktails, milk punch, egg and milk, milk soups, milk hot, and milk iced." The network of establishments multiplied quickly, apparently bankrolled by a single company. When Lord Astor went to the League of Nations Mixed Committee on Nutrition meeting in 1937, he boasted that more than six hundred had opened in England in just the previous two years.[28]

British men ran the risk of becoming "victims of taunts" when patronizing the milk bar. As the *Times* of London put it, such a teetotaling sort of establishment appeared "fusty and effeminate" to beer-swigging British types. But milk drinking capitalized on another association, particularly powerful after the First World War: the tie to American dietary habits. "The Americans are great milk drinkers," reported a writer for the *Sunday Post* in London. "They average a quart a day. The men will go into a bar and order a glass of milk without a blush, drink it, and walk out again with their heads held high." If British readers failed to believe this, then they only had to measure the impressive height and might of their former colonial subjects; why, they were walking advertisements for the stuff. As historian Francis McKee put it, the "example of America and all the New World's vigour, drive and wealth" gave the milk bar a popular appeal in the Old World, which in turn helped to give milk a certain manly flavor.[29]

Australia adopted the milk bar as its own, too, proving the protean nature of milk as a transplanted commodity. "As typical of Australia as the pub is of England or the drug store of America," explained the *Times,* the milk bar "changes like a chameleon with the character and development of the country, because it lives on what the country wants." By the 1960s, in its down-under incarnation, the milk bar would offer the latest high-tech machines ("whirring, restless mixers, hissing, fervent espresso machines")

Milk Bar, Central Railway Station, Sidney, Australia, 1946. Dairy bars proliferated in many Western nations in the decades after World War II, when ice cream floats and milk shakes offered indulgences with a wholesome profile. (State Records, New South Wales)

and assorted milk shakes, sweets, chocolates, espresso, and soft drinks. Men and women, girls and boys of different ages congregated there; "always it is democratic." "Here in Canberra," reported a correspondent, "diplomatists and departmental heads may be seen with builders, bus drivers, and businessmen." Owned by second-generation Greeks or Italians, milk bars operated as "social levellers" in Australian society.[30]

A history of milk's fate during this era is incomplete if we look at only business and advertising. In fact, the farm politics of the Great Depression reconfigured milk's history for the rest of the twentieth century. The early years of the Depression witnessed an unprecedented food glut; stockpiles, lacking buyers, accumulated in Europe and the United States. At first, farmers reduced acreage, slaughtered livestock, and restricted imports. But governments soon calculated that unemployment, wage cuts, and shorter hours had reduced purchasing power. Consumers, it was found, were satisfying hunger with cheaper foods like bread and sugar, not milk. In the United States, the decline in milk consumption was so pronounced that the income of dairy farmers in 1932 was "nearly half that of 1929."[31]

"Farmers panicked, and in many cases violence erupted as frustrated farmers confronted milk dealers and the public." Milk boycotts flared up across the country, organized from the producers' side, in an effort to force up prices. The resulting Agricultural Adjustment Act in 1933 brought the federal government into a direct role in setting dairy prices.[32]

What made this second milk crisis different from the first was a new concept on the horizon of global food history: surplus production. In light of the obvious deprivation of a great part of the American population, the presence of such bounty stirred up emotions on all sides. Estimates suggested that 42 percent of American families received incomes below $1,000 ($13,208 by today's standards) and consumed far less nutritious commodities than they desired. The concept quickly became a lightning rod for debate. "The present market of surpluses of certain protective foods — dairy products, leafy vegetables, and vitamin-C-rich foods — are surpluses from the commercial standpoint but not from the standpoint of nutritional needs," the U.S. Department of Agriculture's *Yearbook of Agriculture* asserted in 1940. Beneath decorous pronouncements about nutrition lay an argument for and against the redistribution of income. Programs promoting milk and dairy consumption could broaden markets and raise the level of living standards of low-income families. But on the other hand, the government would be directing revenue to farmers in order to feed the poor. In the United States, this spelled socialism. "It is not a question of how to 'soak' the rich," proponents quietly pleaded. "It is a question of what distribution of income will enable us to use our productive resources most effectively under a system of private property and production for profit." Opponents reacted fiercely to the imposition of price controls on milk, pointing to Fascist Italy as a prime example of a free market gone bad.[33]

A body called the Surplus Commodities Corporation, established in 1935, helped the United States steer around "a complete overhaul of the distributive system." By purchasing farm commodities that appeared threatened by hard times, the government made basic products like butter, cheese, and evaporated milk available at reduced prices to needy families. Fluid milk presented a particular problem, however, because of its perishability. Only price supports paid directly to farmers could sustain production at levels determined to be adequate for the nation's health requirements. Price supports would prompt further debate as milk became one of the most regulated — and debated — commodities on the market.[34]

More was at stake than simply farmers' incomes. By this time, it was clear to governments throughout the Western world that the dairy industry encompassed enormous business enterprises involved in processing and distributing milk and milk products. The scaffolding that supported modern milk, from expensive farm and pasteurization equipment to custom-made bottles and milk delivery trucks, was here to stay. And now the new knowledge of nutrition provided powerful logic for expansion, not contraction: the whole world should be given a chance to purchase the abundance of milk that the developed world was capable of producing. "Essentially, the impasse is due to the tendency of modern production to outrun consumption," a report for the U.S. Department of Agriculture explained. "The obvious remedy, then, is to make consumption keep up with production." As one economist put it, "The black plague of the twentieth century is under-consumption."[35]

School lunches and school milk programs made perfect sense as ready recipients of farm bounty everywhere, and an added moral imperative loomed: the Depression was taking a particular toll on the resources of poor families. Milk and dairy products offered an opportune blank screen on which the war between socialist and capitalist ideologies could be projected. On the European side of the Atlantic, the determination to redistribute income was explicit. "The first objective of any Government in devising schemes affecting the national food-supply," wrote Sir John Boyd Orr, celebrated author of *Food, Health, and Income* (1936), "should be to ensure that all classes of the community would be able to obtain a diet adequate for health."[36] Now, with the "newer knowledge of nutrition" — McCollum's words became a mantra among policy specialists — the plea for government intervention sounded a clarion call. "One thing is certain," wrote J. C. Drummond and Anne Wilbraham in their pioneering sociological study of English food practices in 1939. "A means must be found of bringing these essential foods within the reach of the poorest section of the community."[37]

School meals had always depended on a mixture of voluntary and state effort, so milk in schools appeared first within the penumbra of charitable acts of kindness. Laws appeared first at the state level, allocating funds that enabled local boards of education to provide milk for schoolchildren, often supplying it free of charge for those in particular need. In California, for example, "every child in school is supplied with milk during school hours which is drunk from special bottles by means of straws, the supply being free in the case of children whose parents are not in a position to pay." Eventually, national laws pointed the way toward universal provision. In Britain, a law in 1934 saw that each child in elementary school received a third of a pint for a penny a day, and with the Education Act of 1944, every child in every school, state or private, received free milk supplied by the state. The United States passed a National School Lunch Act, signed by President Truman in 1946, but the nutritional objectives of the program were muted by the larger goal of serving agricultural interests.[38]

Of course, many more markets for milk lay unexploited around the world. All dairy-rich Western countries were, in some measure, plying other countries with their products already, so the question was only one of how to cast the export net wider. The saturation of markets during the 1930s alerted advanced industrialized countries to the needs of the developing world, which were now perceived as undeveloped purchasing power. A taste for the dairy had yet to be cultivated in Africa, Asia, and South Asia. Here was the next frontier of milk history.

Hindu visions of an ocean of milk proved to be prophetic of the last quarter of the twentieth century. During that time, Indians launched the most ambitious dairy development scheme in the history of the world, enabling India to become today's largest producer of milk. Emphasis needs to be placed on the shift to development, for India was already producing a great deal of milk at the local level. The subcontinent boasts one of the oldest milk cultures in existence, abounding in folk tales and myths illustrating the goodness of rich, white dairy foods. The Asian water buffalo, largely ignored by Western countries wedded to the cow, remained a highly valued producer of fat-rich milk. But centuries of colonial rule had inhibited rural development, and the availability of milk and milk products was still limited by small-scale production and traditional methods and equipment when new plans emerged in the 1970s.

Operation Flood, launched in 1970 by the National Dairy Development Board headed by Verghese Kurien, aimed to empower milk producers and enhance production across India. The program stands as evidence of several historical developments converging on Indian soil: India's struggle for independence from Western surpluses of powdered milk, the growing popularity of dairy cooperatives as the means of organizing rural producers, not just in India but worldwide, and the recognition of the importance of indigenous food programs in postcolonial nations. Nevertheless, as much as dairy development made good sense in India in the decades after independence was achieved in 1947, the stages of Operation Flood unfurled with anything but natural ease. If it hadn't been for extravagant idealism and "bottom-up" power dynamics, two familiar features of milk history, the entire project would have foundered.[39]

The years of the Second World War laid the groundwork for Operation Flood. In its final decade, the British presence in India affected the career path of a precocious science student named Verghese Kurien. As the son of a civil surgeon in British India, Kurien enjoyed all the advantages of an elite class of colonial subjects, including a compulsory graduate education in the United States. Against his wishes (at least according to his autobiography), Kurien was sent to Michigan State University, where he was assigned to a program in dairy engineering. He graduated having studied nuclear physics, his personal choice, informed by the momentous events of 1945, which suggested to him that this was going to be a cutting-edge field. But upon return to India, he was forced to repay his educational support by supervising dairy operations in Anand, an agricultural district in the poverty-stricken area of Gujarat.[40]

The British government was trying to rectify an embarrassing situation in India, one that was superficially similar to the predicament of European cities a half century earlier: the urban milk supply in Bombay (now Mumbai) was undrinkable. As Kurien tells the story, sometime around 1942, "Britishers stationed in Bombay" were sickened by bad milk. "A sample of this milk was sent for testing to a laboratory in London since the British would not trust any Indian laboratory. There was a one-line response from London. It said quite simply: 'The milk of Bombay is more polluted than the gutter water of London.'" Bureaucratic measures followed: a milk commissioner was appointed and the government went to work constructing a scheme to provide fresh, safe milk to the city inhabitants.[41]

The language of this narrative betrayed the prejudices and presumptions of the old colonial era. "Britishers" would soon inspire a milk rebellion through their negative example. The source of milk nearest to

Bombay, the Kaira district, already enjoyed a reputation for its many dairy products. Linked by a railroad line built in the 1860s, the region became the home of a number of butter and casein factories managed by an array of entrepreneurs, including interlopers from England, Germany, and New Zealand. The British government shrewdly tapped the Indian owner of a butter factory, Pestonjee Edulji, as the chosen agent of Bombay's milk supply. The fact that he was an Indian was somewhat promising. But the way Pestonjee and his company, Polson Dairy, went about doing business sensitized the farmers of Gujarat to an important fact: the value laden in their milk supply was being diverted from farmers and consumers, the two most worthy beneficiaries of the local industry.

Pestonjee had promised the British that he would procure, pasteurize, and transport fresh milk the 220 miles to Bombay, which he planned to do by wrapping milk cans in gunny sacks chilled by water. But before this could happen, the factory owner spelled out the precise conditions under which he would take on the job: the British government would give the Polson company all the expensive capital equipment it needed; it would pay Pestonjee an additional processing charge for his effort; and it would protect the whole enterprise by a law that compelled villages around Anand to sell their milk to no one but Polson. Such were typical colonial monopoly practices, which Indian farmers found intolerable in the new climate of independence after 1947.[42]

It seemed unlikely that farmers in this agriculturally rich district would dare to sever their financial ties to the market, given their utter dependence on the proceeds from milk sales in the 1940s. But anticolonial resentments inspired new alliances during the decade of world war and the struggle for autonomy from Britain. The farmers of Gujarat reasoned that a milk boycott against Polson promised to place control of their resources into their own hands. As one account put it, they figured they would "be better off if they drank their milk themselves" than to continue to turn over supplies to Polson. Led by Tribhuvandas Patel, the weight of their numbers and the logic of their arguments were exquisitely timed. In 1946, their boycott won them the right to organize state-sponsored cooperatives in order to market their own milk.[43]

At the time of the boycott, Patel recognized Kurien, then employed by the Indian government, as a valuable ally. Using his trademark powers of persuasion, Patel drew the engineer into the cause, and Kurien became technological adviser to Anand's dairy farmers. The move turned out to be a brilliant one, given the way the two organizers combined their personal philosophies. Patel's sensitivity to village and caste politics and Kurien's alertness to multiple levels of technical expertise went forward into a mine-field of local antagonisms. Farmers would have been justified in suspecting that their milk might land in the hands of processors attached to multi-national corporations. This would bring little return to the district in exchange for its white gold. But a different scenario slowly evolved: while Patel negotiated with farmers, Kurien negotiated with suppliers of equip-ment and personnel, each with the same central goal in mind: democratic dairy cooperatives and Indian leadership and labor would protect their milk project from exploitation by outside interests.

Cooperatives, Kurien asserted plainly in his memoir, were "tailor-made for the dairy industry."[44] Locally based organizations enabled small producers to pool relatively small quantities of milk and gain the benefits of shared services such as veterinarian assistance, technical advice, and expensive capital equipment purchased for large-scale operations. More-over, these arrangements meshed with the new political project of inde-pendent India. "What use is democracy in Delhi if we do not have demo-cratic institutions at the grass-roots level?" Kurien asked. Together, he and Patel organized the dairy farmers of Kaira District into a milk producers' union and established an independent processing plant as a competitor to the Polson Dairy. Calling themselves "Amul"—from the Sanskrit word *amulya,* meaning "priceless"—the organization made history as one of the most important indigenous industries to appear in the early decades of independence.[45]

Amul became the organizational model for all dairying in India in the 1960s, and in 1970, its cooperatives inspired the inception of Operation Flood. As Kurien relates the story in his memoir, *I Too Had a Dream,* the way India's dairy producers took on the Goliath of the developed world's milk corporations is thrilling in its own right. Kurien's allusion to Martin Luther King may seem hubristic to those who think a dairy scheme has little in common with a movement for fundamental civil rights. But a closer look at the obstacles that lay before Indian dairy farmers suggests that the analogy isn't wholly misplaced. Prejudice against rural people in a nation focused on its urban population and still hampered by adherence to caste was painfully apparent at the national level. Conflict at the local level also threatened to destroy the kind of cooperation that was crucial for dairy cooperatives. And at a supra-national level, the likelihood of triumph for an industry fueled by poor villagers with one or two buffaloes per household seemed like a long shot. Rather than invoking the image of a flood, planners would have been within their rights to design a dam, one that would protect against surplus milk shipped from Europe, Australia, and North America. How was Operation Flood to succeed, if outsiders saw the potential of markets in India?

By the 1960s, Europe and America had more milk than anyone could possibly market. "Milk lakes" and "butter mountains" had become bywords in the European Economic Community in the early 1960s, not as visions of an ideal world, but as signs of a significant, even alarming overproduction of dairy products. The reasons for such excess flowed from the dictates of a competitive, highly capitalized industry: facing a market with constantly falling prices, farmers were driven to invest heavily in new methods and technology, even if they could hardly afford to do so, to produce as much milk as they could. Genetic improvements in the cow population, offering higher milk yields per cow, had come about through careful breeding. This was made possible by the 1930s through the identification of high-quality semen and artificial insemination. New feed concentrates enabled farmers to further enhance productivity. Such imperatives drew European farmers into deeper debt, even as increased supplies forced down prices. Making matters worse, the demand for milk was declining, prompted by two ominous developments: a low rate of population growth, which meant fewer children and smaller families; and consumer wariness about the healthiness of dairy products (the fear of cholesterol was now in the air), which marked milk with a new stigma.[46]

So why shouldn't Europeans press their surplus milk products on India, if India's poverty and hunger called for it? No such arrangement was imaginable without an influx of foreign personnel, dictates, and an inevitable (and most likely negative) impact on native industries. The secret to deriving benefit from European surpluses lay in a delicately crafted choreography that could sidestep a surrender to European management and power. Kurien came up with a way for this to happen. The first phase of Operation Flood used financing from the World Bank to purchase European dairy surpluses, called "dairy aid" by the legions of policy experts who became witness to the master plan. In the form of skim milk powder and butter oil (known in the bewildering forest of Indian acronyms as SMP and BO), these products would be reconstituted and offered for sale at low prices in Indian cities, and so they would help to build up demand for milk and butter.[47] In subsequent phases of Operation Flood, such imports would diminish and domestic-produced dairy products would take their place, timed to increase alongside consumer demand. To succeed, the plan had to reorganize production as well as consumption.

Kurien concentrated on communicating the power of egalitarianism implanted in the cooperative dairy system. In Amul's approach, even the everyday chore of delivering milk became a revolutionary act, at least in social terms: according to the rules, dairy cooperators were to queue up, first come, first served, with their contributions to the district plants. In a nation still struggling against caste, the queue unsettled old attitudes. Some wealthy farmers resisted, yet they had no choice. Depositing milk with the cooperative brought Indians together in a new way.

Tensions arose repeatedly at each of the three stages of Operation Flood. Initial suspicions gave way to a concerted campaign by critics to bring down the project with bad press. Indian fears were legitimate: Third World nations understood that multinational corporations could hobble native industries in the same way that colonizing powers had done for centuries. "It is only logical that nobody will invest money in another country unless they hope to take more money out than they brought in," Kurien pointed out in his autobiography. His struggle with Nestlé and Glaxo showed how hard it was to "discipline" foreign capital. International funding of Operation Flood also led to legitimate suspicions that a Western model of development was being imposed on Indian villages. The country lacked infrastructure, and poor villagers wanted a means of a subsistence, not a weak entrée to a cash economy.[48]

Plenty of problems were percolating through Operation Flood. At the village level, dairy democracy was hardly viable if the primary people involved in tending the animals—namely, women—could not be involved in local organizations. Yet the roles of women were restricted by custom and authority. Farms of a few acres might include one or two buffaloes or cows, cared for and milked by the women of the household. Male labor was often limited to cutting fodder, or simply carrying milk to the cooperative. As one young man interviewed in 2001 put it, "As if the milking will be done by me!" He added, "If my father ever found me doing the work of milking, he would immediately sell all the cows." Access to information about lactation, gestation, bovine diseases, and sanitary methods was of paramount importance to those who were in regular contact with milk-producing animals. But according to cooperative rules, only one member of a household could register as a livestock recipient and owner, so the rank-and-file membership of India's thousands of dairy cooperatives was largely male.[49]

The power of milk to spread power: this was the hope of Operation Flood. Theoretically, at least, the project promoted the education of everyone involved in dairying, so village women were invited to take advantage of short courses of instruction at processing plants. The basics of animal nutrition for pregnant buffaloes, the science of live sperm and artificial insemination—nothing was beyond the curriculum of the dairy cooperative. Yet few women could afford a weekend away from home; still fewer would be willing to listen to a male manager lecture about the reproductive process. More women at the level of village organizers, science instructors, and factory managers could confront these problems. This was a tall order to fill, but by 1980, Indian women activists had begun organizing programs for women's development, especially for women in scheduled castes who were poor. Women extension supervisors, trained to spread information and encourage involvement, slowly changed the face of local dairying organizations. With the number of women cooperatives growing, milk output also grew.[50]

A kind of cosmic battle between the sexes erupts from between the lines of Kurien's narrative of Operation Flood. His meeting with the queen of the Netherlands was more than a humorous anecdote ending in a ribald joke. The queen's escort that day was Margaret Alva, a prominent Indian feminist and member of Parliament. According to his account, at the end of a friendly discussion with the queen, Alva abruptly said, "Your majesty, whatever you might think of him, I think Dr Kurien is an MCP!" She then turned to the "male chauvinist pig" himself, Kurien, and explained, "Do you see the crest of the NDDB [National Dairy Development Board]? It is a bull. It should, in fact, have been a cow. After all, NDDB is about dairy development. Doesn't this prove my point?" Kurien's response—"No bull, no milk!"—elicited laughter from the queen, but it didn't deny the fact that patriarchal rules and customs erased the presence of females, human and bovine, from the picture of Indian dairy development. Later phases of reform attempted to address this problem, given the need to recognize and engage as many working women as possible.[51]

It is much easier to discern the successful liberation movement involving Indian buffalo milk in the early years of Operation Flood. It didn't take long for the managers of Amul to realize that a deficit of cow's milk was going to undermine their plans for dairying in Gujarat. A different liquid, buffalo milk, constituted their primary resource. Because of the seasonality of supplies, they then confronted the important technological issue: could excess production of this very different kind of milk, which curdled at different temperatures and had different properties of casein and sugar, be converted to dried and condensed canned milk? Kurien paid a visit to Nestlé headquarters in Switzerland and consulted engineers of every kind. No one in the developed world thought it possible; or rather, as Kurien points out, no one would admit it was possible because processed buffalo milk was of no use to the developed world. But with equipment donated by UNICEF and some technical input from others, the Cinderella of milk transformed into liquid gold for India.[52]

After Operation Flood completed its final phase in 1996, Indian milk production continued to grow. More than 75,000 dairy cooperatives now supply the nation with milk, butter, and cream, helping to reduce hunger and alleviate poverty. Kurien's "White Revolution" enabled the nation to surpass the world's top producer: in 1998, India turned out 81.6 million tons of milk, compared with 78.3 million tons produced in America. By 2006–7, India's output had reached more than 110 million tons.[53] Production is still on the rise, which isn't surprising, given that domestic consumption of milk products remains a priority for the national government. Buffalo milk constituted 15 percent of total world milk production in 2004, compared with 9 percent a decade earlier. Indian producers continue to turn out a substantial portion of the world's supply, which will probably remain the case as long as "the Indian consumer's preference for high fat milk and milk products continues."[54]

Having begun with the myth of the churning of the ocean, perhaps the upward curve of Indian milk production is a good place to end this particular thread of milk history. The lines of the story, hardly complete, are worth reflecting on as a kind of counterpoint to the history of Western milk production. Advancing at a particular stage of Indian history, the pressure to make milk a democratic commodity, controlled by ordinary producers, created a much different model of business. Without idealizing these arrangements, we can see the relationship between milk's small-scale and local identity reflected in the shape of economic and social organizations springing up around the product. Much more remains to be said about the expansion of milk production and consumption into other areas of the globe, where different circumstances are shaping distinctive ways of producing and using the product. The example of India reminds us that the history of milk can be as malleable as milk itself.

CHAPTER FOURTEEN

Milk Today

Look at any very old volume on dairy cows and you'll find a section referring to something called the "escutcheon" or "milk mirror." The second term originated with a maverick specialist in cow anatomy living in the countryside of Bordeaux in the 1820s. François Guenon, a gardener with little formal education, determined that the posterior of the cow's udder foretold the animal's potential as a producer of milk. "One day," he recalled, "as I was whiling away the time in cleaning and scratching my poor old companion, I noticed that a sort of bran or dandruff detached itself in considerable quantities from certain spots on her hind parts, formed by the meeting of the hair, as it grew in opposite directions; which spots I have since called *ears,* from the resemblance they often bear to the bearded ears or heads of wheat or rye." Guenon studied and compared the udders of cows in his own pasture to those around his region and eventually established a system of classification according to the size and placement of marks on the "milk mirror." After a rigorous trial, the Agricultural Society of Bordeaux proclaimed Guenon's method "infallible" and awarded him a gold medal in 1837. Predicting a future "race of animals" with "none but Cows and Bulls of the first order," the society distributed news of this peculiar hind-end variety of cow phrenology. The method made its way across Europe to America; it circulates even today, recommended by an organic farming magazine and a farmer's blog.[1]

Modern dairying depends on sophisticated equipment and vigilant monitoring systems.
This herringbone milking parlor at Holly Green Farm in Bledlow, Buckinghamshire,
serves 480 Holstein cows, each producing roughly 2,325 gallons of milk per year.
(Photo courtesy of Neil and Jane Dyson, Holly Green Farm)

How unlike the science of selection belonging to the twenty-first century. Using artificial insemination, many dairy farmers rely on a collection
of tiny, colored straws full of semen, ordered from a glossy catalogue
picturing varieties of attractive female bovines. The best Holstein-Friesian
cows can produce eight to ten gallons of milk a day. (This is roughly the
same amount as prize animals of the 1860s.) The latest advance, "sexed
semen," now offers farmers the ability to select for (female) gender so that
they can avoid the birth of male calves, which raise little money when sold
as future meat. The more milkers, the better, at least in theory.[2]

But today's heifers have reached maturity at the time of the lowest milk
prices in recent memory: in July 2009, 100 pounds of milk brought only
$11.30 in the United States, compared with as much as $19.30 a year
earlier. Many farmers are accepting a price below what it takes to produce
milk. Forced into competition against large industrial dairy farms pow-
ered by state-of-the-art technology, small farmers have been dropping out
over the past two decades. Low supermarket prices for milk have meant

that even industrial dairy farms are now threatened. One California farmer reported that he was adding new, sexed-semen heifers only as replacements for less productive cows in his megaherd of 4,200. Dairy cow populations have declined in many countries; in the United Kingdom, their numbers fell by 20 percent between 1996 and 2006. The European Union has agreed to gradually withdraw price supports for dairy farmers, while the United States offers meager aid in the current crisis. Once again, the debate over milk's high cost of production and low market price is at an impasse.[3]

Is contemporary milk in trouble? To address this question, Guenon's mirror can be turned to another use, as a gloss on how milk itself can reflect cultural preoccupations and deep anxieties in our own era. Milk has always stood for more than just milk, and its path has always depended on a supporting cast summoned by its association with children and health. Two widely publicized scandals of the late twentieth century have played a big part in shaping how we think about milk today, even if young consumers (always the target of advertising) remain unaware of their occurrence. In the 1970s and 1980s, a vehement campaign against the marketing of infant food in developing countries put milk and milk processors in a negative spotlight. And in the following decade, the use of recombinant bovine somatotrophin, or rBST, a hormone designed to boost milk production in cows, pushed milk and the hidden dangers of modern food into mainstream media discussions. With its power to whip up emotion, milk became the flagship of public sentiment about many other food issues in contemporary society.

Chroniclers of the protest against infant formula marketing look back to 1939, when Cicely Williams, a British doctor working in Singapore, made a prescient speech to the local Rotary Club. Angered by what she saw in her clinic for rich and poor mothers alike, Williams gave her presentation a rather blunt title: "Milk and Murder." "If your lives were embittered as mine is, by seeing day after day this massacre of the innocents by unsuitable feeding," she intoned, "then I believe you would feel as I do that misguided propaganda on infant feeding should be punished as the most criminal form of sedition, and that those deaths should be regarded as murder." The president of the group happened to be the head of the Nestlé Company of Singapore, one of the world's largest manufacturers of infant formula. Did club members shift uneasily in their seats that day?[4]

This was not the first time that infants of poor or uninformed mothers were threatened by commercial substitutes for breast milk. Only fifty years earlier, babies fed on canned condensed milk and reconstituted powdered milk had developed rickets and other deficiency diseases owing to the lack of certain vitamins in processed milk. By the 1930s, the newer age of nutritional knowledge had produced infant formula of a much more sophisticated chemical composition, designed to mimic breast milk in convincing ways. Yet as the French chemist Dumas had warned in the 1870s, scientists tend to exaggerate their own powers. Critics of manufactured food pointed out that formula could not provide the antibodies present in breast milk, vital in the first six months of life, especially in poor nations. And because water supplies in developing countries were not always safe, infant formula could become instantly deadly.

Making such risks more objectionable were the methods companies used to promote sales. By the 1960s, hospitals in the Third World were being targeted with a steady stream of free samples and misleading instructional material (sometimes in the wrong language), which made its way into the hands of uninformed mothers. Even before new mothers left the hospital, their supplies of breast milk had diminished owing to the immediate introduction of commercial substitutes. Cicely Williams's efforts were interrupted by World War II, and her later career was devoted to many other aspects of maternal health. Protest against the role of infant formula in infant malnutrition would have to wait until a new cultural consensus reconsidered the impact of Western business on the postcolonial world.

Milk activism took off during the 1970s, when anti-poverty programs spread an awareness of deep disparities in global wealth. War on Want, a British nongovernmental organization, circulated a report in 1974 called *The Baby Killer,* which attributed widespread infant malnutrition in the Third World to the use of formula. The text was translated in Switzerland later that year as *Nestlé Kills Babies,* and word began to spread. By April 1976, a congregation of Catholics from Dayton, Ohio, appropriately (for those who know the history of beliefs about milk) named Sisters of the Precious Blood, got into the act. As shareholders of Bristol-Myers, the congregation decided to file a lawsuit against the company for misleading investors about its infant formula business in the Third World. And a year later, another group, INFACT, organized an international boycott of Nestlé products, setting off the largest international protest against a multinational corporation in history.[5]

The campaign against Nestlé grew into an international network aimed at transparency and accountability in business. Investigative journalism, legislative hearings, and lawsuits proliferated in Europe and the United States. UNICEF and the World Health Organization played a leading role in advancing principles for the marketing of commercial infant food. In 1981, both organizations endorsed the International Code of Marketing Breast-milk Substitutes, which prohibited dumping free supplies on hospitals and sending white-coated sales representatives on deceptive promotional visits in developing countries. Every nation of the World Health Assembly, except the United States, signed the agreement. Gradually, firms agreed to comply with new standards, and in 1984, Nestlé signed on and the boycott ended. But by 1988, violations incited NGOs to resume action. Appeals continue today, as infant food companies repeatedly insist on their right to freedom of speech in marketing.[6]

The goal of milk purity entered a distinctly modern phase of its history in the 1990s, when the United States became the battleground between a corporation and critics of a drug stimulant of milk production in cows.[7] The use of recombinant bovine somatotrophin (rBST), also known as recombinant bovine growth hormone (rBGH), in dairy herds meant that commercial milk supplies were contaminated with the drug, along with antibiotics and pus excreted by cows with mastitis. The ensuing conflict, pitting dairy scientists and consumer advocates against the Monsanto Corporation, acquainted the American public with a new range of capabilities of laboratory science. As Lisa H. Weasel points out in her riveting account of the conflict, this was the "first time that a food derived from the use of a genetically engineered drug had been allowed to enter the food system." Though Americans generally believe that they have not been touched by the genetic foods controversy currently roiling Europe, the problems of rBST stand as a perfect example of the debate.[8]

When the Monsanto Corporation obtained FDA approval for "Posilac" in 1993, critics had already begun a campaign at the federal level against its use. The effects of rBST on cows were obvious: the drug caused a high incidence of infection in the udders of treated cows, which required infusions of antibiotics. Both substances were detectable in the milk obtained from treated herds. Less obvious was the impact on consumers:

studies showed raised levels of insulin-like growth factor I (IGF-I), which was associated with growth disorders and cancer. Monsanto nevertheless pressed on with promotion and sales of rBST, while covertly squelching product-labeling efforts by states like Maine and Vermont. European dairy farmers, knowing that consumers would rebel, recoiled from using the hormone. The European Union decided to ban its use in October 1999, and Canada followed later that year. But American farmers and lawmakers held fast to a policy of business as usual, while lone critics attempted to publicize disturbing data challenging the safety of rBST.[9]

Opponents of rBST found other ways of mounting pressure against the sale of treated milk. While Monsanto continued to fight the labeling of treated cows' milk in state courts, grassroots efforts aimed their information at the general public. Led by Rick North, a retired director of the Oregon chapter of the American Cancer Society, and Martin Donohoe, a member of Physicians for Social Responsibility, a growing number of concerned milk-drinkers made their own preferences clear to milk processors. North and Donohoe came up with the idea of handing out pre-addressed postcards to the general public, and over six thousand people happened to mail theirs to Tillamook, a cooperative dairy in Oregon, which is also the second largest maker of chunk cheese in the United States. The company changed its mind about use of the drug and by 2005, Tillamook became "rBST-free." Not long after that, Starbucks and Chipotle (then owned by McDonald's) signed on. By August 2008, the struggle seemed to be over, when "Monsanto issued a press release announcing its intention to divest its rBST operations and product."[10]

Milk purity became the by-word of the milk industry from the late 1990s. Consumers demonstrated real concern over what was in their milk, proven by the fact that organic milk became one of the fastest-growing food items on the market. Even industrial dairy farms, which keep cows in close confinement and prohibit them from grazing in pastures (not exactly what you would call wholesome), turned to producing an organic version of the product. Milk became separated into two camps, often housed on opposite sides of the store: low-priced, large-packaged "supermarket" milk and higher-priced organic or boutique milk.

In most developed countries, milk drinking (and milk use) has been in decline since the 1970s, though in Finland, the highest per capita consumer of liquid milk in the world when its figures were last available in

2003, consumption reached an impressive peak of nearly 48 gallons per person. Iceland, Ireland, and Sweden were not far behind. (In these settings, it's worth noting that liquid milk is consumed in fairly large quantities with cereal.) By comparison, United States milk drinkers consumed a mere 21 gallons per capita in 2008, roughly equal to French and Iranian levels.[11] More than half the world's milk supply continues to be dedicated to liquid sales. Schools remain an important channel of milk resources: in 2008, the United States "Special Milk Program" distributed 85 million half pints to children.[12] The latest wave of advertising campaigns has been aimed at teenagers, the foremost buyers of novelty drinks, in the hope of selling milk as a nutritious avenue to health and weight loss. And milk drinks don't have to be called "milk," after all. In Japan, where teenagers drink more bottled coffee drinks than cola, milk products inhabit the vending machine as well as the café.[13] As dairy federations work on guessing the latest trend (the "wheyvolution" is one possibility in the next decade), expansion in China and Latin America indicate that the spread of milk is far from over.[14]

So how does history change our thinking about milk today? Does it help to know more about milk's past? In teaching food history, I've learned to savor the moments late in each semester when students become self-conscious about their analytical distance from the material we've covered. It's as though they're using a Google Earth tool to move closer and farther from their subjects of study, empathizing as they zoom in and reasoning with more detachment as they move out. With a close lens, it's possible for us to feel kinship with Isis worshipers or to enjoy listening to the clang of cymbals around *comos* drinkers. We can see the point of it all and even draw parallels between their rituals and ours. From a greater distance, we can appreciate the technological prowess that brought cheap imported food to poor consumers in Europe, irreversibly altering diets and expectations. But food facts in our own age seldom generate such dispassionate observations and analysis. Cheap milk, for example, may be sold at the expense of small local farmers and its composition may not be what buyers seek in such an exceptional commodity. Beginning with an understanding that milk is exceptional, there are ways that history can guide us through the maelstrom of food-related information in the media today.

From a middle distance, it's possible for us to see the contemporary product trapped in a web of contradictory forces. Now, as before, big corporations, farming interests, and some nutritional experts promote supermarket milk with unembarrassed zeal and self-interest. The history of milk shows how the product itself can act as its own agent in this predicament. Posed against corporate giants, raw milk, fresh and unpasteurized — the ultimate anti-corporate drink — has become the touchstone of a new phase of a movement to promote natural food. For one reason or another, raw milk makes news headlines almost every month. In the United States, over half a million Americans were purchasing the liquid by 2008. The number will most likely increase as more states (so far, twenty-eight) bow to pressure to relax laws and as many family farms turn to raw milk as a way of staying solvent. (The fresh farm product can cost between eight and twelve dollars a gallon, more than twice the current price of an average grocery store container in the United States.) Protest against Canadian milk laws boiled over at the nation's capital in 2008, when Michael Schmidt, a farmer near Toronto, staged a hunger strike against the treatment he was receiving at the hands of the government. Consumers there have learned to buy into "herd shares" as a way of getting around laws that limit how much raw milk dairy farmers are allowed to sell.[15]

Potent, precious, inviolable: these are themes that have recurred through many years of milk history, so we can locate ourselves in the timeline of milk believers (and nonbelievers) who may have invested hope and passion in milk's power. At a recent meeting of Slow Food Boston, panelists discussed two types of food sensibility that have found common cause in milk: consumers who view milk as a "miracle food" and a larger segment of the population that worries over "who controls our food." Both sides of the issue warrant some discussion.

Corby Kummer's recent clear-minded report in the *Atlantic,* "Pasteurization Without Representation," proved how perfectly contradictory the information on raw milk can be. Readers learned how, according to advocates, raw milk "prevents allergies and promotes a healthy immune system" and pasteurized milk has caused populations to miss out on a wide spectrum of health benefits (such as vitamins and enzymes) offered by the real thing. But online readers were also able to click on links that directed them to the FAQ page of the Centers for Disease Control and Prevention, which denied most of these claims, along with another link to the U.S.

Food and Drug Administration's "Question and Answer" page on the subject. There, the reader discovered that raw milk is an "inherently dangerous" host to at least ten deadly pathogens and "should not be consumed by anyone at any time for any purpose." Perspectives on raw milk either prove we're hostages of industrialized food that no longer delivers the nutrition it should, or else shore up our historical sense of how the government ensures a safe food supply for large modern populations, give or take a few nutrients.[16]

The wish for a miracle food, a single commodity or element that might serve as an elixir of immortality or just a solution to the ailment that is bothering us this month: the yearning is not surprising, nor is it unique to modernity. Milk's long legacy of association with divine powers and maternal effusions makes it a strong candidate for the part. Its broad range of nutrients (like the fact or not, milk is very nutritious) means that advocates can claim its central role in both exceptional growth in children and "slimming" diets for adults. (A quick mental comparison with meat, which holds no claim to completeness or purity, underscores milk's greater suitability as an ideal food.) Yet milk may lead us down the path to what Michael Pollan has called "the age of nutrition," with its recent lamentable obsession of viewing food through the lens of its chemical compounds, minerals, and vitamins. "It was in the 1980s," he argues, "that food began disappearing from the American supermarket, gradually to be replaced by 'nutrients,' which are not the same thing." The last decades of the twentieth century witnessed an alteration in the way modern Western countries, particularly the United States, approached food purchases. As populations grew more affluent, health-conscious, and, in many cases, weight-conscious, consumers came to regard medical and scientific data as primary guides for eating practices. Even health food advocacy, paradoxically, has resulted in the micromanagement of nutrients, a strategy that moves us away from a sensible enjoyment of what we eat and drink.[17]

Commonsense milk advocates simply see milk as a thoroughly proven asset to the food system, legitimated by centuries and even millennia of consumption. Convictions run deep in this regard: such arguments flourish in countries with historic dairying populations, such as those in northern Europe and Switzerland, along with the United States. As Stuart Patton, a leading American expert in dairy science, sees it, consumers in the United States take for granted their lives of plenty, their desires, and "lots of good food, including milk." Patton makes the reader reflect on what every vegan knows: milk is just about everywhere in the Western food system, not to mention in plastics, paints, and other compounds. (This latter fact is apparently little known. At the 2010 "Strolling of the Heifers" in Brattleboro, Vermont, for example, many parade viewers responded to the "Milk Paint" float with audible surprise.) "To many people, milk has become like a public utility, resembling such things as water, electricity, gas, and so on," Patton observed. "We need to know about milk."[18]

As a dairy scientist, Patton was broadminded in choosing to include chapters on breast milk and breast-feeding in his coverage, arguing for the biological utility of all milk in cell metabolism, bone growth, and the maintenance of health. But as an advocate of "nature's perfect food," his argument was hardly dispassionate. Perspectives on human biology and dietary practices clearly vary from one culture to the next. Not all food systems see a seamless continuum from breast to udder, and from udder to technologically sophisticated dairy production. And enough is known now about the biological and health outcomes of non-Western food systems to argue that "recommended daily allowances" stipulated by modern Western states aren't infallible. As anthropologist Andrea Wiley points out, "much of the world's population consumes well below the U.S. RDA for calcium without apparent detriment." So milk is not a mandatory food for everyone.[19]

You don't have to be an enemy of milk to wonder about the absence of traditional dairying and milk consumption in many parts of the globe. As the historical pattern of production shows in the preceding pages, northern Europeans who were particularly endowed with dairying traditions carried those foodways elsewhere. And with the development of industrial means of producing and transporting milk, along with massive surpluses of the product, businesses have searched for new markets in hitherto untapped regions of the globe. Large Western food corporations are now sponsoring milk sales and industries in Asia, where milk drinking by adults was once highly unusual and even contradictory to a biological predisposition to lactose intolerance (or, more precisely, lactase impersistence).[20]

Breast milk is a universal food because all infants possess the capability to digest it. Its unique sugar, lactose, is broken down by the enzyme lactase, present in the small intestine at birth. But lactase production begins to decline at weaning in most individuals, and only particular populations boast large numbers of adults with lactase activity (called lactase persistence). Along with northern Europe, other areas of the globe showing lactase persistence include the Middle East, the Arabian Peninsula, portions of sub-Saharan Africa, and South Asia. For the rest of the world population, even where milk drinking has become an adopted way of life, lactose intolerance is a common condition. Fortunately for some who have it, the symptoms experienced after drinking milk can be mild or even nil. Yogurt is often well tolerated because of the natural presence of digestion-aiding bacteria, and some cheeses contain very little lactose. But liquid milk consumption can bring on nausea, cramping, gas, and diarrhea for many adults, including many who live in the dairy-saturated continents of Europe and North America. For them, the contradiction between the dictates of typical nutritional guidelines of modern Western nations and physical comfort must be painfully apparent.[21]

The historian's perspective sheds some valuable light on the implications of these developments. The longstanding cultural meanings of milk exert powerful force over how people perceive its value. As early chapters of this book showed, milk invoked singular associations with supernatural and maternal powers. More recently, its characteristics have drawn from the prestige of science and medicine. Each phase of milk history reveals a changing image of potentialities. As Andrea Wiley has pointed out in her study of milk in China, the liquid "may undergo radical transformations from a marginal or disliked food to one positively associated with modernity and wealth, health and strength." And these markers exert their own magnetism over what foods people crave and consume.[22]

This certainly seems to be the case in China. Studies of fluid milk sales show "record growth over the last decade," particularly in cities and among wealthier consumers who frequent modern grocery stores. According to Wiley, "a fifteen-fold increase in milk consumption" has taken place over the past forty years, by far the greatest increase in any growing market in the world today. The product that seems to be enjoying the most popularity, next to the obligatory household container of liquid milk, is yogurt. This isn't terribly surprising, given that flavored milk drinks and ever more cleverly packaged yogurts generate new sales in Europe and North America, too. Consider that a single Emmi plant in Bern, Switzerland, was turning out roughly 82 million cups of yogurt in 2008. The number of items produced by a single modern dairy plant can also be staggeringly diverse. In the case of the Emmi Bernese facility, which transplants its factory organization and technology to locations around the world, including to North America, the number of different milk-based food products hovered around 450. This highlights just how much flexibility and market potential resides in the middle zone of processing.[23]

Marketing dairy products in China and other frontiers of the global dairy industry relies on the association between milk and growth, along with success in sports. In fact, Wiley shows, milk's alleged power to induce growth continues to inspire dietary studies in non-Western populations. ("Alleged" because studies prove inconclusive, mainly demonstrating milk's magic in undernourished populations.) So as historians, we can trace a line from the milk research of John Boyd Orr and Elmer V. McCollum to studies of school milk programs in Beijing, Malaysia, and Kenya.[24]

Even while milk spreads into new regions of the globe, its positive aura in North America has begun to fade just a little. If a recent edition of a popular nutrition textbook is any measure, milk's main claim to goodness is as breast milk; nowhere else in the text are its benefits trumpeted or enforced.[25] Common nutritional and medical practice now recognizes biological diversity among patients, along with declining lactase capability in aging adult populations. Even the term "lactase deficiency," which implies "presence" as normal, is subjected to critique. So, according to Dr. Mitchell Charap of New York University School of Medicine, budding doctors are no longer taught to urge everyone to drink milk. Calcium and vitamin D remain important to all diets, but patients are given the choice of how to consume them. As our own pediatrician advised us when our daughters proved to be lactose intolerant, "See if you can get them to like bok choy."[26]

My favorite illustration of how milk has suffered a shift in thinking in the past thirty years is in the case of dietary recommendations for stomach ulcers. Until the 1980s, milk remained a crucial component of the notorious "bland diet" imposed on sufferers of gastritis and ulcers. Coating the stomach with milk was believed to provide comfort and protection, just as it did for Thomas Carlyle in the 1840s, despite the fact that milk's slightly acidic profile was adding to the already existing acid bath of the stomach. The other qualities of milk — its soothing and pure nature, its association with maternal care — were implicit in this homely and very old prescription. Doctors schooled in milk history might have hypothesized that the remedy carried a placebo effect. All of this changed after 1982, when Australian researchers Dr. J. Robin Warren and Dr. Barry J. Marshall identified *Helicobacter pylori,* a bacteria that they proved causes stomach ulcers and plays a role in stomach cancer. Physicians and researchers stubbornly resisted believing the Australians' claims, but eventually the resounding success of antibiotics in eliminating ulcers provided sufficient clinical proof. A medical paradigm shifted, replacing milk with prescription medicine in the treatment of a common stomach ailment.[27]

This discussion has brought us a long way from the pastures haunted by François Guenon. Yet some aspects of milk and dairies have not changed all that much. Most dairy farms, apart from the industrial megafarms in the southwestern and western United States, remain relatively small in size and largely family-run. True, Guenon would be astonished to witness the transfer of milk from a farm's holding tank to the stainless steel compartment of a modern refrigerated milk truck. But then the driver might demonstrate something Guenon would recognize as happily familiar: some haulers are required to smell their sample, not just for spoilage, but for other odors, too. In some areas of the United States, dairies are catering to buyers' wishes for a product similar to what their "mothers used to buy" by feeding cows mainly on grass and keeping their delivery circuits within a close range of the milking parlor itself. Home delivery is on the rise in the United States and Europe. Nostalgia for milk's bygone days plays a part in shaping its appeal today.[28]

Perhaps the most surprising aspect of dairying is its job as placeholder for a vision of a vanishing agrarian past. A vista of cows grazing on expansive green pastures is what we think of when we imagine the countryside. As Christoph Grosjean-Sommer, communications director of the Bern office of Swiss Milk Producers, told me, dairy farmers and their cows make the landscape look like it should. Even after all the hard science of the twentieth century, milk remains more magical than the sum of its nutrients.[29]

Introduction

1. Catherine Bertenshaw and Peter Rowlinson, "Exploring Stock Managers' Perceptions of the Human-Animal Relationship on Dairy Farms and an Association with Milk Production," *Anthrozoös* 22 (2009): 59–69. Yields were measured over ten months' lactation. Thanks to Marc Abrahams, the editor and co-founder of *Annals of Improbable Research,* for sending me this article.

2. Peter J. Kuznick, "Losing the World of Tomorrow: The Battle over the Presentation of Science at the 1939 New York World's Fair," *American Quarterly* 46 (1994): 360.

3. Frederick J. Simoons, "The Determinants of Dairying and Milk Use in the Old World: Ecological, Physiological, and Cultural," in *Food, Ecology, and Culture: Readings in the Anthropology of Dietary Practices,* ed. J. R. K. Robson (New York, 1980), 83–91; "white blood" from Philip Franz von Siebold writing on Japan in the late 1700s. Thanks to Merry White for this reference.

4. E. Parmalee Prentice, *American Dairy Cattle, Their Past and Future* (New York, 1942), 95, 100, 114–15; *The Catholic Encyclopedia,* ed. Charles George Herbermann et al. (New York, 1913), 9:153.

5. Peter J. Richerson, Robert Boyd, Joseph Henrich, "Gene-Culture Co-Evolution in the Age of Genomics," *Proceedings of the National Academy of Science* 107, supplement 2 (May 11, 2010): 8985.

6. Harriet Friedmann, "What on Earth Is the Modern World System? Foodgetting and Territory in the Modern Era and Beyond," *Journal of World-Systems Research* 6, no. 2 (Summer–Fall 2000): 480–515; Philip McMichael, "A Food Regime Genealogy," *Journal of Peasant Studies* 36, no. 1 (January 2009): 139–69.

7. Alfred W. Crosby, *Ecological Imperialism: The Biological Expansion of Europe, 900–1900* (Cambridge, 2004), 48.

8. G. F. Fussell, *The English Dairy Farmer, 1500–1900* (London, 1966); E. Melanie DuPuis, *Nature's Perfect Food: How Milk Became America's Drink* (New York, 2002); Peter J. Atkins, *Liquid Materialities: A History of Milk, Science, and the Law* (Ashgate, 2010); Stuart Patton, *Milk: Its Remarkable Contribution to Human Health and Well-Being* (New Brunswick, 2004); Anne Mendelson, *Milk: The Surprising Story of Milk Through the Ages* (Knopf, 2008).

Chapter One: Great Mothers and Cows of Plenty

1. R. E. Witt, *Isis in the Ancient World* (Baltimore, 1971), 288n.23.

2. Joanna Williams, "The Churning of the Ocean of Milk: Myth, Image, and Ecology," in *Indigenous Vision: Peoples of India Attitudes to the Environment,* ed. Geeta Sen (New Delhi, 1992), 145–55; Chitrita Banerji, "How the Bengalis Discovered *Chhana* and Its Delightful Offspring," in *Milk: Beyond the Dairy,* ed. Harlan Walker (Totnes, 2000), 48–59.

3. Quote from *Sibylline Oracles,* 3:744–49, cited in Richard A. Freund, "What Happened to the Milk and Honey? The Changing Symbols of Abundance of the Land of Israel in Late Antiquity," in *"A Land Flowing With Milk and Honey": Visions of Israel from Biblical to Modern Times,* ed. Leonard J. Greenspoon and Ronald A. Simkins (Omaha, 2001), 35.

4. Gail Corrington, "The Milk of Salvation: Redemption by the Mother in Late Antiquity and Early Christianity," *Harvard Theological Review* 82 (1989): 393–420; Hildreth York and Betty Schlossman, "'She Shall Be Called Woman': Ancient Near Eastern Sources of Imagery," *Women's Art Journal* 2 (1981–82): 37–41.

5. Witt, *Isis,* 164–84.

6. Ibid., pl. 13, p. 80.

7. Rebecca Zorach, *Blood, Milk, Ink, and Gold: Abundance and Excess in the French Renaissance* (Chicago, 2005), chap. 3, points to many examples in sixteenth-century France.

8. F. Sokolowski, "A New Testimony on the Cult of Artemis of Ephesus," *Harvard Theological Review* 58 (1965): 427–31.

9. York and Schlossman, "'She Shall Be Called Woman,'" 40; Corrington, "Milk of Salvation," 412.

10. York and Schlossman, "'She Shall Be Called Woman,'" 39; Witt, *Isis,* 148; Pamela C. Berger, *The Goddess Obscured: Transformation of the Grain Protectress from Goddess to Saint* (Boston, 1985), 45–46.

11. Corrington, "Milk of Salvation," 397, 412.

12. Witt, *Isis,* 210, cites the *Pyramid Texts* on this point.

13. Phyllis Pray Bober, *Art, Culture, and Cuisine: Ancient and Medieval Gastronomy* (Chicago, 1999), 39–41, 87, 91, 340n.20.

14. Sandra Ott, "Aristotle Among the Basques: The 'Cheese Analogy' of Conception," *Man*, n.s., 14 (1979): 699–711.

15. Michael Abdalla, "Milk and Its Uses in Assyrian Folklore," in *Milk: Beyond the Dairy*, ed. Walker, 9–18; Helen King, *Hippocrates' Woman: Reading the Female Body in Ancient Greece* (London, 1998), 155. Boiled asses' milk enjoyed the greatest acclaim, allegedly relieving ailments like epilepsy, paralysis, irregular bowel movements, and "dry cholera." *Hippocrates on Diet and Hygiene*, trans. John Renote (London, [1952]), 135, 138, 157–58.

16. Londa Schiebinger, "Why Mammals Are Called Mammals: Gender Politics in Eighteenth-Century Natural History," *American Historical Review* 98 (1993): 394–95.

17. King, *Hippocrates' Woman*, 25–29.

18. Ibid., 49, 71, 143, 154–55. In difficult cases of illness, last-ditch remedies might be designed and administered in public. In one such instance, the liver of a turtle, extracted from a living specimen, was ground up with human breast milk.

19. Geoffrey Stephen Kirk, *Myth: Its Meaning and Functions in Ancient and Other Cultures* (Cambridge, 1970), 135–38.

20. *The Odyssey*, trans. Robert Fagles (New York, 1996), book 9, 276–81.

21. For a brilliant analysis, see Brent D. Shaw, "'Eaters of Flesh, Drinkers of Milk': The Ancient Mediterranean Ideology of the Pastoral Nomad," *Ancient Society* 13–14 (1982–83): 5–31.

22. *The History of Herodotus*, 4 vols., ed. George Rawlinson (London, 1862), 3:1–2.

23. Piero Camporesi, *The Anatomy of the Senses: Natural Symbols in Medieval and Early Modern Italy*, trans. Allan Cameron (Cambridge, 1997), 47.

24. See detail from Tell El-Obaid relief, 2900–2650 B.C.E.; Marten Stol, "Milk, Butter, and Cheese," *Bulletin on Sumerian Agriculture* 7 (1993): 100; Oliver E. Craig, "Dairying, Dairy Products, and Milk Residues: Potential Studies in European Prehistory," in *Food, Culture, and Identity in the Neolithic and Early Bronze Age*, ed. Mike Parker Pearson (Oxford, 2003), 89.

25. Keith Ray and Julian Thomas, "In the Kinship of Cows: The Social Centrality of Cattle in the Earlier Neolithic of Southern Britain," in *Food, Culture, and Identity*, ed. Parker Pearson, 40.

26. Heather Pringle, "Neolithic Agriculture: The Slow Birth of Agriculture," *Science* 282 (1998): 1446–89; Parker Pearson, ed., *Food, Culture, and Identity*, 11; Andrew Sherratt, "Cash Crops Before Cash: Organic Consumables and Trade," *The Prehistory of Food: Appetites for Change*, ed. Chris Gosden and Jon Hather (London, 1999), 15–34.

27. Ray and Thomas, "In the Kinship of Cows," 38; also inconvenient is the fact that the original "land of milk and honey" was Egypt, not Israel. Freund, "What Happened to the Milk and Honey?"

28. Stol, "Milk, Butter, and Cheese," 104.

29. Ibid., 105, 107.

30. Fiona Marshall, "Origin of Specialized Pastoral Production in East Africa," *American Anthropologist* 92 (1990): 873–94; Andrew Sherratt, "The Secondary Exploitation of Animals in the Old World," *World Archaeology* 15 (1983): 95.

31. Sally Grainger, "Cato's Roman Cheesecakes: The Baking Techniques," in *Milk: Beyond the Dairy,* ed. Walker, 168–77.

Chapter Two: Virtuous White Liquor in the Middle Ages

1. Details of Dame Alice's holiday celebration and everyday hospitality are drawn from *The Household Book of Dame Alice de Bryene, of Acton Hall, Suffolk, September 1412 to September 1413,* trans. M. K. Dale, ed. Vincent B. Redstone (Ipswich, 1984). Records for New Year's Day appear on p. 28. See also Ffiona Swabey, "The Household of Alice de Bryene, 1412–13," in *Food and Eating in Medieval Europe,* ed. Martha Carlin and Joel T. Rosenthal (London, 1998), 134–44; Christopher Dyer, *Standards of Living in the Later Middle Ages: Social Change in England c. 1200–1520* (Cambridge, 1989).

2. Kathleen Biddick, *The Other Economy: Pastoral Husbandry on a Medieval Estate* (Berkeley, 1989), 42–43; Maryanne Kowaleski, *Local Markets and Regional Trade in Medieval Exeter* (Cambridge, 1995), 286–87.

3. *Household Book,* 28; B. H. Slicher van Bath, *The Agrarian History of Western Europe,* A.D. *500–1850,* trans. Olive Ordish (London, 1963), 182, 335.

4. Christopher Dawson, ed., *Mission to Asia* (Toronto, 1980), vii.

5. *The Mission of Friar William of Rubruck,* ed. Peter Jackson, trans. David Morgan (London, 1990), 59.

6. Ibid., 75–77.

7. John of Plano Carpini, "History of the Mongols," in *Mission to Asia,* ed. Dawson, 8, 11. Carpini's text, according to Dawson, "was by far the most widely known of all the early accounts of the Mongols" (p. 2).

8. *Mission of Friar William of Rubruck,* 79, 98.

9. Ibid., 81–82, 99.

10. Ibid., 209, 222, 237, 240, 253.

11. Henry Serruys, *Kumiss Ceremonies and Horse Races* (Wiesbaden, 1974), 1.

12. Quoted in ibid., 2.

13. *Mission of Friar William of Rubruck,* 104, 254.

14. James France, *Medieval Images of Saint Bernard of Clairvaux* (Kalamazoo, 2007), 207, 209.

15. Brian Patrick McGuire, *The Difficult Saint: Bernard of Clairvaux and His Tradition* (Kalamazoo, 1991), 198.

16. Marina Warner, *Alone of All Her Sex: The Myth and Cult of the Virgin Mary* (London, 1976), 200.

17. Caroline Walker Bynum, *Jesus as Mother: Studies in the Spirituality of the High Middle Ages* (Berkeley, 1982), 167.

18. St. Bernard, *Commentary on "The Song of Songs,"* Sermon 9.

19. St. Bernard of Clairvaux, *The Glories of the Virgin Mother* (1867), 107.

20. Phyllis Pray Bober, "The Hierarchy of Milk in the Renaissance, and Marsilio Ficino on the Rewards of Old Age," in *Milk: Beyond the Dairy,* ed. Harlan Walker (Totnes, 2000), 96.

21. Medieval medical authorities recommended that mothers nurse their own children not just on these grounds, but to ensure that good food and minimal amounts of intoxicating beverage would contribute toward making "quality" milk. But judging from assertiveness of noble mothers when they *did* nurse, we may assume that other forces (quite possibly their husbands) pressed them to hire wetnurses. Mary Martin McLaughlin, "Survivors and Surrogates: Children and Parents from the Ninth to the Thirteenth Centuries," in *The History of Childhood,* ed. Lloyd deMause (New York, 1974), 116.

22. Richard S. Storrs, *Bernard of Clairvaux: The Times, The Man, and His Work* (New York, 1901), 149.

23. Bynum, *Jesus as Mother,* 122.

24. Clarissa W. Atkinson, *The Oldest Vocation: Christian Motherhood in the Middle Ages* (Ithaca, 1991).

25. Caroline Walker Bynum, *Holy Feast and Holy Fast: The Religious Significance of Food to Medieval Women* (Berkeley, 1987), 122.

26. Ibid., 131, 125–26.

27. Agnes B. C. Dunbar, *A Dictionary of Saintly Women,* 2 vols. (London, 1904), 1:132–35.

28. France, *Medieval Images,* 212–13.

29. For an extensive discussion of the place of Mary in medieval European culture, including the "Marianization" of devotion, see Miri Rubin, *Mother of God: A History of the Virgin Mary* (New Haven, 2009); Warner, *Alone,* 200.

30. Susan Signe Morrison, *Women Pilgrims in Late Medieval England* (London, 2000), 27–35.

31. Phyllis Pray Bober, *Art, Culture, and Cuisine: Ancient and Medieval Gastronomy* (Chicago, 1999), 215.

32. Bernard's new Cistercian order forbade all cheeses, eggs, milk, and even fish, except on special occasions, in an attempt to curb gourmandizing tastes in the monasteries. The mellifluous monk would have seen no contradiction between his raptures over celestial milk and his banishment of the product from his order's diet. Antoni Riera-Melis, "Society, Food, and Feudalism," in *Food: A Culinary History from Antiquity to the Present,* ed. Jean-Louis Flandrin, Massimo Montanari, and Albert Sonnenfeld, trans. Clarissa Botsford, Arthur Goldhammer, et al. (New York, 2000), 260–63; Bober, *Art,* 217.

33. Kees de Roest, *The Production of Parmigiano-Reggiano Cheese: The Force of an Artisanal System in an Industrialised World* (Assen, 2000), 21.

34. See, for example, the discussion by Maguelonne Toussaint-Samat, *A History of Food,* trans. Anthea Bell (Oxford, 1992), 116–17.

35. Riera-Melis, "Society, Food, and Feudalism," 265, 300; Slicher van Bath, *Agrarian History,* 283.

36. Roest, *Production,* 21.

37. Giovanni Boccaccio, *The Decameron,* trans. Guido Waldman (New York, 1993), Eighth Day, story three. A *denier* was a French coin of small denomination.

38. Roest, *Production,* 19.

39. Ibid., 20, 31n.1.

40. Exceptions were often made for eggs and sometimes cheese; according to *Genesis* 3:17–18, the proscription given to Adam forbade eating creatures who set four feet upon the cursed ground. Monasteries seem to have regarded cheese as acceptable food on fast days.

41. Mikhail Bakhtin, *Rabelais and His World,* trans. Hélène Iswolsky (Cambridge, Mass., 1968), 298.

42. Quoted in Carlo Ginzburg, *The Cheese and the Worms: The Cosmos of a Sixteenth-Century Miller,* trans. John and Anne Tedeschi (Baltimore, 1980), 83.

Chapter Three: The Renaissance of Milk

1. The letter is reproduced in part in Mary Ella Milham's excellent Introduction in *Platina, On Right Pleasure and Good Health,* intro., ed., and trans. Mary Ella Milham (Tempe, 1998), 15.

2. Platina, *On Right Pleasure,* book 1, dedication.

3. Ken Albala, *Eating Right in the Renaissance* (Berkeley, 2002), 15.

4. I am indebted to Ken Albala's helpful summaries of milk and the digestive process in ibid., chap. 2, and "Milk: Nutritious and Dangerous," in *Milk: Beyond the Dairy,* ed. Harlan Walker (Totnes, 2000), 19–30.

5. Platina, *On Right Pleasure,* book 2, p. 16. Zeroing in on the source itself, Platina even recommended eating udders, which promised plentiful nourishment according to how abundant they were in furnishing milk (p. 259).

6. Ibid., 365, 367.

7. Milham, "Introduction," 12, 51.

8. Albala, *Eating Right,* 55.

9. Platina, *On Right Pleasure,* 371, 393, 395.

10. Ibid., 335, 339, 365; Roy Strong, *Feast: A History of Grand Eating* (Orlando, 2002), 84.

11. Platina, *On Right Pleasure,* 123, 161, 297, 365, 393–97.

12. Ibid., 159, 381, 383.

13. Ibid., 363, 103.

14. Platina was well over fifty when he participated in a gang attack on a friend's house, based on a suspicion that a servant there had designs on his mistress (37–38). Mary Ella Milham also mentions his reputation for amorous adventures, suggesting that this "may explain the extraordinary number of passages in *De*

honesta voluptate where he records the good or bad effects of various foods upon *Venus*" (ibid., p. 18).

15. Albala, *Eating Right,* 125–26; Irma Naso, *Formaggii Del Medioevo: La "Summa lacticiniorum" di Pantaleone da Confienza* (Torino, 1990). Thanks to my daughter, Emma Gilmore, for giving me a copy of this book.

16. Samuel Kline Cohn, Jr., *The Black Death Transformed: Disease and Culture in Early Renaissance Europe* (London, 2002), 242–43.

17. I am indebted to the excellent critical edition of Marsilio Ficino's *Three Books on Life,* trans. and annotated by Carol V. Kaske and John R. Clark (Binghamton, 1989), 107, 131, 239. [Hereafter cited as *On Life.*]

18. Cited in Kaske and Clark, "Introduction," *On Life,* 43.

19. Ficino, *On Life,* book 1, chap. 4.

20. Ibid., 121. The latter description is taken from the *Aeneid,* 4:700–701.

21. Ficino, *On Life,* book 1, chap. 5, 119, chap. 6, 121.

22. Ibid., book 1, chaps. 5–6.

23. Ibid., book 1, 159.

24. Ibid., book 2, chap. 11.

25. Platina, *On Right Pleasure,* 189; Layinka M. Swinburne, "Milky Medicine and Magic," in *Milk: Beyond the Dairy,* ed. Walker, 339.

26. Ficino, *On Life,* book 1, chap. 10, 135.

27. Cereta's letters have been transcribed, translated, and edited by Diana Robin, *Collected Letters of a Renaissance Feminist* (Chicago, 1997) [hereafter cited as Cereta, *Letters*]. To Veronica di Leno, September 5, 1485, *Letters,* 35.

28. Albert Rabil, Jr., *Laura Cereta, Quattrocento Humanist* (Binghamton, 1981).

29. Cereta, *Letters,* 27, 82.

30. Ibid., 199.

31. Milham, "Introduction," 18–19.

32. Ficino, *On Life,* 213.

33. Cereta, *Letters,* 115–16.

34. "On Entering into the Bonds of Matrimony," February 3, 1486, Cereta, *Letters,* 70.

35. "A Topography and a Defense of Epicurus," December 12, 1487, ibid., 115–22.

36. Cereta, *Letters,* 117–19.

37. Ibid., 119.

38. Ibid.

39. Pierre Joubert, *Erreurs Populaires* (1587), cited in Ken Albala, *Eating Right,* 39.

40. Thomas Cogan, *The Haven of Health* (1584), iii.

41. Ibid., 154–55.

42. Samuel Pepys relates the story of John Caius in his diary entry for November 22, 1667. For numerous examples of the use of breast milk (obtained directly

from the breast) for emergency cures, see Marylynn Salmon, "The Cultural Signif-icance of Breastfeeding and Infant Care in Early Modern England and America," *Journal of Social History* 28 (1994): 247–69; Albala, *Eating Right*, 75.

43. Jean-Louis Flandrin, "Distinction Through Taste," in *A History of Private Life*, ed. Philippe Ariès and Georges Duby (Cambridge, Mass., 1989), 3:289; Stephen Mennell, *All Manners of Food: Eating and Taste in England and France from the Middle Ages to the Present*, 2nd ed. (Chicago, 1996), 73–74.

44. Albala, "Milk: Nutritious and Dangerous," 28.

45. Mennell, *All Manners of Food*, 84.

46. Thomas Moffett, *Healths Improvement: or, Rules Comprizing and Discovering the Nature, Method, and Manner of Preparing All Sorts of Food Used in This Nation* (London, 1655), 128; J. C. Drummond and Anne Wilbraham, *The Englishman's Food* (London, 1958), 154.

47. Moffett, *Healths Improvement*, 125. Moffett died in 1604.

Chapter Four: Cash Cows and Dutch Diligence

1. William Aglionby, *The Present State of the United Provinces of the Low-Countries* (London, 1669), 222, 224–25.

2. "Dutch miracle" is the term used to describe the remarkable transformation of the Dutch economy in the late sixteenth and early seventeenth centuries. See Jonathan Israel, *The Dutch Republic* (Oxford, 1995), 307.

The Republic of the United Provinces was made up of seven northern Dutch provinces, including Holland, that won their independence from Spain between 1568 and 1609. The republic constituted a powerful Protestant nation led by a strong merchant class. The southern Dutch provinces, understood as part of the expression "the Low Countries" at this time, remained under Spanish rule until 1713; they included the province of Flanders and the duchy of Brabant (men-tioned later in this chapter), an area hospitable to Catholics during this century of religious warfare. "Holland," in reference to the Protestant region, and "the Netherlands," as an inclusive and somewhat vague umbrella category for the whole region, were often invoked as general terms for the Dutch provinces.

3. Steven C. A. Pincus, "From Butterboxes to Wooden Shoes: The Shift in English Popular Sentiment from Anti-Dutch to Anti-French in the 1670s," *The Historical Journal* 38 (1995): 338.

4. Quoted in Simon Schama, *Embarrassment of Riches: An Interpretation of Dutch Culture in the Golden Age* (New York, 1987), 265; John Dryden, *Amboyna*, II, i, 38, quoted in Robert Markley, *The Far East and the English Imagination, 1600–1730* (Cambridge, 2006), 165.

5. Aglionby, *Present State of the United Provinces*, 213–14, 328–29.

6. Ibid., 232–33.

7. Ibid., 224, 233.

8. Jan de Vries, *The Dutch Rural Economy in the Golden Age, 1500–1700* (New

Haven, 1974), 96, 107; Jan de Vries and Ad van der Woude, *The First Modern Economy* (Cambridge, 1997), 59.

9. Schama, *Embarrassment*, 169.

10. John Ray, *Observations Topographical, Moral, & Physiological Made in a Journey to Part of the Low-Countries* (London, 1673), 51.

11. Ibid., 85.

12. M. Schoockius, *Exercitatio academica de aversatione casei* (Groningen, 1658), quoted in Josua Bruyn, "Dutch Cheese: A Problem of Interpretation," *Simiolus: Netherlands Quarterly for the History of Art* 24 (1996): 204–5.

13. Ray, *Observations*, 51.

14. Ibid., 50.

15. B. H. Slicher van Bath, *The Agrarian History of Western Europe, A.D. 500–1850* (London, 1963), 183.

16. Ibid., 84.

17. De Vries, *Dutch Rural Economy*, 156, 161.

18. Ibid., 123, 133.

19. Ibid., 151; Slicher van Bath, *Agrarian History*, 257, 260.

20. De Vries, *Dutch Rural Economy*, 150–52.

21. Ibid., 141–42.

22. Slicher van Bath, *Agrarian History*, 284, 335; De Vries, *Dutch Rural Economy*, 144.

23. De Vries and van der Woude, *First Modern Economy*, 124, 144, 186, 210; De Vries, *Dutch Rural Economy*, 285.

24. Schama, *Embarrassment*, 151.

25. Ibid., 159–60.

26. Ibid., 152, 174–75.

27. De Vries, *Dutch Rural Economy*, 221, 123, 219.

28. N. R. A. Vroom, *A Modest Message as Intimated by the Painters of the "Monochrome Banketje"* (Schiedam, 1980), 14–19.

29. Svetlana Alpers, *The Art of Describing: Dutch Art in the Seventeenth Century* (Chicago, 1983).

30. E. de Jongh, *Still-Life in the Age of Rembrandt* (Auckland, 1982), 65–69.

31. Schama, *Embarrassment*, 188.

32. De Vries and van der Woude, *First Modern Economy*, 204.

33. Everett E. Edwards, "Europe's Contribution to the American Dairy Industry," *Journal of Economic History* 9 (1949): 73.

34. De Vries and van der Woude, *First Modern Economy*, 200, 211.

35. *A Discours of Husbandrie Used in Brabant and Flanders* (London, 1650), v, 10–11.

36. Ibid., 3, 13.

37. A. R. Michell, "Sir Richard Weston and the Spread of Clover Cultivation," *Agricultural History Review* 22 (1974): 161.

38. Ibid., 160.

Chapter Five: A Taste for Milk and How It Grew

1. September 4, 1666, *The Diary of Samuel Pepys,* ed. Robert Latham and William Matthews, 11 vols. (Berkeley, 1972), 7:274.

2. Ibid., January 30, 1662/3, May 18, 1664, April 25, 1669, 9:533-34.

3. Anne Mendelson, *Milk: The Surprising Story of Milk Through the Ages* (New York, 2008), 21.

4. Nils-Arvid Bringéus, "A Swedish Beer Milk Shake," in *Milk and Milk Products,* ed. Patricia Lysaght (Edinburgh, 1994), 140-50.

5. Mary Agnes Hickson, *Ireland in the Seventeenth Century* (1884), 75.

6. Daniel Defoe, *A Tour Through the Whole Island of Great Britain,* 7th ed., 4 vols. (1769), 4:204, 337.

7. Quoted in John Burnett, *Liquid Pleasures: A Social History of Drinks in Britain* (London, 1999), 29.

8. G. S. Rousseau, "Mysticism and Millenarianism: 'Immortal Dr. Cheyne,'" in *Millenarianism and Messianism in English Literature and Thought, 1650-1800,* ed. Richard H. Popkin (Leiden, 1988), 87; D. P. Walker, *The Decline of Hell* (Chicago, 1964), 157; John Sekora, *Luxury: The Concept in Western Thought, Eden to Smollett* (Baltimore, 1977), 78-79.

9. Sekora, *Luxury,* 78-79; Walker, *The Decline of Hell,* 255-56.

10. John Pordage, *Theologia mystica* (1683); Jane Lead, *A Fountain of Gardens* (London, 1797), 2:201-2; Madam [Jeanne] Guyon, *The Worship of God in Spirit and Truth* (Bristol, 1775), 57.

11. Benjamin Franklin, *The Autobiography of Benjamin Franklin,* ed. Leonard W. Labaree (New Haven, 1964), 63, 87; Thomas Tryon, *The Merchant, Citizen, and Country-man's Instructor; or, A Necessary Companion for all People* (London, 1701), 91; George Cheyne, *Observations Concerning the Nature and Due Method of Treating the Gout* (London, 1720), 97; Roy Porter and G. S. Rousseau, *Gout: The Patrician Malady* (New Haven, 1998).

12. Anita Guerrini, "Cheyne, George (1671/2-1743)," in *Oxford Dictionary of National Biography* (Oxford, 2004), www.oxforddnb.com/view/article/5258 (accessed November 15, 2007); "The Author's Case" appended to *The English Malady,* 326.

13. Rousseau, "Mysticism and Millenarianism," 105-6.

14. Quoted in Jeremy Schmidt, *Melancholy and the Care of the Soul: Religion, Moral Philosophy, and Madness* (Burlington, 2007), 117.

15. Cheyne, *The English Malady,* 49-50; see also i-ii.

16. Ibid., 50.

17. Rousseau, "Mysticism and Millenarianism," 105-6; Roy Porter, "Introduction," *George Cheyne: The English Malady (1733)* (London, 1991), xxvi-xxxii.

18. Cheyne, "The Case of the Author," in *The English Malady* (1733), 328-30.

19. Rousseau, "Mysticism and Millenarianism," 117.

20. Antonia [*sic*] Bourignon, *An Admirable Treatise of Solid Vertue, Unknown to*

the Men of this Generation (London, 1699), 34–35; *The Light of the World* (London, 1696), xix.

21. Rousseau, "Mysticism and Millenarianism," 124.

22. Cheyne, "The Case of the Author," 335–45. For Tryon's advice on milk, see *The Good House-wife Made a Doctor* (London, 1692) and *The Merchant, Citizen, and Country-Man's Instructor* (London, 1701).

23. *The Letters of Dr. George Cheyne to the Countess of Huntingdon,* ed. Charles F. Mullett (San Marino, 1940), 2, 5, 11, 16, 21.

24. Letter from Selina, Countess of Huntingdon, to Theophilus, Earl of Huntingdon (March 30, 1732), in *In the Midst of Early Methodism: Lady Huntingdon and Her Correspondence,* ed. John R. Tyson with Boyd S. Schlenther (Lanham, 2006), 27.

25. *Letters of Dr. George Cheyne,* 26–27, 39.

26. Ibid., 43, 54.

27. Ibid., 37, 58.

28. John Wesley, *Primitive Physick* (London, 1747), 42.

29. Rebecca Spang, *The Invention of the Restaurant: Paris and Modern Gastronomic Culture* (Cambridge, Mass., 2000), 53, 269.

30. Arthur M. Wilson, *Diderot: The Testing Years, 1713–1759* (New York, 1957), 232; Anne C. Vila, *Enlightenment and Pathology: Sensibility in the Literature and Medicine of Eighteenth-Century France* (Baltimore, 1998), 102ff.

31. Alison McNeil Kettering, *The Dutch Arcadia: Pastoral Art and Its Audience in the Golden Age* (Montclair, 1983), 5.

32. Michael Preston Worley, *Pierre Julien: Sculptor to Queen Marie-Antoinette* (New York, 2003), 77–79.

33. Carolin C. Young, "Marie Antoinette's Dairy at Rambouillet," *Antiques,* October 1, 2000.

Chapter Six: Milk Comes of Age as Cheese

1. "Milk's leap" originated in Clifton Fadiman, *Any Number Can Play* (New York, 1957), 105; on "Suffolk bang," see William and Hugh Raynbird, *On the Agriculture of Suffolk* (London, 1849), 288.

2. Deborah Valenze, "The Art of Women and the Business of Men: Women's Work and the Dairy Industry, c. 1740–1840," *Past and Present,* no. 130 (1991): 142–69; G. E. Fussell, "Eighteenth-Century Traffic in Milk Products," *Economic History* 3 (1937); William Ellis, *Agriculture Improv'd,* 4 vols. (London, 1745), 2:95–98.

3. George Eliot, *Adam Bede* (Harmondsworth, 1980), 189; Rosemary Ashton, "Evans, Marian [George Eliot] (1819–1880)," in *Oxford Dictionary of National Biography* (Oxford, 2004); online ed., May 2008, www.oxforddnb.com/view/article/6794 (accessed 13 June 2010).

4. John Billingsley, *General View of the Agriculture of Somerset* (Bath, 1797), 44.

5. Adrian Henstock, "Cheese Manufacture and Marketing in Derbyshire and North Staffordshire, 1670–1870," *Derbyshire Archeological Journal* 89 (1969): 36; William Ellis, *Modern Husbandman*, 4 vols. (London, 1744), 3:62.

6. Ivy Pinchbeck, *Women Workers and the Industrial Revolution, 1750–1850* (1930; reprinted, Totowa, N.J., 1968), 14–15.

7. William Marshall, *The Rural Economy of Gloucestershire*, 2 vols. (Gloucester, 1789), 1:263, 2:104–5.

8. *A Letter from Sir Digby Legard, Bart., To the President and Vice Presidents of the Society of Agriculture for the East Riding of Yorkshire* [Beverly, 1770], 4, 6.

9. *Letters and Papers on Agriculture, Planting, &c.,* Bath and West of England Society (Bath, 1792). The letter is dated February 1778.

10. Marshall, *Rural Economy of Gloucestershire*, 2:186.

11. Marshall had quarrels with Young's methodology: Young looked at counties, while Marshall preferred to examine whole regions. But in the end, apparently, Young's social and literary flair won the coveted appointment to the directorship of the Board of Agriculture, the body that had been Marshall's idea.

12. Marshall, *Rural Economy of Gloucestershire*, 2:186.

13. Ibid., 184–85.

14. Quoted in G. E. Fussell, *The English Dairy Farmer, 1500–1900* (London, 1966), 211.

15. Marshall, *Rural Economy of Gloucestershire*, 1:297–99.

16. They also were known to put cheeses not yet paid for into storage before weighing them and then paying farmers by the pound. (Over time, cheeses lost weight through evaporation.) *Petition from Cheshire*, 1737.

17. Henstock, "Cheese Manufacture," 39–42; G. E. Fussell, "The London Cheesemongers of the Eighteenth Century," *Economic Journal* (Supplement), Economic History Series, no. 3 (London, 1928): 394–98.

18. Loyal Durand, Jr., "The Migration of Cheese Manufacture in the United States," *Annals of the Association of American Geographers* 42 (1952): 265.

19. Josiah Twamley, *Dairying Exemplified, or the Business of Cheese-making* (Warwick, 1784), 10–11.

20. Valenze, "Art of Women, Business of Men," 154–55.

21. Twamley, *Dairying Exemplified*, 70.

22. Ibid., 74–75.

23. Ibid., 75.

24. Jeremy Barlow, *The Enraged Musician: Hogarth's Musical Imagery* (Burlington, 2005), 206; P. J. Atkins, "The Retail Milk Trade in London, c. 1790–1914," *Economic History Review*, n.s., 33 (1980): 523; Sean Shesgreen, *Images of the Outcast: The Urban Poor in the Cries of London* (Manchester, 2002), 109–10, 174.

25. Quoted in Charles Phythian-Adams, "Milk and Soot: The Changing Vocabulary of a Popular Ritual in Stuart and Hanoverian London," in *The Pursuit of Urban History*, ed. Derek Fraser and Anthony Sutcliffe (London, 1983), 96.

26. Ibid., 99–104.

27. Tobias Smollett, *Humphry Clinker* (London, 1985), 153–54.

28. July 15, 1666, *The Diary of Samuel Pepys,* ed. Robert Latham and William Matthews, 11 vols. (Berkeley, 1983), 7:207.

Chapter Seven: An Interlude of Livestock History

1. "Letter of John Pory, 1619, to Sir Dudley Carleton," in *Narratives of Early Virginia, 1606–1625,* ed. Lyon Gardiner Tyler (New York, 1909), 283–84.

2. "An Account of the Colony of the Lord Baron of Baltimore, 1633," in *Narratives of Early Maryland, 1633–1684,* ed. Clayton Colman Hall (New York, 1910), 9.

3. John Hammond, *Leah and Rachel, or, the Two Fruitfull Sisters Virginia and Mary-land* [1656], in *Narratives of Early Maryland, 1633–1684,* ed. Hall, 291–92.

4. G. A. Bowling, "The Introduction of Cattle into Colonial North America," *Journal of Dairy Science* 25 (1942): 140. For an excellent account of animals in early America, see Virginia DeJohn Anderson, *Creatures of Empire: How Domestic Animals Transformed Early America* (Oxford, 2004).

5. "An Account of the Colony of the Lord Baron of Baltimore, 1633," 9. In his *Natural History of North-Carolina* [Dublin, 1737], John Bricknell recorded "large and delightful Islands" among the rivers of that area, inhabited by "large Stocks of Cattle and Deer, but scarce any Wild Beasts, and few Beasts of Prey" (p. 42). Probably related to bison or ebu, these wild beasts acted as animate indicators of the rich potential of the continent.

6. Edmund S. Morgan, *American Slavery, American Freedom: The Ordeal of Colonial Virginia* (New York, 1975), 109, 136–37.

7. *Travels and Works of Captain John Smith,* ed. Edward Arber, 2 parts (Edinburgh, 1910), 2:595.

8. Ibid., 886; James E. McWilliams, *A Revolution in Eating: How the Quest for Food Shaped America* (New York, 2005), 124.

9. Bowling, "The Introduction of Cattle," 140; Wesley N. Laing, "Cattle in Seventeenth-Century Virginia," *Virginia Magazine of History and Biography* 67 (1959): 143–64.

10. William Cronon, *Changes in the Land: Indians, Colonists, and the Ecology of New England* (New York, 1983), 132–35. The same was true for property laws and cattle in the Chesapeake area; *The Journal of John Winthrop,* ed. Richard S. Dunn and Laetitia Yeandle (Cambridge, Mass., 1996), 72, 81.

11. Edmund Berkeley and Dorothy S. Berkeley, "Another 'Account of Virginia' By the Reverend John Clayton," *Virginia Magazine of History and Biography* 76 (October 1968): 419. The letter was written to the eminent scientist Robert Boyle.

12. "Milk Sickness (Tremetol Poisoning)," *Cambridge World History of Human Disease,* ed. Kenneth R. Kiple (Cambridge, 1993), 880–83; Walter J. Daly, "The 'Slows': The Torment of Milk Sickness on the Midwest Frontier," *Indiana Magazine of History* 102 (2006): 29–40.

13. Bowling, "The Introduction of Cattle," 136–39. Scandinavians spent only a brief period colonizing Delaware, but during that time they contributed high-quality dairy cows to the genetic stew that was to become the North American dairy hybrid.

14. Cronon, *Changes in the Land*, 52, 130.

15. Sarah Kemble Knight, *The Journal of Madame Knight: A Woman's Treacherous Journey by Horseback from Boston to New York in the Year 1704* (1825; repr., Bedford, Mass., 1992), 58.

16. *The Diary of Elizabeth Drinker*, ed. Elaine Forman Crane (Boston, 1991), 244.

17. Diary of Ann Hume Shippen Livingston, January 2, 1784, in *Nancy Shippen Her Journal Book*, ed. Ethel Armes (Philadelphia, 1935), 169, 200; Sarah F. McMahon, "A Comfortable Subsistence: The Changing Composition of Diet in Rural New England, 1620–1840," *The William and Mary Quarterly*, 3rd ser., 42 (1985): 38.

18. Quoted in McWilliams, *Revolution in Eating*, 84.

19. Susannah Carter, *The Frugal Housewife* (New York, 1803), 205.

20. Amelia Simmons, *American Cookery* (Hartford, 1796), 32. The recipe was in many ways similar to that in Hannah Wooley's seventeenth-century cookbook.

21. McWilliams, *Revolution in Eating*, 173.

22. Letter from Abigail Smith Adams to John Adams, September 24, 1777, in *Familiar Letters of John Adams and Abigail Adams During the Revolution*, ed. Charles Francis Adams (New York, 1876), 313.

23. Charles Hitchcock Sherill, *French Memories of Eighteenth-Century America* (1915), 198.

24. [May 31, 1636], *The Journal of John Winthrop*, 95; Cronon, *Changes in the Land*, 72, 95, 141.

25. Winthrop quoted in Cronon, *Changes in the Land*, 141.

26. Loyal Durand, Jr., "The Migration of Cheese Manufacture in the United States," *Annals of the Association of American Geographers* 42 (1952): 263–82.

27. Richard H. Steckel, "Nutritional Status in the Colonial American Economy," *The William and Mary Quarterly*, 3rd ser., 56, no. 1 (January 1999): 38, 44, 46.

28. J. Hector St. John de Crèvecoeur, *Letters from an American Farmer* (London, 1971), letter 3, "What Is an American?" (1782), 64.

Chapter Eight: Milk in the Nursery, Chemistry in the Kitchen

1. Marilyn Yalom, *A History of the Breast* (New York, 1997), 150; for the history of infant feeding, the best general work is still Valerie Fildes, *Breasts, Bottles, and Babies: A History of Infant Feeding* (Edinburgh, 1986); for the United States, see Jacqueline H. Wolf, *Don't Kill Your Baby: Public Health and the Decline of Breastfeeding in the Nineteenth and Twentieth Centuries* (Columbus, 2001); Rima Apple, *Mothers and Medicine: A Social History of Infant Feeding, 1890–1950* (Madison, 1989).

2. Fildes, *Breasts, Bottles, and Babies,* 228.

3. Ibid., 264–65, 288–92.

4. *Jennie June's American Cookery Book* (New York, 1870), 302–3.

5. Fildes, *Breasts,* 290; "pan and spoon" from Thomas Trotter, "A View of the Nervous Temperament" (1807), in *Radical Food: The Culture and Politics of Eating and Drinking, 1780–1830,* 3 vols., ed. Timothy Morton (London, 2000), 3:643. Trotter was an advocate of breast-feeding but conceded that artificial feeding could work if "milk in the best perfection" was used.

6. Fildes, *Breasts,* 290.

7. Brouzet cited in ibid., 265.

8. Jean-Jacques Rousseau, *Émile,* trans. Barbara Foxley (1762; repr., London, 1974), 13.

9. Nancy Senior, "Aspects of Infant Feeding in Eighteenth-Century France," *Eighteenth-Century Studies* 16 (1983): 380; Ann Taylor Allen, *Feminism and Motherhood in Germany, 1800–1914,* n.p., cited in E. Melanie DuPuis, *Nature's Perfect Food: How Milk Became America's Drink* (New York, 2002), 51–52.

10. Senior, "Aspects," 384.

11. Fildes, *Breasts,*

12. Ibid., 268–71, 274.

13. Ann F. La Berge, "Medicalization and Moralization: The Crèches of Nineteenth-Century Paris," *Journal of Social History* 25 (1991): 74.

14. The figure is from Robert M. Hartley, *An Historical, Scientific, and Practical Essay on Milk as an Article of Human Sustenance* (1842), cited also in DuPuis, *Nature's Perfect Food,* 49.

15. DuPuis, *Nature's Perfect Food,* 54–55.

16. Wolf, *Don't Kill Your Baby,* 9–41.

17. William H. Brock, *The Norton History of Chemistry* (New York, 1992), 65.

18. Louis Rosenfeld, "William Prout: Early 19th-Century Physician-Chemist," *Clinical Chemistry* 49, no. 4 (2003): 699–705; W. H. Brock, "Prout, William (1785–1850)," in Oxford Dictionary of National Biography, ed. W. G. H. Matthews and Brian Harrison (Oxford, 2004), www.oxforddnb.com/view/article/22845 (accessed November 3, 2009).

19. *On the Ultimate Composition of Simple Alimentary Substances; With Some Preliminary Remarks on the Analysis of Organized Bodies in General* (London, 1827), 5.

20. *Chemistry, Meteorology, and the Function of Digestion Considered with Reference to Natural Theology* (London, 1834), 481–82.

21. Ibid., 478–79.

22. T. B. Mepham, "'Humanising Milk': The Formulation of Artificial Feeds for Infants, 1850–1910," *Medical History* (1993): 227–28; William H. Brock, *Justus von Liebig: The Chemical Gatekeeper* (Cambridge, 1997), 230–32.

23. Robert M. Hartley, *An Historical, Scientific, and Practical Essay on Milk* (New York, 1842; repr., New York, 1977), 74–75.

24. Ibid., 232n; Peter J. Atkins, "London's Intra-Urban Milk Supply, circa

1790–1914," *Transactions of the Institute of British Geographers,* n.s., 2 (1977): 384–85, 387, 395.

25. Hartley, *Essay on Milk,* 110, 139, 233–34, 242–43.

26. Ibid., 123–26.

27. Ibid., 105–6, citing Prout's *Chemistry, Meteorology, and the Function of Digestion Considered with Reference to Natural Theology* [479–80].

28. John Burnett, *Plenty and Want: A Social History of Food in England, 1815 to the Present Day,* 3rd ed. (Abingdon, 1989), 89–91.

29. The latter publication outlined the chemical dynamics of baking and brewing. Accum also sold simple chemistry sets for the hobbyist. Brian Gee, "Accum, Friedrich Christian (1769–1838)," in *Oxford Dictionary of National Biography* (Oxford, 2004), www.oxforddnb.com/view/article/56 (accessed 15 December 2009).

30. Accum, *Culinary Chemistry* (London, 1821), iv, 3.

31. William H. Brock, *Justus von Liebig: The Chemical Gatekeeper* (Cambridge, 1997), chaps. 8–10. See also Brock, *History of Chemistry,* 194–207.

32. Brock, *Justus von Liebig,* 220. On domestic management, see Laura Schapiro, *Perfection Salad: Women and Cooking at the Turn of the Century* (Berkeley, 1986).

33. See Mark Finlay's entertaining treatment of these subjects in "Quackery and Cookery: Justus Von Liebig's Extract of Meat and the Theory of Nutrition in the Victorian Age," *Bulletin of the History of Medicine* 66 (1992): 404–18, and "Early Marketing of the Theory of Nutrition: The Science and Culture of Liebig's Extract of Meat," in *The Science and Culture of Nutrition, 1840–1940* (Amsterdam, 1995), 48–74.

34. Brock, *Justus von Liebig,* 238–43; Finlay, "Quackery," 409; Finlay, "Early Marketing," 58–60.

35. Brock, *Justus von Liebig,* 243–45.

36. Colin A. Russell, *Edward Frankland: Chemistry, Controversy, and Conspiracy in Victorian England* (Cambridge, 1996), 180–82; Colin A. Russell, "Frankland, Sir Edward (1825–1899)," in *Oxford Dictionary of National Biography* (Oxford, September 2004); online edition, October 2006 www.oxforddnb.com/view/article/10083 (accessed January 13, 2008).

37. Russell, *Edward Frankland,* 182.

38. Brock, *Justus von Liebig,* 246.

39. Rima Apple, *Mothers and Medicine: A Social History of Infant Feeding, 1850–1950* (Madison, 1987).

40. Brock, *Justus von Liebig,* 246–47.

41. Rima D. Apple, "'Advertised by Our Loving Friends': The Infant Formula Industry and the Creation of New Pharmaceutical Markets, 1870–1910," *Journal of the History of Medicine and Allied Sciences* 41 (1986): 7.

42. H. Lebert, *A Treatise on Milk and Henri Nestlé's Milk Food, for the Earliest Period of Infancy and in Later Years* (Vevey, 1881), 13–14, 21.

43. Ibid., 14–15.

44. Quoted in Apple, "The Infant Formula Industry," 10.

45. J. A. B. Dumas, "The Constitution of Blood and Milk," *Philosophical Magazine and Journal of Science,* 4th series, 42 (1871): 129–38. The American nutritionist E. V. McCollum was the first to point out the significance of Dumas's observations, in "Who Discovered Vitamins?" *Science* 118 (November 1953): 632.

Chapter Nine: Beneficial Bovines and the Business of Milk

1. Letter from Eliza Newton Woolsey Howland to Joseph Howland, May 7, 1862, in *Letters of a Family During the War for the Union, 1861–1865,* ed. Georgeanna M. W. Bacon (1899), 1:338.

2. On Civil War accounts of hardtack, see James I. Robertson, Jr., *Soldiers Blue and Gray* (Charleston, 1988), 68–69. As late as 1948, a Connecticut resident reported ownership of an antique jar of Borden's biscuit, still "free from discoloration or odor." Joe B. Frantz, "Gail Borden as a Businessman," *Bulletin of the Business Historical Society* 22 (1948): 126n.

3. Frantz, "Gail Borden," 126.

4. Roberta C. Hendrix, "Some Gail Borden Letters," *Southwestern Historical Quarterly* 51 (1947): 133; Frantz, "Gail Borden," 124.

5. Hendrix, "Some Gail Borden Letters," 133–36.

6. Ibid., 134–36; Mark Finlay, "Early Marketing of the Theory of Nutrition: The Science and Culture of Liebig's Extract of Meat," in *The Science and Culture of Nutrition, 1840–1940* (Amsterdam, 1995), 54.

7. Clarence B. Wharton, *Gail Borden, Pioneer* (San Antonio, 1941), 182. Wharton's volume, with its school-font format, is the closest thing to a true biography of Borden, but its language and style are too adulatory to make it completely reliable.

8. Alan Davidson, *The Oxford Companion to Food* (Oxford, 1999), 130–31; Accum, *Culinary Chemistry,* 213–14.

9. Wharton, *Gail Borden,* 180–93; see also Otto F. Hunziker, *Condensed Milk and Milk Powder,* 4th ed. (La Grange, Ill., 1926), 4–5.

10. Hendrix, "Some Gail Borden Letters," 136.

11. Frantz, "Gail Borden as a Businessman," 128; Wharton, *Gail Borden,* 197–98; Hunziker, *Condensed Milk,* 4.

12. Robertson, *Soldiers Blue and Gray,* 65–66; John D. Billings, *Hardtack and Coffee: Or, the Unwritten Story of Army Life* (Boston, 1887), 124–25; William H. Brock, *Justus von Liebig: The Chemical Gatekeeper* (Cambridge, 1997), 242.

13. Billings, *Hardtack and Coffee,* 125.

14. Frantz, "Gail Borden as a Businessman," 128–30.

15. J. C. Drummond and Anne Wilbraham, *The Englishman's Food* (London, 1958), 302.

16. R. E. Hodgson, "The Dairy Industry in the United States," in *Dairying*

Throughout the World, International Dairy Federation Monograph (Munich, 1966), 17; F. W. Baumgartner, *The Condensed Milk and Milk Powder Industries* (Kingston, Ont., 1920), 8–9. By 1900, the cow population in the United States had reached 16,292,000, when the human population stood at roughly 76 million.

17. Baumgartner, *Condensed Milk,* 3n.

18. "Originator of Phrase, 'From Contented Cows,' Tells History of World-Famous Slogan," *The Carnation* 12 (May–June 1932): 24. Thanks to Jean Thomson Black for a copy of this article and for sharing information about her paternal grandmother with me.

19. Jean Heer, *Nestlé: 125 Years, 1866–1991* (Vevey, 1991), 50, 52.

20. Ibid., 53, 58.

21. Heer, *Nestlé,* 68; Baumgartner, *Condensed Milk,* 3–4.

22. Thomas Fenner, "Die Berneralpen Milchgesellschaft: Ein internationales Unternehmen im Herzen des Emmentals," in *Das Emmental-Ansichten einer Region,* ed. Fritz Von Gunten (Münsingen, Switzerland, 2006), 8–10. Thanks to Thomas Fenner for sending me a copy of this article and for his generous advice on Swiss milk history.

23. Baumgartner, *Condensed Milk,* 10–11.

24. Thanks to Merry White for information on Western dietary influences and milk-drinking in Japan; Baumgartner, *Condensed Milk,* 7; Christopher T. C. Faung, "The Dairy Industry in Taiwan," *Dairying Throughout the World,* 64.

25. Fussell, *English Dairy Farmer,* 290–98; *Journal of the Royal Agricultural Society,* 2nd ser., 8 (1872): 103–57.

26. Hunziker, *Condensed Milk,* 548; T. A. B. Corley, "Horlick, Sir James (1844–1921)," in *Oxford Dictionary of National Biography,* ed. H. C. G. Matthews and Brian Harrison (Oxford, 2004), www.oxforddnb.com/view/article/39011 (accessed November 1, 2009); www.horlicks.com/ind/ba_history.html (accessed March 3, 2004).

27. Hunziker, *Condensed Milk,* 548.

28. Eric J. Hobsbawm, *The Age of Capital, 1848–1875* (London, 1975), 208–9; Sidney W. Mintz, *Sweetness and Power: The Place of Sugar in Modern History* (New York, 1985), 208.

29. Francesco Chiapparino, "Milk and Fondant Chocolate and the Emergence of the Swiss Chocolate Industry at the Turn of the Twentieth Century," in *Food and Material Culture: Proceedings of the Fourth Symposium of the International Commission for Research into European Food History,* ed. Martin R. Schärer and Alexander Fenton (Edinburgh, 1998), 330–44; Alan Davidson, *The Oxford Companion to Food* (Oxford, 1999), 179.

30. Chiapparino, "Milk and Fondant Chocolate," 338.

31. Davidson, *Oxford Companion to Food,* 392–93; Margaret Visser, *Much Depends on Dinner* (New York, 1986), 306; Paul G. Heineman, *Milk* (Philadelphia, 1921), 631. For Ben & Jerry's, see www.xsbusiness.com/Business/Ben-And-Jerry.html.

32. Heineman, *Milk,* 631–32.

33. Visser, *Much Depends on Dinner,* chap. 9.

34. Eric J. Hobsbawm, *The Age of Capital, 1848–1875* (London, 1975), 17 and chap. 10, and also Hobsbawm, *The Age of Empire, 1875–1914* (London, 1987). By 1900, the United States finally joined western Europe in having a larger proportion of its population living in cities and towns; in 1920, only 30 percent of the population lived on farms. R. Douglas Hurt, *Problems of Plenty: The American Farmer in the Twentieth Century* (Chicago, 2002), 9; Maryanna S. Smith, comp., *Chronological Landmarks in American Agriculture,* U.S. Department of Agriculture, Agriculture Information Bulletin no. 425 (Washington, D.C., 1980), 46.

35. "The Dairy Industry in Denmark," in *Dairying Throughout the World,* 166; T. R. Pirtle, *History of the Dairy Industry* (Chicago, 1926; repr., 1973), 272, 276.

Chapter Ten: Milk in an Age of Indigestion

1. Thomas Carlyle to Jane Welsh Carlyle, Scotsbrig, August 7, 1843; Thomas Carlyle to Jane Welsh Carlyle, Scotsbrig, [September 11, 1847]; Jane Welsh Carlyle to Mary Austin, Chelsea, January 2, 1857; Thomas Carlyle to John A. Carlyle, Chelsea, May 1, 1857. The couple distrusted city milk suppliers: "To give real money for imaginary milk is a thing I will not consent to," Carlyle grumbled from Chelsea in a letter to his mother. Thomas Carlyle to Margaret A. Carlyle, Chelsea, July 6, 1834. All letters are from *The Carlyle Letters Online* [CLO], 2007, http://carlyleletters.org. I am grateful to Merry White for pointing me to Carlyle as a subject of milk research.

2. Last three quotations from www.malcolmingram.com/CARLYLEA.HTM.

3. *The Herald of Health* (1875), vol. 25–26, p. 85.

4. Sabine Merta, "Karlsbad and Marienbad: The Spas and Their Cures," in *The Diffusion of Food Culture in Europe from the Late Eighteenth Century to the Present Day,* ed. Derek J. Oddy and Lydia Petranova (Prague, 2005), 152–63; Derek J. Oddy, *From Plain Fare to Fusion Food* (Woodbridge, 2003), 33. I owe thanks to Professor Oddy for his advice on the topic of dyspepsia in relation to milk history.

5. For "Another Remarkable Cure of Dyspepsia," see *Provincial Freeman,* April 4, 1857; John H. Winslow, *Darwin's Victorian Malady* (Philadelphia, 1971), esp. 58–74.

6. Andrew Combe, *The Physiology of Digestion,* 10th ed., ed. James Coxe (Edinburgh, 1860), 14; Thomas K. Chambers, *Digestion and Its Derangements* (London, 1856), 193.

7. Andrew Combe, *The Physiology of Digestion,* 9th ed., ed. James Coxe (Edinburgh, 1849), 112; *The Lancet,* 1862; Mary Tyler Peabody Mann, *Christianity in the Kitchen* (1857), quoted in Sidonia C. Taupin, "'Christianity in the Kitchen' or A Moral Guide for Gourmets," *American Quarterly* 15 (1963): 86.

8. Taupin, "'Christianity in the Kitchen,'" 86–88.

9. Combe, *Physiology of Digestion,* 9th ed., 101.

10. Ibid., 76, 111–12, 115.

11. S. O. Habershon, *Pathological and Practical Observations on Diseases of the Abdomen,* 2nd ed. (London, 1862), 214–17.

12. Philip Karell, M.D., "On the Milk Cure," *Edinburgh Medical Journal* 12 (1866): 98–102.

13. Ibid., 103–4.

14. Ibid., 104–5.

15. Ibid., 111. It was sometimes the case that tuberculosis would disappear on its own accord; in this instance, improved nutrition may have enhanced the ability of the autoimmune system to do its work. F. B. Smith, *The Retreat of Tuberculosis* (London, 1988).

16. Chambers, *Digestion and Its Derangements,* 363.

17. Ibid., 1, 193.

18. Karell, "Milk Cure," 103; Merta, "Karlsbad and Marienbad," 153.

Chapter Eleven: Milk Gone Bad

1. John Spargo, *The Common Sense of the Milk Question* (New York, 1908), 88; Henry E. Alvord, "Dairy Products at the Paris Exposition of 1900," *United States Department of Agriculture, Seventeenth Annual Report of the Bureau of Animal Industry* (Washington, 1901), 201–2, 219.

2. Peter J. Atkins lists a fairly complete roster of diseases communicated by milk, including the following: infectious hepatitis, anthrax, botulism, cholera, listeriosis, salmonellosis, shigellosis, staphylococcal gastroenteritis, and several others. See P. J. Atkins, "White Poison? The Social Consequences of Milk Consumption, 1850–1930," *Social History of Medicine* 5 (1992): 216.

3. Boston physician Milton J. Rosenau offers a complete discussion of early-twentieth-century pasteurization methods in *The Milk Question* (Boston, 1912), 185–230. See also Stuart Patton, *Milk* (New Brunswick, 2004), 192.

4. Richard A. Meckel, *Save the Babies: American Public Health Reform and the Prevention of Infant Mortality, 1850–1929* (Ann Arbor, 1998), 81; Michael French and Jim Phillips, *Cheated not Poisoned? Food Regulation in the United Kingdom, 1875–1938* (Manchester, 2000), 175–84.

5. Bruno Latour, *The Pasteurization of France* (Cambridge, 1988), 21, 25, 36.

6. Rachel G. Fuchs, *Poor and Pregnant in Paris: Strategies for Survival in the Nineteenth Century* (New Brunswick, 1992), 61; Deborah Dwork, *War Is Good for Babies and Other Young Children: A History of the Infant and Child Welfare Movement in England, 1898–1918* (London, 1987), 4.

7. Dwork, *War Is Good for Babies,* 6; Spargo, *Common Sense,* 5.

8. Deborah Dwork, "The Milk Option: An Aspect of the History of the Infant Welfare Movement in England, 1898–1908," *Medical History* 31 (1987): 52.

9. Meckel, *Save the Babies,* 18–19. It was generally acknowledged that differences pertained according to social class. In London, according to one investiga-

tor, "of the children born in the best part of the town, one-fifth die before they attain the fifth year; of the children born in the worst, one-half die before they attain the fifth year" (p. 19).

10. Dwork, "The Milk Option," 52; Spargo, *Common Sense*, 157; Meckel, *Save the Babies*, 38.

11. Spargo, *Common Sense*, 39.

12. G. F. McCleary, *Infantile Mortality and Infants Milk Depots* (London, 1905), 38–39.

13. Dwork, *War Is Good for Babies*, 57–58, 95–97; Spargo, *Common Sense*, 40.

14. Spargo, *Common Sense*, 39; Dwork, *War Is Good for Babies*, 54.

15. *British Medical Journal*, April 25, 1903, 974–75; Dwork, *War Is Good for Babies*, 64.

16. *British Medical Journal*, April 25, 1903, 976; Dwork, *War Is Good for Babies*, 101–4.

17. Nathan Straus, "How the New York Death Rate Was Reduced," *The Forum* (November 1894), reprinted in Lina Straus, *Disease in Milk, The Remedy Pasteurization: The Life and Work of Nathan Straus*, 2nd ed. (New York, 1917), 180–83; Meckel, *Save the Babies*, 78.

18. Meckel, *Save the Babies*, 41; Rosenau, *Milk Question*, 187.

19. Straus, *Disease in Milk*, 180–83.

20. *British Medical Journal*, April 25, 1903, 973; Dwork, *War Is Good for Babies*, 105.

21. Mabel Potter Daggett, "Women: The Larger Housekeeping," *World's Work* (May–October 1912), repr. in *Public Women, Public Words: A Documentary History of American Feminism*, ed. Dawn Keetley and John Charles Pettigrew, 3 vols. (Madison, 1997–2002), 2:126–28; Dorothy Worrell, *The Women's Municipal League of Boston: A History of Thirty-Five Years of Civic Endeavor* [Boston, 1943], 18–20.

22. Nancy Woloch, *Women and the American Experience*, 3rd ed. (Boston, 2000), 297–306; Seth Koven and Sonya Michel, "Womanly Duties: Maternalist Politics and the Origins of Welfare States in France, Germany, Great Britain, and the United States, 1880–1920," *American Historical Review* 95 (1990): 1107.

23. Worrell, *Women's Municipal League of Boston*, xiii, 12–15.

24. Daggett, "Women: The Larger Housekeeping," 129.

25. J. H. M. Knox, "The Claims of the Baby in the Discussion of the Milk Question," *Charities and the Commons* 16 (1906): 492.

26. Meckel, *Save the Babies*, 85.

27. Sonya Michel and Robyn Rosen, "The Paradox of Maternalism: Elizabeth Lowell Putnam and the American Welfare State," *Gender and History* 4 (1992): 369.

28. Ibid., 364–86. Though Elizabeth Lowell Putnam's views were never as outspokenly prejudiced as those of her brother (his persecution of homosexuals at Harvard and his campaign to block the appointment of Louis D. Brandeis to the

U.S. Supreme Court ultimately brought notoriety to the Lowell family), in her final decades she, too, expressed distinctively conservative views. She was a vigorous opponent of renewal of the Sheppard-Towner Act, which provided health and welfare support for mothers across the nation.

29. Elizabeth Lowell Putnam, "Report of the Committee on Milk to the Women's Municipal League of Boston" (1909), 1–2 (section titled "Lucubrations on Milk"); Schlesinger Library, Elizabeth Lowell Putnam Papers [hereafter SL ELPP], MC 360, box 3, folder 44.

30. Spargo, *Common Sense,* 94–104.

31. Putnam, "Report," 2.

32. Michel and Rosen, "Paradox of Maternalism," 369; Rosenau, *Milk Question,* 191. See also Daniel Block, "Saving Milk Through Masculinity: Public Health Officers and Pure Milk, 1880–1930," *Food and Foodways* 13 (2005): 115–34.

33. Massachusetts State Department of Health, *Report of the Special Milk Board* (Boston, 1916), 199.

34. Ibid., 225.

35. Rosenau, *Milk Question,* 186, 189.

36. Letter of March 9, 1911, SL ELPP, MC 360, box 4, folder 49.

37. Massachusetts State Department of Health, *Report of the Special Milk Board,* 225.

38. *Is Loose Milk a Health Hazard?* Report of the Milk Commission, Department of Health, New York City (New York, 1931), 39.

39. "Pay Rate or Get No Milk," *Boston Daily Globe,* April 28, 1910.

40. Rosenau, *Milk Question,* 192; "Milk Order in Force June 15," *Boston Daily Globe,* April 27, 1910; "Milk Men Get Ready," *Boston Daily Globe,* April 29, 1910; Adel P. den Hartog, "Serving the Urban Consumer: The Development of Modern Food Packaging with Special Reference to the Milk Bottle," *Food and Material Culture,* ed. Martin R. Schärer and Alexander Fenton (Vevey, 1998), 248–67.

41. "Farmers Say Milk Should Be Higher," *Hartford Courant,* May 11, 1910.

42. "Advantage to Boston," *Boston Daily Globe,* May 14, 1910.

43. "Milk War Over," *Boston Daily Globe,* June 8, 1910.

44. Putnam's favored regulations, collected under the Ellis Bill, were never passed. Her preference and that of the Women's Municipal League would have been to place the administration of milk regulations under the State Board of Health, rather than in the hands of the Department of Agriculture. Worrell, *Women's Municipal League,* 15.

45. Alan Czaplicki, "'Pure Milk Is Better than Purified Milk': Pasteurization and Milk Purity in Chicago, 1908–1916," *Social Science History* 31 (2007): 411; F. B. Smith, *The Retreat of Tuberculosis* (London, 1988), 183; French and Phillips, *Cheated not Poisoned,* 158–84; Summerskill quoted on 161.

46. E. Melanie DuPuis, *Nature's Perfect Food* (New York, 2002), 88; Czaplicki, "'Pure Milk,'" 420. According to Czaplicki, the problem demonstrates perfectly the "'garbage can' model of organizational decision making," at least in Chicago,

the first American city to legally enforce the pasteurization of cow's milk. There, the right decisions were made because they were closest at hand and an expedient consensus could be reached.

47. Jacqueline H. Wolf, *Don't Kill Your Baby: Public Health and the Decline of Breastfeeding in the Nineteenth and Twentieth Centuries* (Columbus, 2001); Harvey Levenstein, "'Best for Babies' or 'Preventable Infanticide'? The Controversy over Artificial Feeding of Infants in America," *Journal of American History* 70 (1983): 75–94.

48. Spargo, *Common Sense,* 45; McCleary, *Infantile Mortality,* 38. McCleary later became principal medical officer to the National Health Insurance Commission in England.

Chapter Twelve: The ABC's of Milk

1. Massachusetts State Department of Health, *Report of the Special Milk Board* (Boston, 1916), 188. Fresh liquid milk proved scarce even in some rural regions in the early years of the century. In England, because of contracts between Midland farmers and London, "tins of Swiss condensed milk were often seen" on "farmhouse tables." G. F. Fussell, *The English Dairy Farmer, 1500–1900* (London, 1966), 315.

2. Richard Perren, *Agriculture in Depression* (Cambridge, 1995), 9–10.

3. R. Douglas Hurt, *Problems of Plenty: The American Farmer in the Twentieth Century* (Chicago, 2002), 10, 49.

4. W. H. Glover, *Farm and College: The College of Agriculture of the University of Wisconsin, A History* (Madison, Wis., 1952), 26, 87.

5. Ibid., 91, 93. The first "Experiment Station" for agricultural research in the U.S. was started at Wesleyan University in 1875. Federal aid became available to such programs beginning in 1887 with the passage of the Hatch Act.

6. Glover, *Farm and College,* 97; Lincoln Steffens, "Sending a State to College: What the University of Wisconsin Is Doing for Its People," *American Magazine* 67 (1909): 349, 353.

7. Glover, *Farm and College,* 126–27, 188–89. The struggle to test and eradicate all tubercular cattle would continue through the 1890s into the postwar period.

8. E. Parmalee Prentice, *American Dairy Cattle: Their Past and Future* (New York, 1942), 155.

9. Elmer V. McCollum, *The Newer Knowledge of Nutrition: The Use of Food for the Preservation of Vitality and Health* (New York, 1918), [vii].

10. Ibid., 67.

11. Edwards A. Park, "Foreword," in Elmer V. McCollum, *From Kansas Farm Boy to Scientist: The Autobiography of Elmer V. McCollum* (Lawrence, Kan., 1964), xii.

12. McCollum, *From Kansas,* 13–15.

13. Ibid., 11, 20, 26.

14. Ibid., 80–111, 115; Harry G. Day, "Elmer Verner McCollum, 1879–1967," National Academy of Sciences Biographical Memoir Series (Washington, D.C., 1974).

15. Glover, *Farm and College*, 114. Evidently, Chittenden had allowed commercial makers of borax to make claims based on his research.

16. Ibid., 112–13; McCollum, *From Kansas*, 116.

17. McCollum, *From Kansas*, 116–17; Glover, *Farm and College*, 133–48, 300–301. Henry was a widely popular dean of the College of Agriculture from 1891 to 1907, reputed to be able to "do almost anything."

18. McCollum, *From Kansas*, 115.

19. Glover, *Farm and College*, 180.

20. Ibid., 92–95.

21. McCollum, *From Kansas*, 119.

22. Fussell, *English Dairy Farmer*, 314.

23. Glover, *Farm and College*, 121–24.

24. McCollum, *From Kansas*, 114.

25. Ibid., 118–21.

26. Ibid., 124–25. This was not the last time McCollum fought for pay for an unpaid female assistant. During his time at Johns Hopkins, he sensed that money was not forthcoming because Elsa Orent, a Russian-born assistant, was Jewish. Margaret W. Rossiter, *Women Scientists in America: Struggles and Strategies to 1940* (Baltimore, 1982), 373n.

27. Day, "McCollum," 273–77; McCollum, *From Kansas*, 133–34.

28. Day, "McCollum," 278. The master's thesis of Cornelia Kennedy was "the first to use 'A' and 'B' to designate the new dietary essentials," but McCollum eventually came to cite a later paper of his own as the origin of letter nomenclature. She later received her Ph.D. at Johns Hopkins, working again with McCollum, and went on to publish thirty-two articles over thirty-three years. The special hardships of women in science are painfully clear in an account of her career by Patricia B. Swan, "Cornelia Kennedy (1881–1969)," *Journal of Nutrition* 124 (1994): 455–60.

29. Glover, *Farm and College*, 303. In 1922, McCollum and a team of researchers at Johns Hopkins announced that they had "captured" a fourth vitamin, which they called vitamin D (*New York Times*, June 19, 1922).

30. Cited in Glover, *Farm and College*, 299.

31. Russell H. Chittenden, "Lafayette Benedict Mendel, 1872–1935," National Academy of Sciences Biographical Memoir Series (Washington, D.C., 1936), 132.

32. McCollum, *From Kansas*, 155–56.

33. Ibid., 165–69.

34. Ibid., 127, 135. He admitted with chagrin that he had overlooked the instances when this had been occurring with the heifer experiment. "As a Kansas farmer I should have detected an important fact," he confessed; the cows were

often eating leaves with seeds when they ate freshly harvested corn. These two ingredients added up to more than the sum of their parts (p. 127).

35. H. H. Dale, "Hopkins, Sir Frederick Gowland (1861–1947)," in *Oxford Dictionary of National Biography*, ed. H. C. G. Matthew and Brian Harrison (Oxford, 2004), www.oxforddnb.com/view/article/33977 (accessed November 2, 2009); McCollum, *Newer Knowledge*, 18.

36. Ibid., 71.

37. McCollum, *From Kansas*, 159–62.

38. Ibid., 165–70; Day, "McCollum," 278, 284, 286.

39. McCollum, *Newer Knowledge*, 150–51.

40. Ibid., 152.

41. McCollum, *From Kansas*, 163–64; Day, "McCollum," 286.

Chapter Thirteen: Good for Everybody in the Twentieth Century

1. Wyn Grant, *The Dairy Industry: An International Comparison* (Aldershot, 1991), 20.

2. I am indebted to Frank Trentmann for my discussion of milk and consumer politics, much of it drawn from his superlative essay, "Bread, Milk, and Democracy: Consumption and Citizenship in Twentieth-Century Britain," in *The Politics of Consumption: Material Culture and Citizenship in Europe and America*, ed. Martin Daunton and Matthew Hilton (Oxford, 2001), 129–64.

3. Ibid., 140–41 and 140n. France was only slightly ahead of Britain, consuming 16 gallons per capita in 1902. See John Burnett, *Plenty and Want*, 158, cited in Trentmann, "Bread, Milk, and Democracy," 140n.34.

4. William Clinton Mullendore, *History of the United States Food Administration, 1917–1919* (Stanford, 1941), 236.

5. Trentmann, "Bread, Milk, and Democracy," 139, 145. As Trentmann puts it, this new consumer politics "was now tied directly to demands for permanent controls and for consumer representation in economic decision-making to ensure greater transparency in pricing mechanisms and profits" (145–46).

6. Ibid., 142.

7. Ibid., 139.

8. *Cooperative News*, December 27, 1919, 1; "The Poor and Milk," *Cooperative News*, November 22, 1919.

9. Mullendore, *History of the United States Food Administration*, 237–39; Harvey A. Levenstein, *Revolution at the Table: The Transformation of the American Diet* (New York, 1988), 109–10; Trentmann, "Bread, Milk, and Democracy," 146.

10. Mullendore, *History of the United States Food Administration*, 237–39; Dorothy Reed Mendenhall, *Milk: The Indispensable Food for Children*, U.S. Department of Labor, Children's Bureau Care of Children Series (Washington, D.C., 1918), 8–9.

11. Mullendore, *History of the United States Food Administration,* 237–38; Trentmann, "Bread, Milk, and Democracy," 143; Edith Whetham, "The London Milk Trade, 1900–1930," in *The Making of the Modern British Diet,* ed. Derek J. Oddy and Derek S. Miller (London, 1976), 73.

12. Levenstein, *Revolution,* 113–17.

13. Mendenhall, *Milk,* 6.

14. Ibid., 7–8.

15. See Ellen Ross, *Love and Toil: Motherhood in Outcast London, 1870–1918* (Oxford, 1993), for a vivid treatment of milk and motherhood in the nineteenth century.

16. Peter Atkins, "School Milk in Britain, 1900–1934," *Journal of Policy History* 19 (2007): 400–401.

17. Atkins, "School Milk," 404; Jon Pollock, "Two Controlled Trials of Supplementary Feeding of School Children in the 1920s," *Journal of the Royal Society of Medicine* 99 (2006): 323–27.

18. Pollock, "Two Controlled Trials," 324; Francis McKee, "Popularisation of Milk as a Beverage in the 1930s," in *Nutrition in Britain: Science, Scientists, and Politics in the Twentieth Century,* ed. David F. Smith (London, 1997), 125–27.

19. *Cooperative News,* November 8, 1919, 12.

20. Sybil Oldfield, "Pye, Edith Mary (1876–1965)," in *Oxford Dictionary of National Biography* (Oxford, 2004), www.oxforddnb.com/view/article/37871 (accessed August 30, 2009).

21. Levenstein, *Revolution,* 154.

22. Trentmann, "Bread, Milk, and Democracy," 147; Whetham, "London Milk Trade," 70.

23. Atkins, "School Milk," 407; Levenstein, *Revolution,* 151–55; "The National Milk Publicity Council: What It Is and What It Aims to Do," *Milk Trade Gazette* (October 1930): 14.

24. "Borden's Evaporated Milk Book of Recipes," n.d.; "Treasured Recipes: Sanitary Farm Dairies," 1938; "Milk Dishes for Modern Cooks," 1939. I am grateful to Merry White for giving me these booklets.

25. "Milk Dishes for Modern Cooks," 3.

26. Ibid., 4.

27. Advertising leaflet in the Wellcome Library Collection [n.d.].

28. *Times* (London), September 4, 1936, February 25, 1937; McKee, "Popularisation," 136–37.

29. McKee, "Popularisation," 137.

30. "Milk Bar as Social Leveller," *Times* (London), December 15, 1964.

31. R. W. Bartlett, *The Price of Milk* (Danville, Ill., 1941), 44–45.

32. Kenneth W. Bailey, *Marketing and Pricing of Milk and Dairy Products in the United States* (Ames, Iowa, 1997), 4.

33. J. P. Cavin, Hazel K. Stiebeling, and Marius Farioletti, "Agricultural Surpluses and Nutritional Deficits," in *Farmers in a Changing World: The Yearbook of*

Agriculture, 1940, United States Department of Agriculture (Washington, D.C., 1940), 337.

34. Ibid., 339-40; Charles Smith, *Britain's Food Supplies* (London, 1940), 272-74; Grant, *The Dairy Industry,* 20-21. At first, the federal government fixed minimum prices for milk sold by both farmers and retailers. Seven months later, regulations of consumer prices were "abandoned because of inability to enforce them." But minimum prices for dairy farmers continued, reinforced by a variety of state laws. Bartlett, *Price of Milk,* 44.

35. Gove Hambridge, "Farmers in a Changing World — A Summary," in *Farmers in a Changing World,* 56. These were the words of Milo Perkins of the U.S. Surplus Marketing Administration.

36. Sir John Boyd Orr, "Agriculture and National Health," in *Agriculture in the Twentieth Century* (Oxford, 1939), 426.

37. Cavin, Stiebeling, and Farioletti, "Agricultural Surpluses," 333.

38. Peter Atkins, "School Milk in Britain, 1900-1934," *Journal of Policy History* 19 (2007): 395-427; in Britain, the "years 1929-34 were the crucial hinge point." See also Peter J. Atkins, "Fattening Farmers or Fattening Children? School Milk in Britain, 1921-1941," *Economic History Review* 58 (2005): 75-76; Susan Levine, *School Lunch Politics* (Princeton, 2009), 34, 42.

39. Martin Doornbos and K. N. Nair, "The State of Indian Dairying: An Overview," in *Resources, Institutions, and Strategies: Operation Flood and Indian Dairying,* ed. M. Doornbos and K. N. Nair (New Delhi, 1990), 11. Egypt, and to a lesser extent Eastern Europe and Italy, also depend on a fair amount of buffalo milk. In 1974, the Food and Agriculture Organization of the United Nations declared the Asian water buffalo "the most neglected animal" across the developing world.

40. Verghese Kurien, *I Too Had a Dream* (New Delhi, 2005), 3-19.

41. Ibid., 11-12.

42. Ibid., 13.

43. Pratyusha Basu, *Villages, Women, and the Success of Dairy Cooperatives in India* (Amherst, N.Y., 2009), 58-61.

44. Kurien points out: "It was little wonder then that in the US, the capital of capitalism, 85 per cent of the dairy industry was cooperative. In New Zealand, Denmark and Holland, 100 per cent and in erstwhile West Germany 95 per cent of the dairy industry was cooperative" (*I Too Had a Dream,* 56).

45. Ibid., 55, 82. At first, the national government made possible an extension of the "Anand model" by legislating into existence a National Dairy Development Board in 1965.

46. The situation only worsened during the following two decades, so that by 1983, Europe was buying only 54 percent of its total butter production and only 10 percent of its total skim milk powder. Martin Doornbos, Frank van Dorsten, Manoshi Mitra, Plet Terhal, *Dairy Aid and Development: India's Operation Flood* (New Delhi, 1990), 53, 54-55.

47. Ibid., 53–54.

48. Kurien, *I Too Had a Dream,* 61. In 1956, even before Operation Flood was under way, Kurien spurned the advances of Nestlé's because Swiss management indicated that "condensed milk was an extremely delicate procedure and they 'could not leave it to the natives to make.'" In the early 1960s, Amul faced down Glaxo after developing an infant formula of its own (59, 69–72).

49. Basu, *Villages, Women,* 197.

50. Marty Chen, Manoshi Mitra, Geeta Athreya, Anila Dholakia, Preeta Law, Aruna Rao, *Indian Women: A Study of their Role in the Dairy Movement* (New Delhi, 1986); Basu, *Villages, Women,* 78–80.

51. Kurien, *I Too Had a Dream,* 148–49; Basu, *Villages, Women,* passim.

52. Kurien, *I Too Had a Dream,* 42–61.

53. Figures through 1998 are from "India: World's Largest Milk Producer," The Indian Dairy Industry, www.indiadairy.com/ind_world_number_one_milk_pro ducer.html. For Indian milk production in 2006–7, "The World Dairy Situation, 2009," *Bulletin of the International Dairy Federation,* no. 438 (2009): 76.

54. *Bulletin of the International Dairy Federation,* no. 391 (2004): 4; "The World Dairy Situation, 2009," 76.

Chapter Fourteen: Milk Today

1. M. Francis [*sic*] Guenon, *A Treatise on Milch Cows,* trans. N. P. Trist (New York, 1854), 43–48; see, e.g., Robert Jennings, *Cattle and Their Diseases* (Philadelphia, 1864), 62–76. Serious recent discussion of the milk mirror can be found in the December 2007 issue of *Acres, USA,* a magazine for organic farmers. See also http://cedarcovefarm.blogspot.com/2007/12/milk-mirror.html.

2. On ordinary Holstein cow output, see David Pazmiño, "A Transition to Success," *Edible Boston,* no. 14 (Fall 2009), 42. On Holstein-Friesians of the 1860s, see the account of a herd in Belmont, Massachusetts, in E. Parmalee Prentice, *American Dairy Cattle* (New York, 1942), 137–38. "From Science, Plenty of Cows but Little Profit," *New York Times,* September 29, 2009.

3. "From Science, Plenty of Cows"; U.K. dairy cow numbers from "Dairy Statistics: An Insider's Guide, 2007," Milk Development Council, 6. Thanks to Neil and Jane Dyson for this material, as well as their hospitality and illuminating tour of Holly Green Farm, Bledlow, Buckinghamshire, on February 7, 2008.

4. The quotation from "Milk and Murder" is featured in Anwar Fazal and Radha Holla, *The Boycott Book,* available at www.theboycottbook.com, 19. Cicely Williams's speech is available in typescript at the Wellcome Library in London.

5. Fazal and Holla, *The Boycott Book,* 19.

6. Ibid., 19, 41–46.

7. An award-winning documentary film, *The Corporation,* made in 1994 by Mark Achbar, publicized the use of rBST in dairy cows and a legal battle between

CNN and journalists involved in exposing the Monsanto Corporation (www
.thecorporation.com/index.cfm).

8. Lisa H. Weasel, *Food Fray: Inside the Controversy over Genetically Modified Food*
(New York, 2009), 161.

9. Ibid., 158–59, 165.

10. Ibid., 166–73, 178.

11. "Liquid Milk Consumption," *Bulletin of the International Dairy Federation,*
no. 391 (2004): 69; "Liquid Milk Consumption," *Bulletin of the International Dairy
Federation,* no. 438 (2009): 97.

12. Special Milk Program Fact Sheet, 2009, available at www.fns.usda.gov/
cnd/Milk/AboutMilk/SMPFactSheet.pdf.

13. Slow Food Boston meeting, March 8, 2009; Interview, Isador Lauber,
Emmi plant manager, Bern, Switzerland, February 18, 2008; Merry White, *Café
Society* (Berkeley, Calif., publication expected in 2011). I am grateful to Kim Hays
for making my meeting with Isador Lauber possible.

14. Boutique milk items, such as vodkas made with milk and maple sap, or
cosmetics made from colostrum, attract a small but steady following. "Bottoms
Up: Three New Vodkas From the Land of Milk and Sap," *New York Times,* October
12, 2005.

15. Pazmiño, "A Transition to Success," 40–43; "A Raw Deal," *The Activ-
ist Magazine,* October 21, 2008, www.activistmagazine.com/index.php?option=
com_content&task=view&id=946&Itemid=143. Thanks to Slow Food Boston's
screening of "Michael Schmidt: Organic Hero or Bioterrorist?" by Norman Lofts,
March 8, 2009, and participants in a panel discussion, especially Terri Lawton,
owner of Oake Knoll Ayrshires farm in Foxborough, Massachusetts.

16. Corby Kummer, "Pasteurization Without Representation," *The Atlantic,*
May 13, 2010, www.theatlantic.com/food/archive/2010/05/pasteurization-
without-representation/56533/. See also the Centers for Disease Control and Prevention's page on the subject: www.cdc.gov/nczved/divisions/dfbmd/diseases/
raw_milk/#legal. The USFDA's Web site concerning raw milk: www.fda.gov/
Food/FoodSafety/Product-SpecificInformation/MilkSafety/ucm122062.htm.

17. Michael Pollan, "Unhappy Meals," *New York Times Magazine,* January 28,
2007, 41. See also his *In Defense of Food: An Eater's Manifesto* (New York, 2008).

18. Stuart Patton, *Milk: Its Remarkable Contribution to Human Health and Well-
Being* (New Brunswick, 2004), 1, 11.

19. Andrea S. Wiley, " 'Drink Milk for Fitness': The Cultural Politics of Human
Biological Variation and Milk Consumption in the United States," *American Anthropologist* 106 (2004): 514.

20. I have been greatly helped by Andrea Wiley's discussion of this subject in
"The Globalization of Cow's Milk Production and Consumption: Biocultural
Perspectives," *Ecology of Food and Nutrition* 46 (2007): 281–312.

21. Ibid., 299–300.

22. Ibid., 282.

23. Junfei Bai, Thomas I. Wahl, and Jill McCluskey, "Fluid Milk Consumption in Urban Qingdao China," *Australian Journal of Agricultural and Resource Economics* 52 (2008): 133–47; Wiley, "Globalization of Cow's Milk," 290; Lauber interview.

24. Wiley, "Globalization of Cow's Milk," 285, 290, 296, 298, 301, 305.

25. See Gordon M. Wardlaw and Jeffrey S. Hampl, *Perspectives in Nutrition,* 7th ed. (Boston, 2007).

26. Dr. Mitchell Charap, personal communication, June 26, 2010.

27. Drs. Marshall and Warren were awarded the Nobel Prize in Physiology or Medicine for 1995. For a riveting account of their work and the resistance to their discoveries, see Terence Monmaney, "Marshall's Hunch," *New Yorker,* September 20, 1993, 64–72.

28. John Burnett, "Got (Good) Milk? Ask the Dairy Evangelist," *Morning Edition,* WBUR, December 10, 2009.

29. Interview, Christoph Grosjean-Sommer, Bern, Switzerland, February 16, 2008.

Primary Sources

"An Account of the Colony of the Lord Baron of Baltamore, 1633." In *Narratives of Early Maryland, 1633–1684,* ed. Clayton Colman Hall (New York, 1910).

Accum, Friedrich. *Culinary Chemistry* (London, 1821).

Aglionby, William. *The Present State of the United Provinces of the Low-Countries* (London, 1669).

Alvord, Henry E. "Dairy Products at the Paris Exposition of 1900." In United States Department of Agriculture Seventeenth Annual Report of the Bureau of Animal Industry (Washington, 1901).

Bernard of Clairvaux, Saint. *The Glories of the Virgin Mother* (1867).

———. *Selected Works,* trans. G. R. Evans (New York, 1987).

Billingsley, John. *General View of the Agriculture of Somerset* (Bath, 1797).

Boccaccio, Giovanni. *The Decameron,* trans. Guido Waldman (New York, 1993).

Bricknell, John. *Natural History of North-Carolina* (Dublin, 1737).

The Carlyle Letters Online [CLO], 2007, http://carlyleletters.org.

Carter, Susannah. *The Frugal Housewife* (New York, 1803).

Cereta, Laura. *Collected Letters of a Renaissance Feminist,* trans. Diana Robin (Chicago, 1997).

Chambers, Thomas K. *Digestion and Its Derangements* (London, 1856).

Cheyne, George. *The English Malady* [1733], ed. Roy Porter (London, 1991).

———. *Observations Concerning the Nature and Due Method of Treating the Gout* (London, 1720).

Cogan, Thomas. *The Haven of Health* (1584).

Combe, Andrew. *The Physiology of Digestion,* 10th ed., ed. James Coxe (Edinburgh, 1860).

Defoe, Daniel. *A Tour Through the Whole Island of Great Britain* (1769).

Diary of Elizabeth Drinker, ed. Elaine Forman Crane (Boston, 1991).

The Diary of Samuel Pepys, ed. Robert Latham and William Matthews (Berkeley, 1972).

Dumas, J. A. B. "The Constitution of Blood and Milk," *Philosophical Magazine and Journal of Science,* 4th series, 42 (1871): 129–38.

Ellis, William. *Agriculture Improv'd,* 4 vols. (London, 1745).

——. *The Modern Husbandman,* 4 vols. (London, 1744).

Ficino, Marsilio. *Three Books on Life,* trans. and annotated by Carol V. Kaske and John R. Clark (Binghamton, 1989).

Franklin, Benjamin. *The Autobiography of Benjamin Franklin,* ed. Leonard W. Labaree (New Haven, 1964).

Guenon, M. Francis. *A Treatise on Milch Cows,* trans. N. P. Trist, with introductory remarks by John S. Skinner (New York, 1854).

Habershon, S. O. *Pathological and Practical Observations on Diseases of the Abdomen,* 2nd ed. (London, 1862).

Hammond, John. *Leah and Rachel, or, the Two Fruitfull Sisters Virginia and Maryland* [1656]. In *Narratives of Early Maryland, 1633–1684,* ed. Clayton Colman Hall (New York, 1910).

Hartley, Robert M. *An Historical, Scientific, and Practical Essay on Milk as an Article of Human Sustenance* (1842).

Hippocrates on Diet and Hygiene, trans. John Renote (London, 1952).

The Household Book of Dame Alice de Bryene, of Acton Hall, Suffolk, September 1412 to September 1413, trans. M. K. Dale (Ipswich, 1984).

Jennie June's American Cookery Book (New York, 1870).

Jennings, Robert. *Cattle and Their Diseases* (Philadelphia, 1864).

The Journal of John Winthrop, ed. Richard S. Dunn and Laetitia Yeandle (Cambridge, Mass., 1996).

Karell, Philip, M.D. "On the Milk Cure." *Edinburgh Medical Journal* 12 (1866): 98–102.

Knight, Sarah Kemble. *The Journal of Madame Knight: A Woman's Treacherous Journey by Horseback from Boston to New York in the Year 1704* (1825; repr., Bedford, Mass., 1992).

Lead, Jane. *A Fountain of Gardens* (London, 1797).

Lebert, H. *A Treatise on Milk and Henri Nestlé's Milk Food, for the Earliest Period of Infancy and in Later Years* (Vevey, 1881).

Letters and Papers on Agriculture, Planting, &c., Bath and West of England Society (Bath, 1792).

Letter from Selina, Countess of Huntingdon, to Theophilus, Earl of Huntingdon, March 30, 1732. In *In the Midst of Early Methodism: Lady Huntingdon and Her Correspondence,* ed. John R. Tyson with Boyd S. Schlenther (Lanham, Md., 2006).

A Letter from Sir Digby Legard, Bart., To the President and Vice Presidents of the Society of Agriculture for the East Riding of Yorkshire (Beverly, 1770).

Letters of a Family During the War for the Union, 1861–1865, ed. Georgeanna M. W. Bacon (1899).

"Letter of John Pory, 1619, to Sir Dudley Carleton." In *Narratives of Early Virginia, 1606–1625,* ed. Lyon Gardiner Tyler (New York, 1909).

The Letters of Dr. George Cheyne to the Countess of Huntingdon, ed. Charles F. Mullett (San Marino, 1940).

Marshall, William. *The Rural Economy of Gloucestershire,* 2 vols. (Gloucester, 1789).

Massachusetts State Department of Health. *Report of the Special Milk Board* (Boston, 1916).

McCleary, G. F. *Infantile Mortality and Infants Milk Depots* (London, 1905).

Mendenhall, Dorothy Reed. *Milk: The Indispensable Food for Children,* U.S. Department of Labor, Children's Bureau, Care of Children Series (Washington, D.C., 1918).

Milk Commission, Department of Health, New York City. *Is Loose Milk a Health Hazard?* (New York, 1931).

Moffett, Thomas. *Healths Improvement: or, Rules Comprizing and Discovering the Nature, Method, and Manner of Preparing All Sorts of Food Used in This Nation* (London, 1655).

"The National Milk Publicity Council: What It Is and What It Aims to Do." *Milk Trade Gazette* (October 1930).

"Originator of Phrase, 'From Contented Cows,' Tells History of World-Famous Slogan." *The Carnation* 12 (May–June 1932): 24.

Orr, Sir John Boyd. "Agriculture and National Health." In *Agriculture in the Twentieth Century* (Oxford, 1939).

Pantaleone da Confienza. *La summa lacticiniorum,* ed. Irma Naso (Torino, 1990).

Platina. *On Right Pleasure and Good Health,* ed. and trans. Mary Ella Milham (Tempe, 1998).

Pordage, John. *Theologia mystica* (1683).

Prout, William. *Chemistry, Meteorology, and the Function of Digestion Considered with Reference to Natural Theology* (London, 1834).

———. *On the Ultimate Composition of Simple Alimentary Substances; with Some Preliminary Remarks on the Analysis of Organized Bodies in General* (London, 1827).

[Putnam, Elizabeth Lowell.] Elizabeth Lowell Putnam Papers, Schlesinger Library, Radcliffe College.

Raynbird, William and Hugh. *On the Agriculture of Suffolk* (London, 1849).

Rosenau, Milton J. *The Milk Question* (Boston, 1912).

Rousseau, Jean-Jacques. *Émile,* trans. Barbara Foxley (1762; repr., London, 1974).

Shippen, Nancy. *Nancy Shippen Her Journal Book,* ed. Ethel Armes (Philadelphia, 1935).

Simmons, Amelia. *American Cookery* (Hartford, 1796).

Smollett, Tobias. *Humphry Clinker* (London, 1985).

Smith, John. *Travels and Works of Captain John Smith,* ed. Edward Arber, 2 parts (Edinburgh, 1910).

Spargo, John. *The Common Sense of the Milk Question* (New York, 1908).

Steffens, Lincoln. "Sending a State to College: What the University of Wisconsin Is Doing for Its People." *American Magazine* 67 (1909): 349, 353.

Tryon, Thomas. *The Good House-wife Made a Doctor* (London, 1692).

———. *The Merchant, Citizen, and Country-man's Instructor; or, A Necessary Companion for all People* (London, 1701).

Twamley, Josiah. *Dairying Exemplified, or the Business of Cheese-making* (Warwick, 1784).

Wesley, John. *Primitive Physick* (London, 1747).

Weston, Richard. *A Discours of Husbandrie Used in Brabant and Flanders* (London, 1650).

William of Rubruck. *The Mission of Friar William of Rubruck,* trans. David Morgan, ed. Peter Jackson (London, 1990).

Secondary Sources

Abdalla, Michael. "Milk and Its Uses in Assyrian Folklore." In *Milk: Beyond the Dairy,* ed. Harlan Walker (Totnes, 2000).

Albala, Ken. *Eating Right in the Renaissance* (Berkeley, 2002).

———. "Milk: Nutritious and Dangerous." In *Milk: Beyond the Dairy,* ed. Harlan Walker (Totnes, 2000).

Alpers, Svetlana. *The Art of Describing: Dutch Art in the Seventeenth Century* (Chicago, 1983).

Anderson, Virginia DeJohn. *Creatures of Empire: How Domestic Animals Transformed Early America* (Oxford, 2004).

Apple, Rima D. "'Advertised by Our Loving Friends': The Infant Formula Industry and the Creation of New Pharmaceutical Markets, 1870–1910." *Journal of the History of Medicine and Allied Sciences* 41 (1986): 7.

———. *Mothers and Medicine: A Social History of Infant Feeding, 1890–1950* (Madison, 1989).

Atkins, Peter J. "Fattening Farmers or Fattening Children? School Milk in Britain, 1921–1941." *Economic History Review* 58 (2005): 75–76.

———. "London's Intra-Urban Milk Supply, Circa 1790–1914." *Transactions of the Institute of British Geographers,* n.s., 2 (1977): 384–85, 387, 395.

———. "The Retail Milk Trade in London, c. 1790–1914." *Economic History Review,* n.s., 33 (1980): 523.

———. "School Milk in Britain, 1900–1934." *Journal of Policy History* 19 (2007): 400–401.

Atkins, Peter J., and Derek J. Oddy, eds. *Food and the City Since 1800* (Aldershot, 2007).

Atkinson, Clarissa W. *The Oldest Vocation: Christian Motherhood in the Middle Ages* (Ithaca, 1991).

Bailey, Kenneth W. *Marketing and Pricing of Milk and Dairy Products in the United States* (Ames, Iowa, 1997).

Bakhtin, Mikhail. *Rabelais and His World,* trans. Hélène Iswolsky (Cambridge, Mass., 1968).

Banerji, Chitrita. "How the Bengalis Discovered *Chhana* and its Delightful Offspring." In *Milk: Beyond the Dairy,* ed. Harlan Walker (Totnes, 2000).

Barlow, Jeremy. *The Enraged Musician: Hogarth's Musical Imagery* (Burlington, 2005).

Bartlett, R. W. *The Price of Milk* (Danville, Ill., 1941).

Basu, Pratyusha. *Villages, Women, and the Success of Dairy Cooperatives in India* (Amherst, N.Y., 2009).

Baumgartner, F. W. *The Condensed Milk and Milk Powder Industries* (Kingston, Ont., 1920).

Berger, Pamela C. *The Goddess Obscured: Transformation of the Grain Protectress from Goddess to Saint* (Boston, 1985).

Biddick, Kathleen. *The Other Economy: Pastoral Husbandry on a Medieval Estate* (Berkeley, 1989).

Billings, John D. *Hardtack and Coffee: Or, the Unwritten Story of Army Life* (Boston, 1887).

Block, Daniel. "Protecting and Connecting: Separation, Connection, and the U.S. Dairy Economy, 1840–2002." *Journal for the Study of Food and Society* 6 (2002): 22–30.

——. "Public Health, Cooperatives, Local Regulation, and the Development of Modern Milk Policy: The Chicago Milkshed, 1900–1940." *Journal of Historical Geography* 35 (2009): 128–53.

——. "Saving Milk Through Masculinity, Public Health Officers and Pure Milk, 1880–1930." *Food and Foodways* 13 (2005): 115–34.

Bober, Phyllis Pray. *Art, Culture, and Cuisine: Ancient and Medieval Gastronomy* (Chicago, 1999).

——. "The Hierarchy of Milk in the Renaissance, and Marsilio Ficino on the Rewards of Old Age." In *Milk: Beyond the Dairy,* ed. Harlan Walker (Totnes, 2000).

Bowling, G. A. "The Introduction of Cattle into Colonial North America." *Journal of Dairy Science* 25 (1942): 140.

Brock, William H. *Justus von Liebig: The Chemical Gatekeeper* (Cambridge, 1997).

——. *The Norton History of Chemistry* (New York, 1992).

Burnett, John. *Liquid Pleasures: A Social History of Drinks in Britain* (London, 1999).

——. *Plenty and Want: A Social History of Food in England, 1815 to the Present Day,* 3rd ed. (Abingdon, 1989).

Bynum, Caroline Walker. *Holy Feast and Holy Fast: The Religious Significance of Food to Medieval Women* (Berkeley, 1987).

———. *Jesus as Mother: Studies in the Spirituality of the High Middle Ages* (Berkeley, 1982).

Camporesi, Piero. *The Anatomy of the Senses: Natural Symbols in Medieval and Early Modern Italy,* trans. Allan Cameron (Cambridge, 1997).

Cavin, J. P., Hazel K. Stiebeling, and Marius Farioletti. "Agricultural Surpluses and Nutritional Deficits." In *Farmers in a Changing World: The Yearbook of Agriculture, 1940,* United States Department of Agriculture (Washington, D.C., 1940).

Chen, Marty, et al. *Indian Women: A Study of their Role in the Dairy Movement* (New Delhi, 1986).

Chiapparino, Francesco. "Milk and Fondant Chocolate and the Emergence of the Swiss Chocolate Industry at the Turn of the Twentieth Century." In *Food and Material Culture: Proceedings of the Fourth Symposium of the International Commission for Research into European Food History,* ed. Martin R. Schärer and Alexander Fenton (East Linton, Scotland, 1998).

Cohn, Samuel Kline, Jr. *The Black Death Transformed: Disease and Culture in Early Renaissance Europe* (London, 2002).

Corrington, Gail. "The Milk of Salvation: Redemption by the Mother in Late Antiquity and Early Christianity." *Harvard Theological Review* 82 (1989): 393–420.

Craig, Oliver E. "Dairying, Dairy Products, and Milk Residues: Potential Studies in European Prehistory." In *Food, Culture, and Identity in the Neolithic and Early Bronze Age,* ed. Mike Parker Pearson (Oxford, 2003).

Cronon, William. *Changes in the Land: Indians, Colonists, and the Ecology of New England* (New York, 1983).

Czaplicki, Alan. "'Pure Milk Is Better Than Purified Milk': Pasteurization and Milk Purity in Chicago, 1908–1916." *Social Science History* 31 (2007): 411.

Dairying Throughout the World. International Dairy Federation Monograph (Munich, 1966).

Davidson, Alan. *The Oxford Companion to Food* (Oxford, 1999).

Dawson, Christopher, ed. *Mission to Asia* (Toronto, 1980).

Day, Harry G. "Elmer Verner McCollum, 1879–1967." National Academy of Sciences Biographical Memoir Series (Washington, D.C., 1974).

den Hartog, Adel P., ed. *Food Technology, Science, and Marketing: European Diet in the Twentieth Century* (East Lothian, 1995).

———. "Serving the Urban Consumer: The Development of Modern Food Packaging with Special Reference to the Milk Bottle." In *Food and Material Culture,* ed. Martin R. Schärer and Alexander Fenton (East Linton, Scotland, 1998).

de Jongh, E. *Still-Life in the Age of Rembrandt* (Auckland, 1982).

de Roest, Kees. *The Production of Parmigiano-Reggiano Cheese: The Force of an Artisanal System in an Industrialised World* (Assen, 2000).

de Vries, Jan. *The Dutch Rural Economy in the Golden Age, 1500–1700* (New Haven: Yale University Press, 1974).

de Vries, Jan, and Ad van der Woude. *The First Modern Economy* (Cambridge, 1997).

Doornbos, Martin, et al. *Dairy Aid and Development: India's Operation Flood* (New Delhi, 1990).

Doornbos, Martin, and Nair, K. N. "The State of Indian Dairying: An Overview." In *Resources, Institutions, and Strategies: Operation Flood and Indian Dairying,* ed. M. Doornbos and K. N. Nair (New Delhi, 1990).

Drummond, J. C., and Wilbraham, Anne. *The Englishman's Food* (London, 1958).

DuPuis, E. Melanie. *Nature's Perfect Food: How Milk Became America's Drink* (New York, 2002).

Durand, Loyal, Jr. "The Migration of Cheese Manufacture in the United States." *Annals of the Association of American Geographers* 42 (1952): 263–82.

Dwork, Deborah. "The Milk Option: An Aspect of the History of the Infant Welfare Movement in England, 1898–1908." *Medical History* 31 (1987): 52.

——. *War Is Good for Babies and Other Young Children: A History of the Infant and Child Welfare Movement in England, 1898–1918* (London, 1987).

Dyer, Christopher. *Standards of Living in the Later Middle Ages: Social Change in England c. 1200–1520* (Cambridge, 1989).

Edwards, Everett E. "Europe's Contribution to the American Dairy Industry." *Journal of Economic History* 9 (1949): 73.

Eliot, George. *Adam Bede* (1859; repr., Harmondsworth, 1980).

Faung, Christopher T. C. "The Dairy Industry in Taiwan." *Dairying Throughout the World*. International Dairy Federation Monograph (Munich, 1966).

Fenner, Thomas. "Die Berneralpen Milchgesellschaft: Ein internationales Unternehmen im Herzen des Emmentals." In *Das Emmental-Ansichten einer Region,* ed. Fritz Von Gunten (Münsingen, Switzerland, 2006).

Fildes, Valerie. *Breasts, Bottles, and Babies: A History of Infant Feeding* (Edinburgh, 1986).

Finlay, Mark. "Early Marketing of the Theory of Nutrition: The Science and Culture of Liebig's Extract of Meat." In *The Science and Culture of Nutrition, 1840–1940* (Amsterdam, 1995).

——. "Quackery and Cookery: Justus Von Liebig's Extract of Meat and the Theory of Nutrition in the Victorian Age." *Bulletin of the History of Medicine* 66 (1992): 404–18.

Flandrin, Jean-Louis. "Distinction Through Taste." In *A History of Private Life,* ed. Philippe Ariès and Georges Duby, vol. 3 (Cambridge, Mass., 1989).

France, James. *Medieval Images of Saint Bernard of Clairvaux* (Kalamazoo, 2007).

Frantz, Joe B. "Gail Borden as a Businessman." *Bulletin of the Business Historical Society* 22 (1948): 126.

French, Michael, and Jim Phillips. *Cheated not Poisoned? Food Regulation in the United Kingdom, 1875–1938* (Manchester, 2000).

Freund, Richard A. "What Happened to the Milk and Honey? The Changing Symbols of Abundance of the Land of Israel in Late Antiquity." In *"A Land Flowing With Milk and Honey": Visions of Israel from Bibilical to Modern Times,* ed. Leonard J. Greenspoon and Ronald A. Simkins (Omaha, 2001).

Fussell, G. E. "Eighteenth-Century Traffic in Milk Products." In *Economic History* 3 (1937).

——. *The English Dairy Farmer, 1500–1900* (London, 1966).

——. "The London Cheesemongers of the Eighteenth Century." *Economic Journal* (Supplement), Economic History Series, 3 (London, 1928): 394–98.

Ginzburg, Carlo. *The Cheese and the Worms: The Cosmos of a Sixteenth-Century Miller,* trans. John and Anne Tedeschi (Baltimore, 1980).

Glover, W. H. *Farm and College: The College of Agriculture of the University of Wisconsin, A History* (Madison, Wis., 1952).

Grainger, Sally. "Cato's Roman Cheesecakes: The Baking Techniques." In *Milk: Beyond the Dairy,* ed. Harlan Walker (Totnes, 2000).

Grant, Wyn. *The Dairy Industry: An International Comparison* (Aldershot, 1991).

Guerrini, Anita. *Obesity and Depression in the Enlightenment: The Life and Times of George Cheyne* (Norman, 2000).

Hambridge, Gove. "Farmers in a Changing World — A Summary." In *Farmers in a Changing World,* United States Department of Agriculture (Washington, D.C., 1940).

Heer, Jean. *Nestlé: 125 Years, 1866–1991* (Vevey, 1991).

Hendrix, Roberta C. "Some Gail Borden Letters." *Southwestern Historical Quarterly* 51 (1947): 133.

Henstock, Adrian. "Cheese Manufacture and Marketing in Derbyshire and North Staffordshire, 1670–1870." *Derbyshire Archeological Journal* 89 (1969): 32–46.

Hobsbawm, Eric J. *The Age of Capital, 1848–1875* (London, 1975).

——. *The Age of Empire, 1875–1914* (London, 1987).

Hunziker, Otto F. *Condensed Milk and Milk Powder,* 4th ed. (La Grange, Ill., 1926).

Hurt, R. Douglas. *Problems of Plenty: The American Farmer in the Twentieth Century* (Chicago, 2002).

Israel, Jonathan. *The Dutch Republic* (Oxford, 1995).

Kamminga, Harmke, and Andrew Cunningham, eds. *The Science and Culture of Nutrition, 1840–1940* (Amsterdam, 1995).

Kettering, Alison McNeil. *The Dutch Arcadia: Pastoral Art and Its Audience in the Golden Age* (Montclair, 1983).

King, Helen. *Hippocrates' Woman: Reading the Female Body in Ancient Greece* (London, 1998).

Kirk, Geoffrey Stephen. *Myth: Its Meaning and Functions in Ancient and Other Cultures* (Cambridge, 1970).

Koven, Seth, and Sonya Michel. "Womanly Duties: Maternalist Politics and the Origins of Welfare States in France, Germany, Great Britain, and the United States, 1880–1920." *American Historical Review* 95 (1990): 1107.

Kowaleski, Maryanne. *Local Markets and Regional Trade in Medieval Exeter* (Cambridge, 1995).

Kurien, Verghese. *I Too Had a Dream* (New Delhi, 2005).

La Berge, Ann F. "Medicalization and Moralization: The Crèches of Nineteenth-Century Paris." *Journal of Social History* 25 (1991): 74.

Laing, Wesley N. "Cattle in Seventeenth-Century Virginia." *Virginia Magazine of History and Biography* 67 (1959): 143–64.

Latour, Bruno. *The Pasteurization of France* (Cambridge, 1988).

Levenstein, Harvey A. "'Best for Babies' or 'Preventable Infanticide'? The Controversy over Artificial Feeding of Infants in America." *Journal of American History* 70 (1983): 75–94.

——. *Revolution at the Table: The Transformation of the American Diet* (New York, 1988).

Levine, Susan. *School Lunch Politics* (Princeton, 2009).

McCollum, Elmer V. *From Kansas Farm Boy to Scientist: The Autobiography of Elmer V. McCollum* (Lawrence, Kan., 1964).

——. *The Newer Knowledge of Nutrition: The Use of Food for the Preservation of Vitality and Health* (New York, 1918).

——. "Who Discovered Vitamins?" *Science* 118 (November 1953): 632.

McGuire, Brian Patrick. *The Difficult Saint: Bernard of Clairvaux and His Tradition* (Kalamazoo, 1991).

McKee, Francis. "Popularisation of Milk as a Beverage in the 1930s." In *Nutrition in Britain: Science, Scientists, and Politics in the Twentieth Century,* ed. David F. Smith (London, 1997).

McLaughlin, Mary Martin. "Survivors and Surrogates: Children and Parents from the Ninth to the Thirteenth Centuries." In *The History of Childhood*, ed. Lloyd deMause (New York, 1974).

McMahon, Sarah F. "A Comfortable Subsistence: The Changing Composition of Diet in Rural New England, 1620–1840." *The William and Mary Quarterly,* 3rd series, 42 (1985): 38.

McWilliams, James E. *A Revolution in Eating: How the Quest for Food Shaped America* (New York, 2005).

Markley, Robert. *The Far East and the English Imagination, 1600–1730* (Cambridge, 2006).

Marshall, Fiona. "Origin of Specialized Pastoral Production in East Africa." *American Anthropologist* 92 (1990): 873–94.

Meckel, Richard A. *Save the Babies: American Public Health Reform and the Prevention of Infant Mortality, 1850–1929* (Ann Arbor, 1998).

Mendelson, Anne. *Milk: The Surprising Story of Milk Through the Ages* (Knopf, 2008).

Mennell, Stephen. *All Manners of Food: Eating and Taste in England and France from the Middle Ages to the Present,* 2nd ed. (Chicago, 1996).

Mepham, T. B. "'Humanising Milk': The Formulation of Artificial Feeds for Infants (1850–1910)." *Medical History* (1993): 227–28.

Merta, Sabine. "Karlsbad and Marienbad: The Spas and Their Cures." In *The Diffusion of Food Culture in Europe from the Late Eighteenth Century to the Present Day,* ed. Derek J. Oddy and Lydia Petranova (Prague, 2005).

Michel, Sonya, and Robyn Rosen. "The Paradox of Maternalism: Elizabeth Lowell Putnam and the American Welfare State." *Gender and History* 4 (1992): 369.

Michell, A. R. "Sir Richard Weston and the Spread of Clover Cultivation." *Agricultural History Review* 22 (1974): 161.

"Milk Sickness (Tremetol Poisoning)." In *Cambridge World History of Human Disease,* ed. Kenneth R. Kiple (Cambridge, 1993).

Mintz, Sidney W. *Sweetness and Power: The Place of Sugar in Modern History* (New York, 1985).

Morgan, Edmund S. *American Slavery, American Freedom: The Ordeal of Colonial Virginia* (New York, 1975).

Morrison, Susan Signe. *Women Pilgrims in Late Medieval England* (London, 2000).

Mullendore, William Clinton. *History of the United States Food Administration, 1917–1919* (Stanford, 1941).

Murcott, Anne. "Scarcity in Abundance: Food and Non-Food." *Social Research* 66 (Spring 1999): 305–39.

Oddy, Derek J. "Food, Drink, and Nutrition." In *Cambridge Social History of Britain,* vol. 2, ed. F. M. L. Thompson (Cambridge, 2008).

———. *From Plain Fare to Fusion Food* (Woodbridge, 2003).

Oddy, Derek J., and Derek S. Miller, eds. *The Making of the Modern British Diet* (London, 1976).

Ott, Sandra. "Aristotle Among the Basques: The 'Cheese Analogy' of Conception." *Man* 14 (1979): 699–711.

Oxford Dictionary of National Biography, ed. H. C. G. Matthew and Brian Harrison (Oxford, 2004).

Patton, Stuart. *Milk: Its Remarkable Contribution to Human Health and Well-Being* (New Brunswick, 2004).

Pazmiño, David. "A Transition to Success." *Edible Boston* 14 (Fall 2009): 42.

Perren, Richard. *Agriculture in Depression* (Cambridge, 1995).

Phythian-Adams, Charles. "Milk and Soot: The Changing Vocabulary of a Popular Ritual in Stuart and Hanoverian London." In *The Pursuit of Urban History,* ed. Derek Fraser and Anthony Sutcliffe (London, 1983).

Pinchbeck, Ivy. *Women Workers and the Industrial Revolution, 1750–1850* (1930; repr., Totowa, N.J., 1968).

Pincus, Steven C. A. "From Butterboxes to Wooden Shoes: The Shift in English Popular Sentiment from Anti-Dutch to Anti-French in the 1670s." *Historical Journal* 38 (1995): 338.

Pirtle, T. R. *History of the Dairy Industry* (Chicago, 1926; repr., 1973).

Pollan, Michael. *In Defense of Food: An Eater's Manifesto* (New York, 2008).

Pollock, Jon. "Two Controlled Trials of Supplementary Feeding of School Children in the 1920s." *Journal of the Royal Society of Medicine* 99 (2006): 323–27.

Porter, Roy, and G. S. Rousseau. *Gout: The Patrician Malady* (New Haven, 1998).

Prentice, E. Parmalee. *American Dairy Cattle: Their Past and Future* (New York, 1942).

Pringle, Heather. "Neolithic Agriculture: The Slow Birth of Agriculture." *Science* 282 (1998): 1446–89.

Rabil, Albert, Jr. *Laura Cereta, Quattrocento Humanist* (Binghamton, 1981).

Ray, John. *Observations Topographical, Moral, & Physiological Made in a Journey to Part of the Low-Countries* (London, 1673)

Ray, Keith, and Julian Thomas. "In the Kinship of Cows: The Social Centrality of Cattle in the Earlier Neolithic of Southern Britain." In *Food, Culture, and Identity,* ed. Mike Parker Pearson (Oxford, 2003).

Riera-Melis, Natoni. "Society, Food, and Feudalism." In *Food: A Culinary History,* ed. Jean-Louis Flandrin, Massimo Montanari, and Albert Sonnenfeld, trans. Clarissa Botsford et al. (New York, 2000).

Robertson, James I., Jr. *Soldiers Blue and Gray* (Charleston, 1988).

Rosenfeld, Louis. "William Prout: Early 19th-Century Physician-Chemist." *Clinical Chemistry* 49 (2003): 699–705.

Rossiter, Margaret W. *Women Scientists in America: Struggles and Strategies to 1940* (Baltimore, 1982).

Rousseau, G. S. "Mysticism and Millenarianism: 'Immortal Dr. Cheyne.'" In *Millenarianism and Messianism in English Literature and Thought, 1650–1800,* ed. Richard H. Popkin (Leiden, 1988).

Rubin, Miri. *Mother of God: A History of the Virgin Mary* (New Haven, 2009).

Russell, Colin A. *Edward Frankland: Chemistry, Controversy, and Conspiracy in Victorian England* (Cambridge, 1996).

Salmon, Marylynn. "The Cultural Significance of Breastfeeding and Infant Care in Early Modern England and America." *Journal of Social History* 28 (1994): 247–69.

Schama, Simon. *Embarrassment of Riches: An Interpretation of Dutch Culture in the Golden Age* (New York, 1987).

Schapiro, Laura. *Perfection Salad: Women and Cooking at the Turn of the Century* (Berkeley, 1986).

Schiebinger, Londa. "Why Mammals Are Called Mammals: Gender Politics in Eighteenth-Century Natural History." *American Historical Review* 98 (1993): 394.

Schlossman, Betty L., and Hildreth York. "'She Shall Be Called Woman': Ancient Near Eastern Sources of Imagery." *Women's Art Journal* 2 (Autumn 1981–Winter 1982): 37–41.

Schmidt, Jeremy. *Melancholy and the Care of the Soul: Religion, Moral Philosophy, and Madness* (Burlington, 2007).

Sekora, John. *Luxury: The Concept in Western Thought, Eden to Smollett* (Baltimore, 1977).

Senior, Nancy. "Aspects of Infant Feeding in Eighteenth-Century France." *Eighteenth-Century Studies* 16 (1983): 380.

Serruys, Henry. *Kumiss Ceremonies and Horse Races* (Wiesbaden, 1974).

Shaw, Brent D. "'Eaters of Flesh, Drinkers of Milk': The Ancient Mediterranean Ideology of the Pastoral Nomad." *Ancient Society* 13–14 (1982–83): 5–31.

Sherill, Charles Hitchcock. *French Memories of Eighteenth-Century America* (1915).

Sherratt, Andrew. "Cash Crops Before Cash: Organic Consumables and Trade." In *The Prehistory of Food: Appetites for Change,* ed. Chris Gosden and Jon Hather (London, 1999).

——. "The Secondary Exploitation of Animals in the Old World." *World Archaeology* 15 (1983): 95.

Shesgreen, Sean. *Images of the Outcast: The Urban Poor in the Cries of London* (Manchester, 2002).

Slicher van Bath, B. H. *The Agrarian History of Western Europe, A.D. 500–1850,* trans. Olive Ordish (London, 1963).

Smith, Charles. *Britain's Food Supplies* (London, 1940).

Smith, F. B. *The Retreat of Tuberculosis* (London, 1988).

Smith, Maryanna S., comp. *Chronological Landmarks in American Agriculture.* U.S. Department of Agriculture, Agriculture Information Bulletin No. 425 (Washington, D.C., 1980).

Spang, Rebecca. *The Invention of the Restaurant: Paris and Modern Gastronomic Culture* (Cambridge, Mass., 2000).

Steckel, Richard H. "Nutritional Status in the Colonial American Economy." *The William and Mary Quarterly,* 3rd Ser., 56, no. 1 (January 1999): 38, 44, 46.

Stol, Marten. "Milk, Butter, and Cheese." *Bulletin on Sumerian Agriculture* 7 (1993): 100.

Strong, Roy. *Feast: A History of Grand Eating* (Orlando, 2002).

Storrs, Richard S. *Bernard of Clairvaux: The Times, the Man, and His Work* (New York, 1901).

Swabey, Ffiona. "The Household of Alice de Bryene, 1412–13." In *Food and Eating in Medieval Europe,* ed. Martha Carlin and Joel T. Rosenthal (London, 1998).

Swinburne, Layinka M. "Milky Medicine and Magic." In *Milk: Beyond the Dairy,* ed. Harlan Walker (Totnes, 2000).

Taupin, Sidonia C. "'Christianity in the Kitchen,' or A Moral Guide for Gourmets." *American Quarterly* 15 (1963): 86.

Teuteberg, Hans J. "The Beginnings of the Modern Milk Age in Germany." In *Food in Perspective: Proceedings of the Third International Conference on Ethnological Food Research,* ed. Alexander Fenton and Trefor M. Owen (Edinburgh, 1981), 283–311.

Toussaint-Samat, Maguelonne. *A History of Food,* trans. Anthea Bell (Oxford, 1992).

Trentmann, Frank. "Bread, Milk, and Democracy: Consumption and Citizenship in Twentieth-Century Britain." In *The Politics of Consumption: Material Culture and Citizenship in Europe and America,* ed. Martin Daunton and Matthew Hilton (Oxford, 2001).

Trentmann, Frank, and Flemming Just, eds. *Food and Conflict in Europe in the Age of the Two World Wars* (Basingstoke, 2006).

Valenze, Deborah. "The Art of Women and the Business of Men: Women's Work and the Dairy Industry, c. 1740–1840." *Past and Present* 130 (1991): 142–69.

Vernon, James. *Hunger: A Modern History* (Cambridge, Mass., 2007).

Vila, Anne C. *Enlightenment and Pathology: Sensibility in the Literature and Medicine of Eighteenth-Century France* (Baltimore, 1998).

Visser, Margaret. *Much Depends on Dinner* (New York, 1986).

Warner, Marina. *Alone of All Her Sex: The Myth and Cult of the Virgin Mary* (London, 1976).

Weasel, Lisa H. *Food Fray: Inside the Controversy over Genetically Modified Food* (New York, 2009).

Wharton, Clarence B. *Gail Borden, Pioneer* (San Antonio, 1941).

Whetham, Edith. "The London Milk Trade, 1900–1930." In *The Making of the Modern British Diet,* ed. Derek J. Oddy and Derek S. Miller (London, 1976).

Wiley, Andrea S. "'Drink Milk for Fitness': The Cultural Politics of Human Biological Variation and Milk Consumption in the United States." *American Anthropologist* 106 (2004): 506–17.

——. "The Globalization of Cow's Milk Production and Consumption: Biocultural Perspectives." *Ecology of Food and Nutrition* 46 (2007): 281–312.

Williams, Joanna. "The Churning of the Ocean of Milk — Myth, Image, and Ecology." In *Indigenous Vision: Peoples of India Attitudes to the Environment,* ed. Geeta Sen (New Delhi, 1992).

Witt, R. E. *Isis in the Ancient World* (Baltimore, 1971).

Wolf, Jacqueline H. *Don't Kill Your Baby: Public Health and the Decline of Breastfeeding in the Nineteenth and Twentieth Centuries* (Columbus, 2001).

Worley, Michael Preston. *Pierre Julien: Sculptor to Queen Marie-Antoinette* (New York, 2003).

Worrell, Dorothy. *The Women's Municipal League of Boston: A History of Thirty-Five Years of Civic Endeavor* (Boston, 1943).

Zorach, Rebecca. *Blood, Milk, Ink, and Gold: Abundance and Excess in the French Renaissance* (Chicago, 2005).

Page numbers in italic type refer to illustrations